STUDENT'S SOLUTIONS MANUAL

BASIC MATHEMATICS

SEVENTH EDITION

STUDENT'S SOLUTIONS MANUAL

BASIC MATHEMATICS

SEVENTH EDITION

Marvin L. Bittinger
Indiana University—Purdue University at Indianapolis

Mervin L. Keedy
Purdue University

Judith A. Penna

ADDISON-WESLEY PUBLISHING COMPANY
Reading, Massachusetts • Menlo Park, California • New York
Don Mills, Ontario • Wokingham, England • Amsterdam • Bonn
Sydney • Singapore • Tokyo • Madrid • San Juan • Milan • Paris

Table of Contents

The author thanks Patsy Hammond for her
excellent typing and Pam Smith for her careful
proofreading. Their skill, patience, and good
humor made the author's work much easier.

STUDENT'S SOLUTIONS MANUAL

BASIC MATHEMATICS

SEVENTH EDITION

Chapter 1
Operations on the Whole Numbers

Exercise Set 1.1

1. 5742 = 5 thousands + 7 hundreds + 4 tens + 2 ones

3. 27,342 = 2 ten thousands + 7 thousands + 3 hundreds + 4 tens + 2 ones

5. 5609 = 5 thousands + 6 hundreds + 0 tens + 9 ones, or 5 thousands + 6 hundreds + 9 ones

7. 2300 = 2 thousands + 3 hundreds + 0 tens + 0 ones, or 2 thousands + 3 hundreds

9. 2 thousands + 4 hundreds + 7 tens + 5 ones = 2475

11. 6 ten thousands + 8 thousands + 9 hundreds + 3 tens + 9 ones = 68,939

13. 7 thousands + 3 hundreds + 0 tens + 4 ones = 7304

15. 1 thousand + 9 ones = 1009

17. 85 = eighty-five

19.

88,000

Eighty-eight thousand ———↑

21.

123,765

One hundred twenty-three thousand, ———↑
seven hundred sixty-five ———↑

23.

7, 754, 211, 577

Seven billion, ———↑
seven hundred fifty-four million, ———↑
two hundred eleven thousand, ———↑
five hundred seventy-seven ———↑

25.

6, 469, 952

Six million, ———↑
four hundred sixty-nine thousand, ———↑
nine hundred fifty-two ———↑

27.

1, 954, 116

One million, ———↑
nine hundred fifty-four thousand, ———↑
one hundred sixteen ———↑

29.

Two million, ———↑
two hundred thirty-three thousand, ———↑
eight hundred twelve ———↑

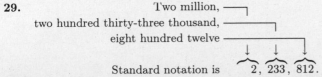

Standard notation is   2, 233, 812.

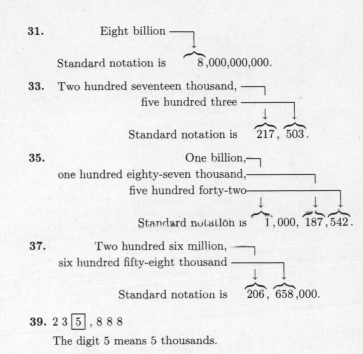

31.   Eight billion ———↑

Standard notation is   8,000,000,000.

33. Two hundred seventeen thousand, ———↑
five hundred three ———↑

Standard notation is   217, 503.

35.   One billion, ———↑
one hundred eighty-seven thousand, ———↑
five hundred forty-two ———↑

Standard notation is   1, 000, 187, 542.

37.   Two hundred six million, ———↑
six hundred fifty-eight thousand ———↑

Standard notation is   206, 658,000.

39. 2 3 $\boxed{5}$, 8 8 8

The digit 5 means 5 thousands.

41. 4 8 8, $\boxed{5}$ 2 6

The digit 5 means 5 hundreds.

43. 8 9, $\boxed{3}$ 0 2

The digit 3 tells the number of hundreds.

45. 8 9, 3 $\boxed{0}$ 2

The digit 0 tells the number of tens.

47. All digits are 9's. Answers may vary. For an 8-digit read-out, for example, it would be 99,999,999. This number has three periods.

Exercise Set 1.2

1.
| They rent 3 videos one week. | They rent 6 videos the next week. | They rent 9 videos in all. |

3 videos +   6 videos   = 9 videos

3.
| Earns $23 one day. | Earns $31 the next day. | The total earned is $54. |

$23   + $31   = $54

5. 325 ft + 325 ft + 325 ft + 325 ft = 1300 ft

7.
```
   3 6 4
 +   2 3
   3 8 7
```
Add ones, add tens, then add hundreds.

9.
```
   1
  1716
 +3482
 ─────
  5198
```
Add ones: We get 8. Add tens: We get 9 tens. Add hundreds: We get 11 hundreds, or 1 thousand + 1 hundred. Write 1 in the hundreds column and 1 above the thousands. Add thousands: We get 5 thousands.

11.
```
   1
   86
  +78
 ────
  164
```
Add ones: We get 14, or 1 ten + 4 ones. Write 4 in the ones column and 1 above the tens. Add tens: We get 16 tens.

13.
```
   1
   99
  + 1
 ────
  100
```
Add ones: We get 10 ones, or 1 ten + 0 ones. Write 0 in the ones column and 1 above the tens. Add tens: We get 10 tens.

15.
```
   1 1
   789
  +111
 ─────
   900
```
Add ones: We get 10 ones, or 1 ten + 0 ones. Write 0 in the ones column and 1 above the tens. Add tens: We get 10 tens, or 1 hundred + 0 tens. Write 0 in the tens column and 1 above the hundreds. Add hundreds: We get 9 hundreds.

17.
```
    1
   909
  +101
 ─────
  1010
```
Add ones: We get 10. Write 0 in the ones column and 1 above the tens. Add tens: We get 1 ten. Add hundreds: We get 10 hundreds.

19.
```
    1
   811
  +390
 ─────
  1201
```
Add ones: We get 1. Add tens: We get 10 tens. Write 0 in the tens column and 1 above the hundreds. Add hundreds: We get 12 hundreds.

21.
```
    1
   356
  +491
 ─────
   847
```
Add ones: We get 7. Add tens: We get 14 tens. Write 4 in the tens column and 1 above the hundreds. Add hundreds: We get 8 hundreds.

23.
```
   1 1
  5093
 +3217
 ─────
  8310
```
Add ones: We get 10. Write 0 in the ones column and 1 above the tens. Add tens: We get 11. Write 1 in the tens column and 1 above the hundreds. Add hundreds: We get 3 hundreds. Add thousands: We get 8 thousands.

25.
```
   1 1
  4825
 +1783
 ─────
  6608
```
Add ones: We get 8. Add tens: We get 10 tens. Write 0 in the tens column and 1 above the hundreds. Add hundreds: We get 16 hundreds. Write 6 in the hundreds column and 1 above the thousands. Add thousands: We get 6 thousands.

27.
```
   1 1 1
   9999
  +6785
 ──────
 16,784
```
Add ones: We get 14. Write 4 in the ones column and 1 above the tens. Add tens: We get 18 tens. Write 8 in the tens column and 1 above the hundreds. Add hundreds: We get 17 hundreds. Write 7 in the hundreds column and 1 above the thousands. Add thousands: We get 16 thousands.

29.
```
  1 1 1
 23,443
+10,989
───────
 34,432
```
Add ones: We get 12. Write 2 in the ones column and 1 above the tens. Add tens: We get 13 tens. Write 3 in the tens column and 1 above the hundreds. Add hundreds: We get 14 hundreds. Write 4 in the hundreds column and 1 above the thousands. Add thousands: We get 4 thousands. Add ten thousands: We get 3 ten thousands.

31.
```
  1 1 1 1
  77,543
 +23,767
────────
 101,310
```
Add ones: We get 10. Write 0 in the ones column and 1 above the tens. Add tens: We get 11 tens. Write 1 in the tens column and 1 above the hundreds. Add hundreds: We get 13 hundreds. Write 3 in the hundreds column and 1 above the thousands. Add thousands: We get 11 thousands. Write 1 in the thousands column and 1 above the ten thousands. Add ten thousands: We get 10 ten thousands.

33.
```
  1 1 1 1
  99,999
 +    112
────────
 100,111
```
Add ones: We get 11. Write 1 in the ones column and 1 above the tens. Add tens: We get 11 tens. Write 1 in the tens column and 1 above the hundreds. Add hundreds: We get 11 hundreds. Write 1 in the hundreds column and 1 above the thousands. Add thousands: We get 10 thousands. Write 0 in the thousands column and 1 above the ten thousands. Add ten thousands: We get 10 ten thousands.

35. Add from the top.

We first add 7 and 9, getting 16; then 16 and 4, getting 20; then 20 and 8, getting 28.

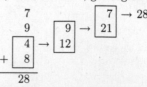

Check by adding from the bottom.

We first add 8 and 4, getting 12; then 12 and 9, getting 21; then 21 and 7, getting 28.

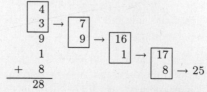

37. Add from the top.

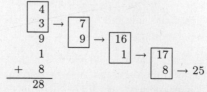

Check:

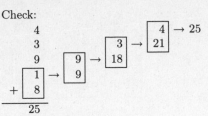
$$4 + 3 + 9 + 1 + 8 = 25$$

39. Add from the top.

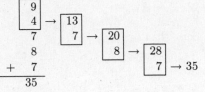
$$9 + 4 + 7 + 8 + 7 = 35$$

Check:

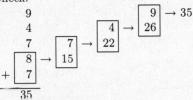
$$9 + 4 + 7 + 8 + 7 = 35$$

41. We look for pairs of numbers whose sums are 10, 20, 30, and so on.

```
23  ───→   23
16  ───→   16
11  ───→   30
18  ──╱→   18
+19 ─╱     ──
87         87
```

43. We look for pairs of numbers whose sums are 10, 20, 30, and so on.

```
45  ───→   70
25  ──╱
36  ───→   80
44  ──╱
+80 ───→   80
230        230
```

45.
```
  1
  2 3
  6 2
+ 4 5
─────
1 3 0
```
Add ones: We get 10. Write 0 in the ones column and 1 above the tens. Add tens: We get 13 tens.

47.
```
  5 1
  3 6
+ 6 2
─────
1 4 9
```
Add ones, then add tens.

49.
```
  2 6
  8 2
+ 6 1
─────
1 6 9
```
Add ones, then add tens.

51.
```
  1 1
  2 0 7
  2 9 5
+ 3 4 0
───────
  8 4 2
```
Add ones: We get 12. Write 2 in the ones column and 1 above the tens. Add tens: We get 14 tens. Write 4 in the tens column and 1 above the hundreds. Add hundreds: We get 8 hundreds.

53.
```
  2 4
  3 2 7
  4 2 8
  5 6 9
  7 8 7
+ 2 0 9
───────
2 3 2 0
```
Add ones: We get 40. Write 0 in the ones column and 4 above the tens. Add tens: We get 22 tens. Write 2 in the tens column and 2 above the hundreds. Add hundreds: We get 23 hundreds.

55.
```
   1 1 1
   2 0 3 7
   4 9 2 3
   3 4 7 1
 + 1 2 4 8
──────────
1 1,6 7 9
```
Add ones: We get 19. Write 9 in the ones column and 1 above the tens. Add tens: We get 17 tens. Write 7 in the tens column and 1 above the hundreds. Add hundreds: We get 16 hundreds. Write 6 in the hundreds column and 1 above the thousands. Add thousands: We get 11 thousands.

57.
```
   1   1
   3 4 2 0
   8 7 1 9
   4 3 1 2
 + 6 2 0 3
──────────
2 2,6 5 4
```
Add ones: We get 14. Write 4 in the ones column and 1 above the tens. Add tens: We get 5 tens. Add hundreds: We get 16 hundreds. Write 6 in the hundreds column and 1 above the thousands. Add thousands: We get 22 thousands.

59.
```
   2 2 3 3 1 1
   5,6 7 8,9 8 7
   1,4 0 9,3 1 2
     8 9 8,8 8 8
 + 4,7 7 7,9 1 0
────────────────
1 2,7 6 5,0 9 7
```

61. $7000 + 900 + 90 + 2 = 7$ thousands $+ 9$ hundreds $+ 9$ tens $+ 2$ ones $= 7992$

63. 4 $\boxed{8}$ 6, 2 0 5

The digit 8 tells the number of ten thousands.

65. One method is described in the answer section in the text. Another method is: $1 + 100 = 101$, $2 + 99 = 101$, ..., $50 + 51 = 101$. Then $50 \cdot 101 = 5050$.

Exercise Set 1.3

1.

$$\begin{pmatrix}\text{Number of} \\ \text{gallons to} \\ \text{start with}\end{pmatrix} \quad \begin{pmatrix}\text{Number of} \\ \text{gallons sold}\end{pmatrix} \quad \begin{pmatrix}\text{Number of} \\ \text{gallons left}\end{pmatrix}$$

2400 gal $-$ 800 gal $=$ $\boxed{}$

3. $7 - 4 = 3$
$\uparrow$
This number gets added (after 3).
$\downarrow$
$7 = 3 + 4$

(By the commutative law of addition, $7 = 4 + 3$ is also correct.)

5. $13 - 8 = 5$
$\uparrow$
This number gets added (after 5).
$\downarrow$
$13 = 5 + 8$

(By the commutative law of addition, $13 = 8 + 5$ is also correct.)

7. $23 - 9 = 14$

This number gets added (after 14).

$$23 = 14 + 9$$

(By the commutative law of addition, $23 = 9 + 14$ is also correct.)

9. $43 - 16 = 27$

This number gets added (after 27).

$$43 = 27 + 16$$

(By the commutative law of addition, $43 = 16 + 27$ is also correct.)

11. $6 + 9 = 15$ $6 + 9 = 15$

This number gets subtracted (moved).   This number gets subtracted (moved).

$6 = 15 - 9$ $9 = 15 - 6$

13. $8 + 7 = 15$ $8 + 7 = 15$

This number gets subtracted (moved).   This number gets subtracted (moved).

$8 = 15 - 7$ $7 = 15 - 8$

15. $17 + 6 = 23$ $17 + 6 = 23$

This number gets subtracted (moved).   This number gets subtracted (moved).

$17 = 23 - 6$ $6 = 23 - 17$

17. $23 + 9 = 32$ $23 + 9 = 32$

This number gets subtracted (moved).   This number gets subtracted (moved).

$23 = 32 - 9$ $9 = 32 - 23$

19. We write an addition sentence first.

Number already sold plus Number yet to be sold is Sales goal

$$190 \quad + \quad \boxed{} \quad = \quad 220$$

Now we write a related subtraction.

$$190 \quad + \quad \boxed{} \quad = \quad 220$$

$$\boxed{} \quad = \quad 220 - 190$$ 190 gets subtracted (moved)

21.
```
  1 6
-   4
-----
  1 2
```
Subtract ones, then subtract tens.

23.
```
  6 5
- 2 1
-----
  4 4
```
Subtract ones, then subtract tens.

25.
```
  8 6 6
- 3 3 3
-------
  5 3 3
```
Subtract ones, subtract tens, then subtract hundreds.

27.
```
  4 5 4 7
- 3 4 2 1
---------
  1 1 2 6
```
Subtract ones, subtract tens, subtract hundreds, then subtract thousands.

29.
```
    7 16
    8̸ 6̸
-   4  7
--------
    3  9
```
We cannot subtract 7 ones from 6 ones. Borrow 1 ten to get 16 ones. Subtract ones, then subtract tens.

31.
```
       11
    5  7̸ 15
    6̸  2̸ 5̸
-   3  2  7
-----------
    2  9  8
```
We cannot subtract 7 ones from 5 ones. Borrow 1 ten to get 15 ones. Subtract ones. We cannot subtract 2 tens from 1 ten. Borrow 1 hundred to get 11 tens. Subtract tens, then subtract hundreds.

33.
```
       2 15
    8  3̸ 5̸
-   6  0  9
-----------
    2  2  6
```
We cannot subtract 9 ones from 5 ones. Borrow 1 ten to get 15 ones. Subtract ones, subtract tens, then subtract hundreds.

35.
```
       7 11
    9  8̸ 1̸
-   7  4  7
-----------
    2  3  4
```
We cannot subtract 7 ones from 1 one. Borrow 1 ten to get 11 ones. Subtract ones, subtract tens, then subtract hundreds.

37.
```
       6 16
    7  7̸ 9
-   2  3 8 7
-----------
    5  3 8 2
```
Subtract ones. We cannot subtract 8 tens from 6 tens. Borrow 1 hundred to get 16 tens. Subtract tens, subtract hundreds, then subtract thousands.

39.
```
         17
      8  7̸ 12
    3 9̸ 8̸ 2̸
-   2 4  8  9
------------
    1 4  9  3
```
We cannot subtract 9 ones from 2 ones. Borrow 1 ten to get 12 ones. Subtract ones. We cannot subtract 8 tens from 7 tens. Borrow 1 hundred to get 17 tens. Subtract tens, subtract hundreds, then subtract thousands.

41.
```
        13
     4 9 3̸ 16
    5̸ 0̸ 4̸ 6̸
-   2 8 5  9
------------
    2 1 8  7
```
We cannot subtract 9 ones from 6 ones. Borrow 1 ten to get 16 ones. Subtract ones. We cannot subtract 5 tens from 3 tens. We have 5 thousands or 50 hundreds. We borrow 1 hundred to get 13 tens. We have 49 hundreds. Subtract tens, subtract hundreds, then subtract thousands.

43.
$$
\begin{array}{r}
\overset{6\ \ 16\ \ 3\ \ 10}{7\ \cancel{6}\ \cancel{4}\ \cancel{0}} \\
-\ 3\ 8\ 0\ 9 \\
\hline
3\ 8\ 3\ 1
\end{array}
$$
We cannot subtract 9 ones from 0 ones. Borrow 1 ten to get 10 ones. Subtract ones, then tens. We cannot subtract 8 hundreds from 6 hundreds. Borrow 1 thousand to get 16 hundreds. Subtract hundreds, then thousands.

45.
$$
\begin{array}{r}
\overset{11\ 15\ 13}{\overset{1\ \ 5\ \ 3\ \ 17}{1\ \cancel{2},\ \cancel{6}\ \cancel{4}\ \cancel{7}}} \\
-\ \ \ 4\ 8\ 9\ 9 \\
\hline
7\ 7\ 4\ 8
\end{array}
$$

47.
$$
\begin{array}{r}
\overset{16\ \ 16}{\overset{5\ \ \cancel{6}\ \ \cancel{6}\ \ 11}{4\ \cancel{6},\ 7\ 7\ \cancel{1}}} \\
-\ 1\ 2,\ 9\ 7\ 7 \\
\hline
3\ 3,\ 7\ 9\ 4
\end{array}
$$

49.
$$
\begin{array}{r}
\overset{7\ 10}{\cancel{8}\ \cancel{0}} \\
-\ 2\ 4 \\
\hline
5\ 6
\end{array}
$$

51.
$$
\begin{array}{r}
\overset{8\ 10}{\cancel{9}\ \cancel{0}} \\
-\ 5\ 4 \\
\hline
3\ 6
\end{array}
$$

53.
$$
\begin{array}{r}
\overset{13}{\overset{\cancel{3}\ \ 10}{1\ \cancel{4}\ \cancel{0}}} \\
-\ \ \ 5\ 6 \\
\hline
8\ 4
\end{array}
$$

55.
$$
\begin{array}{r}
\overset{8\ 10}{6\ \cancel{9}\ \cancel{0}} \\
-\ 2\ 3\ 6 \\
\hline
4\ 5\ 4
\end{array}
$$

57.
$$
\begin{array}{r}
\overset{8\ 10}{\cancel{9}\ \cancel{0}\ 3} \\
-\ 1\ 3\ 2 \\
\hline
7\ 7\ 1
\end{array}
$$

59.
$$
\begin{array}{r}
\overset{2\ \ 9\ \ 10}{2\ \cancel{3}\ \cancel{0}\ \cancel{0}} \\
-\ \ \ 1\ 0\ 9 \\
\hline
2\ 1\ 9\ 1
\end{array}
$$
We have 3 hundreds or 30 tens. We borrow 1 ten to get 10 ones. We then have 29 tens. Subtract ones, then tens, then hundreds, then thousands.

61.
$$
\begin{array}{r}
\overset{7\ \ 9\ \ 18}{6\ \cancel{8}\ \cancel{0}\ \cancel{8}} \\
-\ 3\ 0\ 5\ 9 \\
\hline
3\ 7\ 4\ 9
\end{array}
$$
We have 8 hundreds or 80 tens. We borrow 1 ten to get 18 ones. We then have 79 tens. Subtract ones, then tens, then hundreds, then thousands.

63.
$$
\begin{array}{r}
\overset{8\ 12}{8\ 0\ \cancel{9}\ \cancel{2}} \\
-\ 1\ 0\ 7\ 3 \\
\hline
7\ 0\ 1\ 9
\end{array}
$$
Borrow 1 ten to get 12 ones. Subtract ones, then tens, then hundreds, then thousands.

65.
$$
\begin{array}{r}
\overset{6\ \ 9\ \ 9\ \ 10}{\cancel{7}\ \cancel{0}\ \cancel{0}\ \cancel{0}} \\
-\ 2\ 7\ 9\ 4 \\
\hline
4\ 2\ 0\ 6
\end{array}
$$
We have 7 thousands or 700 tens. We borrow 1 ten to get 10 ones. We then have 699 tens. Subtract ones, then tens, then hundreds, then thousands.

67.
$$
\begin{array}{r}
\overset{7\ \ 9\ \ 9\ \ 10}{4\ \cancel{8},\ \cancel{0}\ \cancel{0}\ \cancel{0}} \\
-\ 3\ 7,\ 6\ 9\ 5 \\
\hline
1\ 0,\ 3\ 0\ 5
\end{array}
$$
We have 8 thousands or 800 tens. We borrow 1 ten to get 10 ones. We then have 799 tens. Subtract ones, then tens, then hundreds, then thousands, then ten thousands.

69. 6, 3 7 5, 6 0 2

The digit 7 means 7 ten thousands.

Exercise Set 1.4

1. Round 48 to the nearest ten.

4 8

↑

The digit 4 is in the tens place. Consider the next digit to the right. Since the digit, 8, is 5 or higher, round 4 tens up to 5 tens. Then change the digit to the right of the tens digit to zero.

The answer is 50.

3. Round 67 to the nearest ten.

6 7
↑

The digit 6 is in the tens place. Consider the next digit to the right. Since the digit, 7, is 5 or higher, round 6 tens up to 7 tens. Then change the digit to the right of the tens digit to zero.

The answer is 70.

5. Round 731 to the nearest ten.

7 3 1
↑

The digit 3 is in the tens place. Consider the next digit to the right. Since the digit, 1, is 4 or lower, round down, meaning that 3 tens stays as 3 tens. Then change the digit to the right of the tens digit to zero.

The answer is 730.

7. Round 895 to the nearest ten.

8 9 5

↑

The digit 9 is in the tens place. Consider the next digit to the right. Since the digit, 5, is 5 or higher, we round up. The 89 tens become 90 tens. Then change the digit to the right of the tens digit to zero.

The answer is 900.

9. Round 146 to the nearest hundred.

1 4 6
↑

The digit 1 is in the hundreds place. Consider the next digit to the right. Since the digit, 4, is 4 or lower, round down, meaning that 1 hundred stays as 1 hundred. Then change all digits to the right of the hundreds digit to zeros.

The answer is 100.

11. Round 957 to the nearest hundred.

9 [5] 7

↑

The digit 9 is in the hundreds place. Consider the next digit to the right. Since the digit, 5, is 5 or higher, round up. The 9 hundreds become 10 hundreds. Then change all digits to the right of the hundreds digit to zeros.

The answer is 1000.

13. Round 9079 to the nearest hundred.

9 0 [7] 9

↑

The digit 0 is in the hundreds place. Consider the next digit to the right. Since the digit, 7, is 5 or higher, round 0 hundreds up to 1 hundred. Then change all digits to the right of the hundreds digit to zeros.

The answer is 9100.

15. Round 2850 to the nearest hundred.

2 8 [5] 0

↑

The digit 8 is in the hundreds place. Consider the next digit to the right. Since the digit, 5, is 5 or higher, round 8 hundreds up to 9 hundreds. Then change all digits to the right of the hundreds digit to zeros.

The answer is 2900.

17. Round 5876 to the nearest thousand.

5 [8] 7 6

↑

The digit 5 is in the thousands place. Consider the next digit to the right. Since the digit, 8, is 5 or higher, round 5 thousands up to 6 thousands. Then change all digits to the right of the thousands digit to zeros.

The answer is 6000.

19. Round 7500 to the nearest thousand.

7 [5] 0 0

↑

The digit 7 is in the thousands place. Consider the next digit to the right. Since the digit, 5, is 5 or higher, round 7 thousands up to 8 thousands. Then change all the digits to the right of the thousands digit to zeros.

The answer is 8000.

21. Round 45,340 to the nearest thousand.

4 5, [3] 4 0

↑

The digit 5 is in the thousands place. Consider the next digit to the right. Since the digit, 3, is 4 or lower, round down, meaning that 5 thousands stays as 5 thousands. Then change all the digits to the right of the thousands digit to zeros.

The answer is 45,000.

23. Round 373,405 to the nearest thousand.

3 7 3, [4] 0 5

↑

The digit 3 is in the thousands place. Consider the next digit to the right. Since the digit, 4, is 4 or lower, round down, meaning that 3 thousands stays as 3 thousands. Then change all the digits to the right of the thousands digit to zeros.

The answer is 373,000.

25.

	Rounded to the nearest ten
7 8 2 8	7 8 3 0
+ 9 7 8 6	+ 9 7 9 0
	1 7, 6 2 0 ← Estimated answer

27.

	Rounded to the nearest ten
8 0 7 4	8 0 7 0
− 2 3 4 7	− 2 3 5 0
	5 7 2 0 ← Estimated answer

29.

	Rounded to the nearest ten
4 5	5 0
7 7	8 0
2 5	3 0
+ 5 6	+ 6 0
3 4 3	2 2 0 ← Estimated answer

The sum 343 seems to be incorrect since 220 is not close to 343.

31.

	Rounded to the nearest ten
6 2 2	6 2 0
7 8	8 0
8 1	8 0
+ 1 1 1	+ 1 1 0
9 3 2	8 9 0 ← Estimated answer

The sum 932 seems to be incorrect since 890 is not close to 932.

33.

	Rounded to the nearest hundred
7 3 4 8	7 3 0 0
+ 9 2 4 7	+ 9 2 0 0
	1 6, 5 0 0 ← Estimated answer

35.

	Rounded to the nearest hundred
6 8 5 2	6 9 0 0
− 1 7 4 8	− 1 7 0 0
	5 2 0 0 ← Estimated answer

37.

	Rounded to the nearest hundred
2 1 6	2 0 0
8 4	1 0 0
7 4 5	7 0 0
+ 5 9 5	+ 6 0 0
1 6 4 0	1 6 0 0 ← Estimated answer

The sum 1640 seems to be correct since 1600 is close to 1640.

39.

	Rounded to the nearest hundred
7 5 0	8 0 0
4 2 8	4 0 0
6 3	1 0 0
+ 2 0 5	+ 2 0 0
1 4 4 6	1 5 0 0 ← Estimated answer

The sum 1446 seems to be correct since 1500 is close to 1446.

41.

	Rounded to the nearest thousand
9 6 4 3	1 0,0 0 0
4 8 2 1	5 0 0 0
8 9 4 3	9 0 0 0
+ 7 0 0 4	+ 7 0 0 0
	3 1,0 0 0 ← Estimated answer

43.

	Rounded to the nearest thousand
9 2,1 4 9	9 2,0 0 0
− 2 2,5 5 5	− 2 3,0 0 0
	6 9,0 0 0 ← Estimated answer

45.

Since 0 is to the left of 17, $0 < 17$.

47.

Since 34 is to the right of 12, $34 > 12$.

49.

Since 1000 is to the left of 1001, $1000 < 1001$.

51.

Since 133 is to the right of 132, $133 > 132$.

53.

Since 460 is to the right of 17, $460 > 17$.

55.

Since 37 is to the right of 11, $37 > 11$.

57.
$$
\begin{array}{r}
^{1\ \ 1\ \ 1\ \ 1} \\
6\ 7,7\ 8\ 9 \\
+\ 1\ 8,9\ 6\ 5 \\
\hline
8\ 6,7\ 5\ 4
\end{array}
$$
Add ones. We get 14. Write 4 in the ones column and 1 above the tens. Add tens: We get 15 tens. Write 5 in the tens column and 1 above the hundreds. Add hundreds: We get 17 hundreds. Write 7 in the hundreds column and 1 above the thousands. Add thousands: We get 16 thousands. Write 6 in the thousands column and 1 above the ten thousands. Add ten thousands: We get 8 ten thousands.

59.
$$
\begin{array}{r}
^{16} \\
^{5\ \ \not{6}\ \ 17} \\
\not{6}\ 7,7\ 8\ 9 \\
-\ 1\ 8,9\ 6\ 5 \\
\hline
4\ 8,8\ 2\ 4
\end{array}
$$
Subtract ones: We get 4. Subtract tens: We get 2. We cannot subtract 9 hundreds from 7 hundreds. We borrow 1 thousand to get 17 hundreds. Subtract hundreds. We cannot subtract 8 thousands from 6 thousands. We borrow 1 ten thousand to get 16 thousands. Subtract thousands, then ten thousands.

61. Using a calculator, we find that the sum is 30,411.

63. Using a calculator, we find that the difference is 69,594.

Exercise Set 1.5

1. Repeated addition fits best in this case.

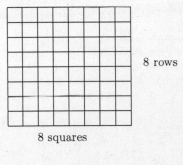

$$32 \cdot \$10 = \$320$$

3. We have a rectangular array.

8 rows

8 squares

$$8 \cdot 8 = 64$$

5. If we think of filling the rectangle with square feet, we have a rectangular array.

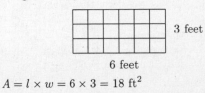

3 feet

6 feet

$$A = l \times w = 6 \times 3 = 18 \text{ ft}^2$$

7.
$$\begin{array}{r} 87 \\ \times 10 \\ \hline 870 \end{array}$$
Multiplying by 1 ten (We write 0 and then multiply 87 by 1.)

9.
$$\begin{array}{r} 2340 \\ \times 1000 \\ \hline 2,340,000 \end{array}$$
Multiplying by 1 thousand (We write 000 and then multiply 2340 by 1.)

11.
$$\begin{array}{r} 4 \\ 65 \\ \times 8 \\ \hline 520 \end{array}$$
Multiplying by 8

13.
$$\begin{array}{r} 2 \\ 94 \\ \times 6 \\ \hline 564 \end{array}$$
Multiplying by 6

15.
$$\begin{array}{r} 2 \\ 509 \\ \times 3 \\ \hline 1527 \end{array}$$
Multiplying by 3

17.
$$\begin{array}{r} 126 \\ 9229 \\ \times 7 \\ \hline 64,603 \end{array}$$
Multiplying by 7

19.
$$\begin{array}{r} 53 \\ \times 90 \\ \hline 4770 \end{array}$$
Multiplying by 9 tens (We write 0 and then multiply 53 by 9.)

21.
$$\begin{array}{r} 2 \\ 3 \\ 85 \\ \times 47 \\ \hline 595 \\ 3400 \\ \hline 3995 \end{array}$$
Multiplying by 7
Multiplying by 40
Adding

23.
$$\begin{array}{r} 2 \\ 640 \\ \times 72 \\ \hline 1280 \\ 44800 \\ \hline 46,080 \end{array}$$
Multiplying by 2
Multiplying by 70
Adding

25.
$$\begin{array}{r} 11 \\ 11 \\ 444 \\ \times 33 \\ \hline 1332 \\ 13320 \\ \hline 14,652 \end{array}$$
Multiplying by 3
Multiplying by 30
Adding

27.
$$\begin{array}{r} 3 \\ 7 \\ 509 \\ \times 408 \\ \hline 4072 \\ 203600 \\ \hline 207,672 \end{array}$$
Multiplying by 8
Multiplying by 4 hundreds (We write 00 and then multiply 509 by 4.)

29.
$$\begin{array}{r} 42 \\ 1 \\ 31 \\ 853 \\ \times 936 \\ \hline 5118 \\ 25590 \\ 767700 \\ \hline 798,408 \end{array}$$
Multiplying by 6
Multiplying by 30
Multiplying by 900
Adding

31.
$$\begin{array}{r} 22 \\ 33 \\ 489 \\ \times 340 \\ \hline 19560 \\ 146700 \\ \hline 166,260 \end{array}$$
← Multiplying by 4 tens (We write 0 and then multiply 489 by 4.)
← Multiplying by 3 hundreds (We write 00 and then multiply 489 by 3.)

33.
$$\begin{array}{r} 11 \\ 244 \\ 377 \\ 133 \\ 4378 \\ \times 2694 \\ \hline 17512 \\ 394020 \\ 2626800 \\ 8756000 \\ \hline 11,794,332 \end{array}$$
Multiplying by 4
Multiplying by 90
Multiplying by 600
Multiplying by 2000
Adding

35.
$$\begin{array}{r} 12 \\ 1 \\ 1 \\ 113 \\ 6428 \\ \times 3224 \\ \hline 25712 \\ 128560 \\ 1285600 \\ 19284000 \\ \hline 20,723,872 \end{array}$$
Multiplying by 4
Multiplying by 20
Multiplying by 200
Multiplying by 3000
Adding

37.
$$\begin{array}{r} 13 \\ 3482 \\ \times 104 \\ \hline 13928 \\ 348200 \\ \hline 362,128 \end{array}$$
Multiplying by 4
Multiplying by 1 hundred (We write 00 and then multiply 3482 by 1.)

39.
$$\begin{array}{r} 2 \\ 4 \\ 5006 \\ \times 4008 \\ \hline 40048 \\ 20024000 \\ \hline 20,064,048 \end{array}$$
Multiplying by 8
Multiplying by 4 thousands (We write 000 and then multiply 5006 by 4.)

41.
$$\begin{array}{r} 23 \\ 34 \\ 5608 \\ \times 4500 \\ \hline 2804000 \\ 22432000 \\ \hline 25,236,000 \end{array}$$
Multiplying by 5 hundreds (We write 00 and then multiply 5608 by 5.)
Multiplying by 4000
Adding

43.

```
    2 1
    3 2
    3 3
    8 7 6
  × 3 4 5
    4 3 8 0    Multiplying by 5
  3 5 0 4 0    Multiplying by 40
2 6 2 8 0 0    Multiplying by 300
3 0 2, 2 2 0    Adding
```

45.

```
      5 5 5
      1 1 1
      1 1 1
      3 3 3
      7 8 8 9
    × 6 2 2 4
      3 1 5 5 6    Multiplying by 4
    1 5 7 7 8 0    Multiplying by 20
  1 5 7 7 8 0 0    Multiplying by 200
4 7 3 3 4 0 0 0    Multiplying by 6000
4 9, 1 0 1, 1 3 6    Adding
```

47.

```
    2 2
    2 2
    5 5 5
  ×   5 5
    2 7 7 5    Multiplying by 5
  2 7 7 5 0    Multiplying by 50
  3 0, 5 2 5    Adding
```

49.

```
    1 1
    2 2
    7 3 4
  × 4 0 7
    5 1 3 8    Multiplying by 7
  2 9 3 6 0 0    Multiplying by 4 hundreds (We write 00
  2 9 8, 7 3 8    and then multiply 734 by 4.)
```

51.

Rounded to
the nearest ten

```
    4 5          5 0
  × 6 7        × 7 0
             3 5 0 0 ← Estimated answer
```

53.

Rounded to
the nearest ten

```
    3 4          3 0
  × 2 9        × 3 0
               9 0 0 ← Estimated answer
```

55.

Rounded to
the nearest hundred

```
    8 7 6          9 0 0
  × 3 4 5        × 3 0 0
             2 7 0, 0 0 0 ← Estimated answer
```

57.

Rounded to
the nearest hundred

```
    4 3 2          4 0 0
  × 1 9 9        × 2 0 0
             8 0, 0 0 0 ← Estimated answer
```

59.

Rounded to
the nearest thousand

```
    5 6 0 8          6 0 0 0
  × 4 5 7 6        × 5 0 0 0
             3 0, 0 0 0, 0 0 0 ← Estimated
                                    answer
```

61.

Rounded to
the nearest thousand

```
    7 8 8 8          8 0 0 0
  × 6 2 2 4        × 6 0 0 0
             4 8, 0 0 0, 0 0 0 ← Estimated
                                    answer
```

63.

```
      1
      2 0
      8 5 0
  + 3 5 0 0
    4 3 7 0
```

Add ones: We get 0. Add tens: We get 7 tens. Add hundreds: We get 13 hundreds. Write 3 in the hundreds column and 1 above the thousands. Add thousands: We get 4 thousands.

65. Round 2 3 4 ⑤ to the nearest ten.

The digit 4 is in the tens place. Since the next digit to the right, 5, is 5 or higher, round 4 tens up to 5 tens. Then change the digit to the right of the tens digit to zero.

The answer is 2350.

Round 2 3 ④ 5 to the nearest hundred.

The digit 3 is in the hundreds place. Since the next digit to the right, 4, is 4 or lower, round down, meaning 3 hundreds stays as 3 hundreds. Then change all digits to the right of the hundreds digit to zeros.

The answer is 2300.

Round 2 ③ 4 5 to the nearest thousand.

The digit 2 is in the thousands place. Since the next digit to the right, 3, is 4 or lower, round down, meaning 2 thousands stays as 2 thousands. Then change all digits to the right of the thousands digit to zeros.

The answer is 2000.

Exercise Set 1.6

1. Think of an array with 4 lb in each row. The cheese in each row will go in a box.

```
176 ÷ 4 = [    ]
```

3. Think of an array with $23,000 in each row. The money in each row will go to a child.

23,000 in each row

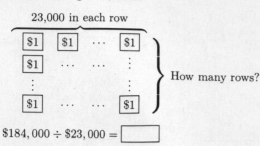

How many rows?

$184,000 \div $23,000 = \boxed{}$

5. $18 \div 3 = 6$ The 3 moves to the right. A related multiplication sentence is $18 = 6 \cdot 3$. (By the commutative law of multiplication, there is also another multiplication sentence: $18 = 3 \cdot 6$.)

7. $22 \div 22 = 1$ The 22 on the right of the ÷ symbol moves to the right. A related multiplication sentence is $22 = 1 \cdot 22$. (By the commutative law of multiplication, there is also another multiplication sentence: $22 = 22 \cdot 1$.)

9. $54 \div 6 = 9$ The 6 moves to the right. A related multiplication sentence is $54 = 9 \cdot 6$. (By the commutative law of multiplication, there is also another multiplication sentence: $54 = 6 \cdot 9$.)

11. $37 \div 1 = 37$ The 1 moves to the right. A related multiplication sentence is $37 = 37 \cdot 1$. (By the commutative law of multiplication, there is also another multiplication sentence: $37 = 1 \cdot 37$.)

13. $9 \times 5 = 45$

Move a factor to the other side and then write a division.

$9 \times 5 = 45$ $9 \times 5 = 45$

$9 = 45 \div 5$ $5 = 45 \div 9$

15. Two related division sentences for $37 \cdot 1 = 37$ are:

$37 = 37 \div 1$ $(37 \cdot 1 = 37)$

and

$1 = 37 \div 37$ $(37 \cdot 1 = 37)$

17. $8 \times 8 = 64$

Since the factors are both 8, moving either one to the other side gives the related division sentence $8 = 64 \div 8$.

19. Two related division sentences for $11 \cdot 6 = 66$ are:

$11 = 66 \div 6$ $(11 \cdot 6 = 66)$

and

$6 = 66 \div 11$ $(11 \cdot 6 = 66)$

21.
```
      5 5
  5 ) 2 7 7
    2 5 0
      2 7
      2 5
        2
```
Think: 2 hundreds ÷ 5. There are no hundreds in the quotient.
Think: 27 tens ÷ 5. Estimate 5 tens.
Think: 27 ones ÷ 5. Estimate 5 ones.

The answer is 55 R 2.

23.
```
      1 0 8
  8 ) 8 6 4
    8 0 0
      6 4
      6 4
        0
```
Think: 8 hundreds ÷ 8. Estimate 1 hundred.
Think: 6 tens ÷ 8. There are no tens in the quotient (other than the tens in 100). Write a 0 to show this.
Think: 64 ones ÷ 8. Estimate 8 ones.

The answer is 108.

25.
```
      3 0 7
  4 ) 1 2 2 8
    1 2 0 0
        2 8
        2 8
          0
```
Think: 12 hundreds ÷ 4. Estimate 3 hundreds.
Think: 2 tens ÷ 4. There are no tens in the quotient (other than the tens in 300). Write a 0 to show this.
Think: 28 ones ÷ 4. Estimate 7 ones.

The answer is 307.

27.
```
      7 5 3
  6 ) 4 5 2 1
    4 2 0 0
      3 2 1
      3 0 0
        2 1
        1 8
          3
```
Think: 45 hundreds ÷ 6. Estimate 7 hundreds.
Think: 32 tens ÷ 6. Estimate 5 tens.
Think: 21 ones ÷ 6. Estimate 3 ones.

The answer is 753 R 3.

29.
```
      7 4
  4 ) 2 9 7
    2 8 0
      1 7
      1 6
        1
```
Think: 29 tens ÷ 4. Estimate 7 tens.
Think: 17 ones ÷ 4. Estimate 4 ones.

The answer is 74 R 1.

31.
```
      9 2
  8 ) 7 3 8
    7 2 0
      1 8
      1 6
        2
```
Think: 73 tens ÷ 8. Estimate 9 tens.
Think: 18 ones ÷ 8. Estimate 2 ones.

The answer is 92 R 2.

33.
```
      1 7 0 3
  5 ) 8 5 1 5
    5 0 0 0
    3 5 1 5
    3 5 0 0
        1 5
        1 5
          0
```
Think: 8 thousands ÷ 5. Estimate 1 thousand.
Think: 35 hundreds ÷ 5. Estimate 7 hundreds.
Think: 1 ten ÷ 5. There are no tens in the quotient (other than the tens in 1700). Write a 0 to show this.
Think: 15 ones ÷ 5. Estimate 3 ones.

The answer is 1703.

35.
```
      9 8 7
  9 | 8 8 8 8
    8 1 0 0
    ‾‾‾‾‾‾‾
      7 8 8
      7 2 0
      ‾‾‾‾‾
        6 8
        6 3
        ‾‾‾
          5
```
Think: 88 hundreds ÷ 9. Estimate 9 hundreds.

Think: 78 ten ÷ 9. Estimate 8 tens.

Think: 68 ones ÷ 9. Estimate 7 ones.

The answer is 987 R 5.

37.
```
         5 2
  7 0 | 3 6 9 2
       3 5 0 0
       ‾‾‾‾‾‾‾
         1 9 2
         1 4 0
         ‾‾‾‾‾
           5 2
```
Think: 369 tens ÷ 70. Estimate 5 tens.

Think: 192 ones ÷ 70. Estimate 2 ones.

The answer is 52 R 52.

39.
```
        2 9
  3 0 | 8 7 5
       6 0 0
       ‾‾‾‾‾
       2 7 5
       2 7 0
       ‾‾‾‾‾
           5
```
Think: 87 tens ÷ 30. Estimate 2 tens.

Think: 275 ones ÷ 30. Estimate 9 ones.

The answer is 29 R 5.

41.
```
        4 0
  2 1 | 8 5 2
       8 4 0
       ‾‾‾‾‾
         1 2
```
Round 21 to 20.

Think: 85 tens ÷ 20. Estimate 4 tens.

Think: 12 ones ÷ 20. There are no ones in the quotient (other than the ones in 40). Write a 0 to show this.

The answer is 40 R 12.

43.
```
           8
  8 5 | 7 6 7 2
       6 8 0 0
       ‾‾‾‾‾‾‾
        |8 7| 2
```
Round 85 to 90.

Think: 767 tens ÷ 90. Estimate 8 tens.

Since 87 is larger than the divisor, the estimate is too low.

```
           9 0
  8 5 | 7 6 7 2
       7 6 5 0
       ‾‾‾‾‾‾‾
             2 2
```
Think: 767 tens ÷ 90. Estimate 9 tens.

Think: 22 ones ÷ 90. There are no ones in the quotient (other than the ones in 90). Write a 0 to show this.

The answer is 90 R 22.

45.
```
              3
  1 1 1 | 3 2 1 9
         3 3 3 0
```
Round 111 to 100.

Think: 321 tens ÷ 100. Estimate 3 tens.

Since we cannot subtract 3330 from 3219, the estimate is too high.

```
             2 9
  1 1 1 | 3 2 1 9
         2 2 2 0
         ‾‾‾‾‾‾‾
           9 9 9
           9 9 9
           ‾‾‾‾‾
               0
```
Think: 321 tens ÷ 100. Estimate 2 tens.

Think: 999 ones ÷ 100. Estimate 9 ones.

The answer is 29.

47.
```
        1 0 5
  8 | 8 4 3
    8 0 0
    ‾‾‾‾‾
      4 3
      4 0
      ‾‾‾
        3
```
Think: 8 hundreds ÷ 8. Estimate 1 hundred.

Think: 4 tens ÷ 8. There are no tens in the quotient (other than the tens in 100). Write a 0 to show this.

Think: 43 ones ÷ 8. Estimate 5 ones.

The answer is 105 R 3.

49.
```
        1 6 0 9
  5 | 8 0 4 7
    5 0 0 0
    ‾‾‾‾‾‾‾
    3 0 4 7
    3 0 0 0
    ‾‾‾‾‾‾‾
        4 7
        4 5
        ‾‾‾
          2
```
Think: 8 thousands ÷ 5. Estimate 1 thousand.

Think: 30 hundreds ÷ 5. Estimate 6 hundreds.

Think: 4 tens ÷ 5. There are no tens in the quotient (other than the tens in 1600). Write a 0 to show this.

Think: 47 ones ÷ 5. Estimate 9 ones.

The answer is 1609 R 2.

51.
```
        1 0 0 7
  5 | 5 0 3 6
    5 0 0 0
    ‾‾‾‾‾‾‾
        3 6
        3 5
        ‾‾‾
          1
```
Think: 5 thousands ÷ 5. Estimate 1 thousand.

Think: 0 hundreds ÷ 5. There are no hundreds in the quotient (other than the hundreds in 1000). Write a 0 to show this.

Think: 3 tens ÷ 5. There are no tens in the quotient (other than the tens in 1000). Write a 0 to show this.

Think: 36 ones ÷ 5. Estimate 7 ones.

The answer is 1007 R 1.

53.
```
           2 2
  4 6 | 1 0 5 8
         9 2 0
         ‾‾‾‾‾
         1 3 8
           9 2
           ‾‾‾
           4 6
```
Round 46 to 50.

Think: 105 tens ÷ 50. Estimate 2 tens.

Think: 138 ones ÷ 50. Estimate 2 ones.

Since 46 is not smaller than the divisor, 46, the estimate is too low.

```
           2 3
  4 6 | 1 0 5 8
         9 2 0
         ‾‾‾‾‾
         1 3 8
         1 3 8
         ‾‾‾‾‾
             0
```
Think: 138 ones ÷ 50. Estimate 3 ones.

The answer is 23.

55.
```
        1 0 7
  3 2 | 3 4 2 5
       3 2 0 0
       ‾‾‾‾‾‾‾
         2 2 5
         2 2 4
         ‾‾‾‾‾
             1
```
Round 32 to 30.

Think: 34 hundreds ÷ 30. Estimate 1 hundred.

Think: 22 tens ÷ 30. There are no tens in the quotient (other than the tens in 100). Write 0 to show this.

Think: 225 ones ÷ 30. Estimate 7 ones.

The answer is 107 R 1.

57.

$$\begin{array}{r} 4 \\ 24\overline{)8880} \\ 9600 \end{array}$$

Round 24 to 20.
Think: 88 hundreds ÷ 20. Estimate 4 hundreds.

Since we cannot subtract 9600 from 8880, the estimate is too high.

$$\begin{array}{r} 38 \\ 24\overline{)8880} \\ 7200 \\ \hline 1680 \\ 1920 \end{array}$$

Think: 88 hundreds ÷ 20. Estimate 3 hundreds.

Think: 168 tens ÷ 20. Estimate 8 tens.

Since we cannot subtract 1920 from 1680, the estimate is too high.

$$\begin{array}{r} 370 \\ 24\overline{)8880} \\ 7200 \\ \hline 1680 \\ 1680 \\ \hline 0 \end{array}$$

Think: 168 tens ÷ 20. Estimate 7 tens.

Think: 0 ones ÷ 20. There are no ones in the quotient (other than the ones in 370). Write a 0 to show this.

The answer is 370.

59.

$$\begin{array}{r} 5 \\ 28\overline{)17,067} \\ 14000 \\ \hline \boxed{30}67 \end{array}$$

Round 28 to 30.
Think: 170 hundreds ÷ 30. Estimate 5 hundreds.

Since 30 is larger than the divisor, 28, the estimate is too low.

$$\begin{array}{r} 608 \\ 28\overline{)17,067} \\ 16800 \\ \hline 267 \\ 224 \\ \hline \boxed{43} \end{array}$$

Think: 170 hundreds ÷ 30. Estimate 6 hundreds.

Think: 26 tens ÷ 30. There are no tens in the quotient (other than the tens in 600.) Write a zero to show this.

Think: 267 ones ÷ 30. Estimate 8 ones.

Since 43 is larger than the divisor, 28, the estimate is too low.

$$\begin{array}{r} 609 \\ 28\overline{)17,067} \\ 16800 \\ \hline 267 \\ 252 \\ \hline 15 \end{array}$$

Think: 267 ones ÷ 30. Estimate 9 ones.

The answer is 609 R 15.

61.

$$\begin{array}{r} 304 \\ 80\overline{)24,320} \\ 24000 \\ \hline 320 \\ 320 \\ \hline 0 \end{array}$$

Think: 243 hundreds ÷ 80. Estimate 3 hundreds.

Think: 32 tens ÷ 80. There are no tens in the quotient (other than the tens in 300). Write a 0 to show this.

Think: 320 ones ÷ 80. Estimate 4 ones.

The answer is 304.

63.

$$\begin{array}{r} 3508 \\ 285\overline{)999,999} \\ 855000 \\ \hline 144999 \\ 142500 \\ \hline 2499 \\ 2280 \\ \hline 219 \end{array}$$

The answer is 3508 R 219.

65.

$$\begin{array}{r} 8070 \\ 456\overline{)3,679,920} \\ 3648000 \\ \hline 31920 \\ 31920 \\ \hline 0 \end{array}$$

The answer is 8070.

67. 7882 = 7 thousands + 8 hundreds + 8 tens + 2 ones

69. We divide 1231 by 42:

$$\begin{array}{r} 29 \\ 42\overline{)1231} \\ 840 \\ \hline 391 \\ 378 \\ \hline 13 \end{array}$$

The answer is 29 R 13. Since 13 students will be left after 29 buses are filled, then 30 buses are needed.

Exercise Set 1.7

1. $x + 0 = 14$

We replace x by different numbers until we get a true equation. If we replace x by 14, we get a true equation: $14 + 0 = 14$. No other replacement makes the equation true, so the solution is 14.

3. $y \cdot 17 = 0$

We replace y by different numbers until we get a true equation. If we replace y by 0, we get a true equation: $0 \cdot 17 = 0$. No other replacement makes the equation true, so the solution is 0.

5.
$$\begin{aligned} 13 + x &= 42 \\ 13 + x - 13 &= 42 - 13 && \text{Subtracting 13 on both sides} \\ 0 + x &= 29 && \text{13 plus } x \text{ minus 13 is } 0 + x. \\ x &= 29 \end{aligned}$$

7.
$$\begin{aligned} 12 &= 12 + m \\ 12 - 12 &= 12 + m - 12 && \text{Subtracting 12 on both sides} \\ 0 &= 0 + m && \text{12 plus } m \text{ minus 12 is } 0 + m. \\ 0 &= m \end{aligned}$$

9.
$$\begin{aligned} 3 \cdot x &= 24 \\ \frac{3 \cdot x}{3} &= \frac{24}{3} && \text{Dividing by 3 on both sides} \\ x &= 8 && \text{3 times } x \text{ divided by 3 is } x. \end{aligned}$$

11. $112 = n \cdot 8$

$\dfrac{112}{8} = \dfrac{n \cdot 8}{8}$ Dividing by 8 on both sides

$14 = n$

13. $45 \times 23 = x$

To solve the equation we carry out the calculation.

$$\begin{array}{r} 4\,5 \\ \times\ 2\,3 \\ \hline 1\,3\,5 \\ 9\,0\,0 \\ \hline 1\,0\,3\,5 \end{array}$$

The solution is 1035.

15. $t = 125 \div 5$

To solve the equation we carry out the calculation.

$$\begin{array}{r} 2\,5 \\ 5\overline{)1\,2\,5} \\ 1\,0\,0 \\ \hline 2\,5 \\ 2\,5 \\ \hline 0 \end{array}$$

The solution is 25.

17. $p = 908 - 458$

To solve the equation we carry out the calculation.

$$\begin{array}{r} 9\,0\,8 \\ -\ 4\,5\,8 \\ \hline 4\,5\,0 \end{array}$$

The solution is 450.

19. $x = 12,345 + 78,555$

To solve the equation we carry out the calculation.

$$\begin{array}{r} 1\,2,3\,4\,5 \\ +\ 7\,8,5\,5\,5 \\ \hline 9\,0,9\,0\,0 \end{array}$$

The solution is 90,900.

21. $3 \cdot m = 96$

$\dfrac{3 \cdot m}{3} = \dfrac{96}{3}$ Dividing by 3 on both sides

$m = 32$

23. $715 = 5 \cdot z$

$\dfrac{715}{5} = \dfrac{5 \cdot z}{5}$ Dividing by 5 on both sides

$143 = z$

25. $10 + x = 89$

$10 + x - 10 = 89 - 10$

$x = 79$

27. $61 = 16 + y$

$61 - 16 = 16 + y - 16$

$45 = y$

29. $6 \cdot p = 1944$

$\dfrac{6 \cdot p}{6} = \dfrac{1944}{6}$

$p = 324$

31. $5 \cdot x = 3715$

$\dfrac{5 \cdot x}{5} = \dfrac{3715}{5}$

$x = 743$

33. $47 + n = 84$

$47 + n - 47 = 84 - 47$

$n = 37$

35. $x + 78 = 144$

$x + 78 - 78 = 144 - 78$

$x = 66$

37. $165 = 11 \cdot n$

$\dfrac{165}{11} = \dfrac{11 \cdot n}{11}$

$15 = n$

39. $624 = t \cdot 13$

$\dfrac{624}{13} = \dfrac{t \cdot 13}{13}$

$48 = t$

41. $x + 214 = 389$

$x + 214 - 214 = 389 - 214$

$x = 175$

43. $567 + x = 902$

$567 + x - 567 = 902 - 567$

$x = 335$

45. $18 \cdot x = 1872$

$\dfrac{18 \cdot x}{18} = \dfrac{1872}{18}$

$x = 104$

47. $40 \cdot x = 1800$

$\dfrac{40 \cdot x}{40} = \dfrac{1800}{40}$

$x = 45$

49. $2344 + y = 6400$

$2344 + y - 2344 = 6400 - 2344$

$y = 4056$

51. $8322 + 9281 = x$

$17,603 = x$ Doing the addition

53. $234 \cdot 78 = y$

$18,252 = y$ Doing the multiplication

55. $58 \cdot m = 11,890$

$\dfrac{58 \cdot m}{58} = \dfrac{11,890}{58}$

$m = 205$

57. $x \cdot 198 = 10,890$

$\dfrac{x \cdot 198}{198} = \dfrac{10,890}{198}$

$x = 55$

59. Two related division sentences for $6 \cdot 8 = 48$ are:

$6 = 48 \div 8$ $\qquad (\ 6 \cdot 8 = 48 \ )$

and

$8 = 48 \div 6$ $\qquad (\ 6 \cdot 8 = 48 \ )$

61.
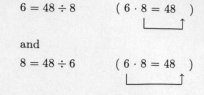

Since 342 is to the right of 339, $342 > 339$.

Exercise Set 1.8

1. a) First we find how many hits Ty Cobb and Pete Rose got together.

Familiarize. We visualize the situation. We let $n =$ the number of hits they got together.

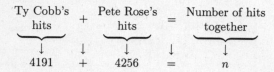

Translate. We write the number sentence that corresponds to the situation:

$4191 + 4256 = n$

Solve. We carry out the addition:

$$\begin{array}{r} \overset{1}{} \\ 4\,1\,9\,1 \\ +\,4\,2\,5\,6 \\ \hline 8\,4\,4\,7 \end{array}$$

Check. We can repeat the calculation. We note that the answer seems reasonable since it is larger than either of the separate hit totals. We can also find an estimated answer by rounding: $4191 + 4256 \approx 4200 + 4300 = 8500 \approx 8447$. The answer checks.

State. Ty Cobb and Pete Rose got 8447 hits together.

Next we find how many more hits Rose got than Cobb.

Familiarize. We visualize the situation. We let $h =$ the number by which Rose's hits exceed Cobb's.

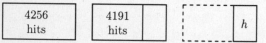

Translate. We see that this is a "take-away" situation. We write the equation:

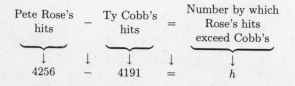

Solve. We carry out the subtraction:

$$\begin{array}{r} \overset{1}{}\,\overset{15}{} \\ 4\,\cancel{2}\,\cancel{5}\,6 \\ -\,4\,1\,9\,1 \\ \hline 6\,5 \end{array}$$

Thus, $65 = h$, or $h = 65$.

Check. We can repeat the calculation. We note that the answer seems reasonable since it is less than the number of Pete Rose's hits. We can add the answer to 4191, the number being subtracted: $4191 + 65 = 4256$. We can also estimate: $4256 - 4191 \approx 4300 - 4200 = 100 \approx 65$. The answer checks.

State. Pete Rose got 65 more hits than Ty Cobb.

b) First we find how many hits Hank Aaron and Stan Musial got together.

Familiarize. We visualize the situation. We let $n =$ the number of hits they got together.

Hank Aaron's hits $+$ Stan Musial's hits $=$ Number of hits together

$3771 \quad + \quad 3630 \quad = \quad n$

Translate. We write the number sentence that corresponds to the situation:

$3771 + 3630 = n$

Solve. We carry out the addition:

$$\begin{array}{r} \overset{1}{}\,\overset{1}{} \\ 3\,7\,7\,1 \\ +\,3\,6\,3\,0 \\ \hline 7\,4\,0\,1 \end{array}$$

Check. We can repeat the calculation. We note that the answer seems reasonable since it is larger than either of the separate hit totals. We can also find an estimated answer by rounding: $3771 + 3630 \approx 3800 + 3700 = 7500 \approx 7401$. The answer checks.

State. Hank Aaron and Stan Musial got 7401 hits together.

Next we find how many more hits Aaron got than Musial.

Familiarize. We visualize the situation. We let $h =$ the number by which Aaron's hits exceed Musial's.

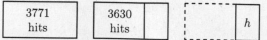

Translate. We see that this is a "take-away" situation. We write the equation:

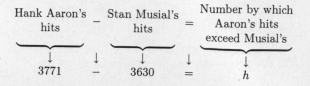

Solve. We carry out the subtraction:

$$\begin{array}{r} 3\,7\,7\,1 \\ -\,3\,6\,3\,0 \\ \hline 1\,4\,1 \end{array}$$

Thus, $141 = h$, or $h = 141$.

Check. We can repeat the calculation. We note that the answer seems reasonable since it is less than the number of Hank Aaron's hits. We can also add the answer to 3630, the number being subtracted: $3630 + 141 = 3771$. We can also estimate: $3771 - 3630 \approx 3800 - 3600 = 200 \approx 141$. The answer checks.

State. Hank Aaron got 141 more hits than Stan Musial.

3. *Familiarize*. We first make a drawing.

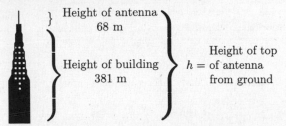

Since we are combining lengths, addition can be used. We let $h =$ the height of the top of the antenna from the ground.

Translate. We translate to the following addition sentence:

$$381 + 68 = h$$

Solve. To solve we carry out the addition.

$$\begin{array}{r} 3\,8\,1 \\ +\,6\,8 \\ \hline 4\,4\,9 \end{array}$$

Check. We can repeat the calculation. We can also find an estimated answer by rounding: $381 + 68 \approx 380 + 70 = 450 \approx 449$. The answer checks.

State. The top of the antenna is 449 m from the ground.

5. *Familiarize*. We first make a drawing.

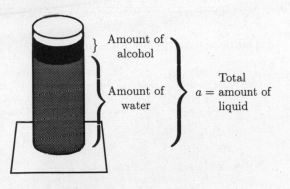

Since we are combining amounts, addition can be used. We let $a =$ the total amount of liquid.

Translate. We translate to the following addition sentence:

$$2340 + 655 = a$$

Solve. To solve we carry out the addition.

$$\begin{array}{r} 2\,3\,4\,0 \\ +\,6\,5\,5 \\ \hline 2\,9\,9\,5 \end{array}$$

Check. We can repeat the calculation. We can also find an estimated answer by rounding: $2340 + 655 \approx 2300 + 700 = 3000 \approx 2995$. The answer checks.

State. A total of 2995 cubic centimeters of liquid was poured.

7. *Familiarize*. This is a multistep problem. We visualize the situation. We let $p =$ the number of megabytes of memory used by the three programs that have been installed and $m =$ the number of megabytes of memory left over.

Word processor: 15 megabytes	Data-base: 6 megabytes	Spread-sheet: 12 megabytes	Remaining memory
p			m
170 megabytes			

Translate. First we write a number sentence that corresponds to the number of megabytes of memory that have been used:

$$15 + 6 + 12 = p$$

Solve. We carry out the addition.

$$\begin{array}{r} 1\,5 \\ 6 \\ +\,1\,2 \\ \hline 3\,3 \end{array}$$

Thus $33 = p$, or $p = 33$.

Now we must find the remaining memory. We have a "take-away" situation. We write the corresponding number sentence.

$$170 - 33 = m$$
$$137 = m \quad \text{Carrying out the subtraction}$$

Check. Since $15 + 6 + 12 + 137 = 170$, the answer checks.

State. There are 137 megabytes of memory left on the computer.

9. *Familiarize*. We visualize the situation. We let $p =$ the number by which the population of the Tokyo-Yokohama area exceeds the population of the New York-northeastern New Jersey area.

NY – NJ population 18,087,521	Excess
	p
Tokyo – Yokohama population 29,971,000	

Translate. This is a "how-much-more" situation. Translate to an equation.

People in New York-north-eastern New Jersey area	plus	How many more people	is	People in Tokyo-Yokohama area
↓	↓	↓	↓	↓
$18,087,521$	$+$	p	$=$	$29,971,000$

Solve. We solve the equation.

$$18,087,521 + p = 29,971,000$$
$$18,087,521 + p - 18,087,521 = 29,971,000 - 18,087,521$$
$$\text{Subtracting 18,087,521 on both sides}$$
$$p = 11,883,479$$

Check. We can add the result to $18,087,521$:

$$18,087,521 + 11,883,479 = 29,971,000.$$

We can also estimate:

$$18,087,521 + 11,883,479 \approx 18,000,000 + 12,000,000 =$$
$$30,000,000 \approx 29,971,000$$

The answer checks.

State. There are $11,883,479$ more people in the Tokyo-Yokohama area.

11. *Familiarize*. This is a multistep problem.

To find the new balance, find the total amount of the three checks. Then take that amount away from the original balance. We visualize the situation. We let $t =$ the total amount of the three checks and $n =$ the new balance.

	$246		
$45	$78	$32	n
	t		n

Translate. To find the total amount of the three checks, write an addition sentence.

Amount of first check	+	Amount of second check	+	Amount of third check	=	Total amount
↓	↓	↓	↓	↓	↓	↓
$45	$+$	$78	$+$	$32	$=$	t

Solve. To solve the equation we carry out the addition.

$$\begin{array}{r} \$\ 4\ 5 \\ 7\ 8 \\ +\ 3\ 2 \\ \hline \$1\ 5\ 5 \end{array}$$

The total amount of the three checks is $155.

To find the new balance, we have a "take-away" situation.

Original balance	−	Amount of checks	=	New balance
↓	↓	↓	↓	↓
$246	$-$	$155	$=$	n

To solve we carry out the subtraction.

$$\begin{array}{r} \$\ 2\ 4\ 6 \\ -\ 1\ 5\ 5 \\ \hline \$\ \ 9\ 1 \end{array}$$

Check. We add the amounts of the three checks and the new balance. The sum should be the original balance.

$$\$45 + \$78 + \$32 + \$91 = \$246$$

The answer checks

State. The new balance is $91.

13. *Familiarize*. We first make a drawing. Repeated addition works well here. We let $n =$ the number of calories burned in 5 hours.

133 calories	133 calories	133 calories	133 calories	133 calories
1	2	3	4	5

Translate. Translate to a number sentence.

Number of hours	times	Number of calories	is	Total calories burned
↓	↓	↓	↓	↓
5		133	$=$	n

Solve. To solve the equation, we carry out the multiplication.

$$5 \cdot 133 = n$$
$$665 = n \quad \text{Doing the multiplication}$$

Check. We can repeat the calculation. We can also estimate: $5 \cdot 133 \approx 5 \cdot 130 = 650 \approx 665$. The answer checks.

State. In 5 hours, 665 calories would be burned.

15. *Familiarize*. We first make a drawing. Repeated addition works well here. We let $x =$ the number of seconds in an hour.

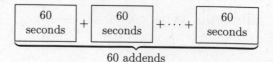

60 seconds	+	60 seconds	$+ \cdots +$	60 seconds

60 addends

Translate. We translate to an equation.

Number of minutes in an hour	times	Number of seconds in a minute	is	Number of seconds in an hour
↓	↓	↓	↓	↓
60		60	$=$	x

Solve. We carry out the multiplication.

$$60 \cdot 60 = x$$
$$3600 = x \quad \text{Doing the multiplication}$$

Check. We can check by repeating the calculation. The answer checks.

State. There are 3600 seconds in an hour.

17. Familiarize. We first make a drawing.

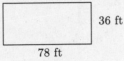
36 ft

78 ft

Translate. Using the formula for area, we have $A = l \cdot w = 78 \cdot 36$.

Solve. Carry out the multiplication.

$$
\begin{array}{r}
7\ 8 \\
\times\ 3\ 6 \\
\hline
4\ 6\ 8 \\
2\ 3\ 4\ 0 \\
\hline
2\ 8\ 0\ 8
\end{array}
$$

Thus $A = 2808$.

Check. We repeat the calculation. The answer checks.

State. The area is 2808 ft^2.

19. Familiarize. We visualize the situation. We let $d =$ the diameter of the earth.

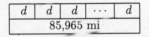

Translate. Repeated addition applies here. The following multiplication corresponds to the situation.

The diameter of Jupiter	is	11	times	the diameter of the earth
↓	↓	↓	↓	↓
85,965	=	11	·	d

Solve. To solve the equation, we divide by 11 on both sides.

$$85,965 = 11 \cdot d$$
$$\frac{85,965}{11} = \frac{11 \cdot d}{11}$$
$$7815 = d$$

Check. To check, we multiply 7815 by 11:

$$11 \cdot 7815 = 85,965$$

This checks.

State. The diameter of the earth is 7815 mi.

21. Familiarize. This is a multistep problem.

We must find the total cost of the 8 suits and the total cost of the 3 shirts. The amount spent is the sum of these two totals.

Repeated addition works well in finding the total cost of the 8 suits and the total cost of the 3 shirts. We let $x =$ the total cost of the suits, and $y =$ the total cost of the shirts.

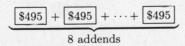
8 addends

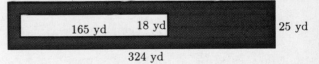
3 addends

Translate. We translate to two equations.

Number of suits	times	Cost per suit	is	Total cost of suits
↓	↓	↓	↓	↓
8	·	495	=	x

Number of shirts	times	Cost per shirt	is	Total cost of shirts
↓	↓	↓	↓	↓
3	·	46	=	y

Solve. To solve these equations, we carry out the multiplications.

$$
\begin{array}{r}
4\ 9\ 5 \\
\times\ \ \ 8 \\
\hline
3\ 0\ 0\ 0
\end{array}
$$ Thus, $x = \$3960$.

$$
\begin{array}{r}
4\ 6 \\
\times\ \ 3 \\
\hline
1\ 3\ 8
\end{array}
$$ Thus $y = \$138$.

We let $a =$ the total amount spent.

Total cost of suits	plus	Total cost of shirts	is	Amount spent
↓	↓	↓	↓	↓
$3960	+	$138	=	a

To solve the equation, carry out the addition.

$$
\begin{array}{r}
\$\ 3\ 9\ 6\ 0 \\
+\ \ \ 1\ 3\ 8 \\
\hline
\$\ 4\ 0\ 9\ 8
\end{array}
$$

Check. We repeat the calculation. The answer checks.

State. The amount spent is $4098.

23. Familiarize. This is a multistep problem.

We must find the area of the lot and the area of the garden. Then we take the area of the garden away from the area of the lot. We let $A =$ the area of the lot and $G =$ the area of the garden.

First, make a drawing of the lot with the garden. The area left over is shaded.

165 yd   18 yd   25 yd

324 yd

Translate. We use the formula for area twice.

$$A = l \cdot w = 324 \cdot 25$$
$$G = l \cdot w = 165 \cdot 18$$

Solve. We carry out the multiplications.

$$
\begin{array}{r}
3\,2\,4 \\
\times\ \ 2\,5 \\
\hline
1\,6\,2\,0 \\
6\,4\,8\,0 \\
\hline
8\,1\,0\,0
\end{array}
$$

$A = 8100$ yd^2

$$
\begin{array}{r}
1\,6\,5 \\
\times\ \ \ 1\,8 \\
\hline
1\,3\,2\,0 \\
1\,6\,5\,0 \\
\hline
2\,9\,7\,0
\end{array}
$$

$G = 2970$ yd^2

To find the area left over we have a "take-away" situation. We let $a =$ the area left over.

Area of lot	minus	Area of garden	is	Area left over
$\downarrow$	$\downarrow$	$\downarrow$	$\downarrow$	$\downarrow$
8100	$-$	2970	$=$	a

To solve we carry out the subtraction .

$$
\begin{array}{r}
8\,1\,0\,0 \\
-\,2\,9\,7\,0 \\
\hline
5\,1\,3\,0
\end{array}
$$

Check. We repeat the calculations. The answer checks.

State. The area left over is 5130 yd^2.

25. *Familiarize*. We first draw a picture. We let $n =$ the number of bottles to be filled.

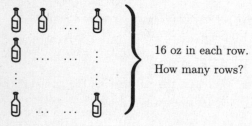

16 oz in each row.
How many rows?

Translate and Solve. We translate to an equation and solve as follows:

$$608 \div 16 = n$$

$$
\begin{array}{r}
3\,8 \\
1\,6\,\overline{\smash{)}\,6\,0\,8} \\
4\,8\,0 \\
\hline
1\,2\,8 \\
1\,2\,8 \\
\hline
0
\end{array}
$$

Check. We can check by multiplying the number of bottles by 16: $16 \cdot 38 = 608$. The answer checks.

State. 38 sixteen-oz bottles can be filled.

27. *Familiarize*. We first draw a picture. We let $y =$ the number rows.

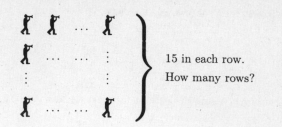

15 in each row.
How many rows?

Translate and Solve. We translate to an equation and solve as follows:

$$225 \div 15 = y$$

$$
\begin{array}{r}
1\,5 \\
1\,5\,\overline{\smash{)}\,2\,2\,5} \\
1\,5\,0 \\
\hline
7\,5 \\
7\,5 \\
\hline
0
\end{array}
$$

Check. We can check by multiplying the number of rows by 15: $15 \cdot 15 = 225$. The answer checks.

State. There are 15 rows.

29. *Familiarize*. We first draw a picture. We let $x =$ the amount of each payment.

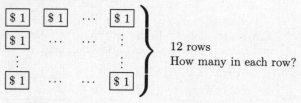

12 rows
How many in each row?

Translate and Solve. We translate to an equation and solve as follows:

$$324 \div 12 = x$$

$$
\begin{array}{r}
2\,7 \\
1\,2\,\overline{\smash{)}\,3\,2\,4} \\
2\,4\,0 \\
\hline
8\,4 \\
8\,4 \\
\hline
0
\end{array}
$$

Check. We can check by multiplying 27 by 12: $12 \cdot 27 = 324$. The answer checks.

State. Each payment is $27.

31. *Familiarize*. We draw a picture. We let $n =$ the number of bags to be filled.

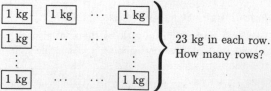

23 kg in each row.
How many rows?

Translate and Solve. We translate to an equation and solve as follows:

$$885 \div 23 = n$$

$$
\begin{array}{r}
3\,8 \\
2\,3\,\overline{\smash{)}\,8\,8\,5} \\
6\,9\,0 \\
\hline
1\,9\,5 \\
1\,8\,4 \\
\hline
1\,1
\end{array}
$$

Check. We can check by multiplying the number of bags by 23 and adding the remainder of 11:

$$23 \cdot 38 = 874$$
$$874 + 11 = 885$$

The answer checks.

State. 38 twenty-three-kg bags can be filled. There will be 11 kg of sand left over.

33. *Familiarize*. This is a multistep problem.

We must find the total price of the 5 coats. Then we must find how many 20's there are in the total price. Let $p =$ the total price of the coats.

To find the total price of the 5 coats we can use repeated addition.

$$\underbrace{\boxed{\$64} + \boxed{\$64} + \boxed{\$64} + \boxed{\$64} + \boxed{\$64}}_{\text{5 addends}}$$

Translate.

Price per coat	times	Number of coats	is	Total price of coats
↓	↓	↓	↓	↓
64	·	5	=	p

Solve. First we carry out the multiplication.

$$64 \cdot 5 = p$$
$$320 = p$$

The total price of the 5 coats is $320. Repeated addition can be used again to find how many 20's there are in $320. We let $x =$ the number of $20 bills required.

$320			
$20	$20	$\cdots$	$20

Translate to an equation and solve.

$$20 \cdot x = 320$$
$$\frac{20 \cdot x}{20} = \frac{320}{20}$$
$$x = 16$$

Check. We repeat the calculation. The answer checks.

State. It took 16 twenty dollar bills.

35. *Familiarize*. To find how far apart on the map the cities are we must find how many 55's there are in 605. Repeated addition applies here. We let $d =$ the distance between the cities on the map.

605 mi			
55 mi	55 mi	$\cdots$	55 mi

Translate.

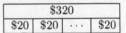

55 mi	times	How many 55's	is	605 mi?
↓	↓	↓	↓	↓
55	·	d	=	605

Solve. We divide on both sides by 55.

$$55 \cdot d = 605$$
$$\frac{55 \cdot d}{55} = \frac{605}{55}$$
$$d = 11$$

Check. We multiply the number of inches by 55: $55 \cdot 11 = 605$. The answer checks.

State. The cities are 11 inches apart on the map.

Familiarize. When two cities are 14 in. apart on the map, we can use repeated addition to find how far apart they are in reality. We let $n =$ the distance between the cities in reality.

Since 1 in. represents 55 mi, we can draw the following picture:

$$\underbrace{\boxed{55 \text{ mi}} + \boxed{55 \text{ mi}} + \cdots + \boxed{55 \text{ mi}}}_{\text{14 addends}}$$

Translate.

Number of inches	times	Number of miles per inch	is	Real distance between cities
↓	↓	↓	↓	↓
14	·	55	=	n

Solve. We carry out the multiplication.

$$14 \times 55 = n$$
$$770 = n \qquad \text{Doing the multiplication}$$

Check. We repeat the calculation. The answer checks.

State. The cities are 770 mi apart in reality.

37. *Familiarize*. This is a multistep problem.

We must first find the total number of ounces of soda. Then we must find how many 16's there are in that total. We let $x =$ the total number of ounces of soda.

To find the total number of ounces of soda we can use repeated addition.

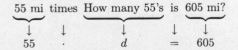

$$\underbrace{\boxed{12} + \boxed{12} + \boxed{12} + \boxed{12} + \boxed{12} + \boxed{12} + \cdots + \boxed{12}}_{\text{640 addends}}$$

Translate.

Number of bottles	times	Ounces per bottle	is	Total number of ounces
↓	↓	↓	↓	↓
640	·	12	=	x

Solve. We carry out the multiplication.

$$640 \cdot 12 = x$$
$$7680 = x \qquad \text{Doing the multiplication}$$

There are 7680 oz of soda to be bottled. We can also use repeated addition to find how many 16's there are in 7680. Let y = this number.

7680 oz			
16 oz	16 oz	$\cdots$	16 oz

Translate to an equation and solve.

$$16 \cdot y = 7680$$
$$\frac{16 \cdot y}{16} = \frac{7680}{16} \qquad \text{Dividing by 16 on both sides}$$
$$y = 480$$

Check. If we multiply 480 by 16 we get 7680, and 7680 ÷ 12 = 640. The answer checks.

State. The soda will fill 480 sixteen-oz bottles.

39. *Familiarize*. This is a multistep problem.

We must find how many 100's there are in 3500. Then we must find that number times 15.

First we draw a picture

One pound			
3500 calories			
100 cal	100 cal	$\cdots$	100 cal
15 min	15 min	$\cdots$	15 min

In Example 11 it was determined that there are 35 100's in 3500. We let t = the time you have to bicycle to lose a pound.

Translate and Solve. We know that bicycling at 9 mph for 15 min burns off 100 calories, so we need to bicycle for 35 times 15 min in order to burn off one pound. Translate to an equation and solve. We let t = the time required to lose one pound by bicycling.

$$35 \times 15 = t$$
$$525 = t$$

Check. Suppose you bicycle for 525 minutes. If we divide 525 by 15, we get 35, and 35 times 100 is 3500, the number of calories that must be burned off to lose one pound. The answer checks.

State. You must bicycle at 9 mph for 525 min, or 8 hr 45 min, to lose one pound.

41. Round 234,562 to the nearest thousand.

$$2\,3\,4,\;\boxed{5}\,6\,2$$
$$\uparrow$$

The digit 4 is in the thousands place. Consider the next digit to the right. Since the digit, 5, is 5 or higher, round 4 thousands up to 5 thousands. Then change all digits to the right of the thousands place to zeros.

The answer is 235,000.

43. *Familiarize*. We visualize the situation. Let d = the distance light travels in 1 sec.

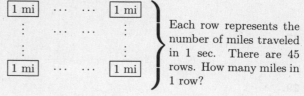

Each row represents the number of miles traveled in 1 sec. There are 45 rows. How many miles in 1 row?

Translate and Solve. We translate to an equation and solve as follows:

$$8,370,000 \div 45 = d$$

$$
\begin{array}{r}
186\,000 \\
45\,\overline{)8,370,000} \\
4\,500\,000 \\
\hline
3\,870\,000 \\
3\,600\,000 \\
\hline
270\,000 \\
270\,000 \\
\hline
0
\end{array}
$$

Check. We can check by multiplying the number of miles traveled in 1 sec by 45 sec:

$$45 \cdot 186,000 = 8,370,000$$

The answer checks.

State. Light travels 186,000 mi in 1 sec.

Exercise Set 1.9

1. Exponential notation for $3 \cdot 3 \cdot 3 \cdot 3$ is 3^4.

3. Exponential notation for $5 \cdot 5$ is 5^2.

5. Exponential notation for $7 \cdot 7 \cdot 7 \cdot 7 \cdot 7$ is 7^5.

7. Exponential notation for $10 \cdot 10 \cdot 10$ is 10^3.

9. $7^2 = 7 \cdot 7 = 49$

11. $9^3 = 9 \cdot 9 \cdot 9 = 729$

13. $12^4 = 12 \cdot 12 \cdot 12 \cdot 12 = 20,736$

15. $11^2 = 11 \cdot 11 = 121$

17. $\begin{aligned} 12 + (6 + 4) &= 12 + 10 \qquad \text{Doing the calculation inside the parentheses} \\ &= 22 \qquad\qquad\;\; \text{Adding} \end{aligned}$

19. $\begin{aligned} 52 - (40 - 8) &= 52 - 32 \qquad \text{Doing the calculation inside the parentheses} \\ &= 20 \qquad\qquad\;\; \text{Subtracting} \end{aligned}$

21. $1000 \div (100 \div 10)$

$\begin{aligned} &= 1000 \div 10 \qquad \text{Doing the calculation inside the parentheses} \\ &= 100 \qquad\qquad\;\; \text{Dividing} \end{aligned}$

23. $\begin{aligned} (256 \div 64) \div 4 &= 4 \div 4 \qquad \text{Doing the calculation inside the parentheses} \\ &= 1 \qquad\qquad\; \text{Dividing} \end{aligned}$

25. $(2+5)^2 = 7^2$ Doing the calculation inside the parentheses

$= 49$ Evaluating the exponential expression

27. $2 + 5^2 = 2 + 25$ Evaluating the exponential expression

$= 27$ Adding

29. $16 \cdot 24 + 50 = 384 + 50$ Doing all multiplications and divisions in order from left to right

$= 434$ Doing all additions and subtractions in order from left to right

31. $83 - 7 \cdot 6 = 83 - 42$ Doing all multiplications and divisions in order from left to right

$= 41$ Doing all additions and subtractions in order from left to right

33. $10 \cdot 10 - 3 \times 4$

$= 100 - 12$ Doing all multiplications and divisions in order from left to right

$= 88$ Doing all additions and subtractions in order from left to right

35. $4^3 \div 8 - 4$

$= 64 \div 8 - 4$ Evaluating the exponential expression

$= 8 - 4$ Doing all multiplications and divisions in order from left to right

$= 4$ Doing all additions and subtractions in order from left to right

37. $17 \cdot 20 - (17 + 20)$

$= 17 \cdot 20 - 37$ Carrying out the operation inside parentheses

$= 340 - 37$ Doing all multiplications and divisions in order from left to right

$= 303$ Doing all additions and subtractions in order from left to right

39. $6 \cdot 10 - 4 \cdot 10$

$= 60 - 40$ Doing all multiplications and divisions in order from left to right

$= 20$ Doing all additions and subtractions in order from left to right

41. $300 \div 5 + 10$

$= 60 + 10$ Doing all multiplications and divisions in order from left to right

$= 70$ Doing all additions and subtractions in order from left to right

43. $3 \cdot (2+8)^2 - 5 \cdot (4-3)^2$

$= 3 \cdot 10^2 - 5 \cdot 1^2$ Carrying out operations inside parentheses

$= 3 \cdot 100 - 5 \cdot 1$ Evaluating the exponential expressions

$= 300 - 5$ Doing all multiplications and divisions in order from left to right

$= 295$ Doing all additions and subtractions in order from left to right

45. $4^2 + 8^2 \div 2^2 = 16 + 64 \div 4$

$= 16 + 16$

$= 32$

47. $10^3 - 10 \cdot 6 - (4 + 5 \cdot 6) = 10^3 - 10 \cdot 6 - (4 + 30)$

$= 10^3 - 10 \cdot 6 - 34$

$= 1000 - 10 \cdot 6 - 34$

$= 1000 - 60 - 34$

$= 940 - 34$

$= 906$

49. $6 \times 11 - (7 + 3) \div 5 - (6 - 4) = 6 \times 11 - 10 \div 5 - 2$

$= 66 - 2 - 2$

$= 64 - 2$

$= 62$

51. $120 - 3^3 \cdot 4 \div (30 - 24) = 120 - 3^3 \cdot 4 \div 6$

$= 120 - 27 \cdot 4 \div 6$

$= 120 - 108 \div 6$

$= 120 - 18$

$= 102$

53. We add the numbers and then divide by the number of addends.

$$(\$64 + \$97 + \$121) \div 3 = \$282 \div 3$$
$$= \$94$$

55. $8 \times 13 + \{42 \div [18 - (6 + 5)]\}$

$= 8 \times 13 + \{42 \div [18 - 11]\}$

$= 8 \times 13 + \{42 \div 7\}$

$= 8 \times 13 + 6$

$= 104 + 6$

$= 110$

57. $[14 - (3 + 5) \div 2] - [18 \div (8 - 2)]$

$= [14 - 8 \div 2] - [18 \div 6]$

$= [14 - 4] - 3$

$= 10 - 3$

$= 7$

59. $(82 - 14) \times [(10 + 45 \div 5) - (6 \cdot 6 - 5 \cdot 5)]$

$= (82 - 14) \times [(10 + 9) - (36 - 25)]$

$= (82 - 14) \times [19 - 11]$

$= 68 \times 8$

$= 544$

61. $4 \times \{(200 - 50 \div 5) - [(35 \div 7) \cdot (35 \div 7) - 4 \times 3]\}$

$= 4 \times \{(200 - 10) - [5 \cdot 5 - 4 \times 3]\}$

$= 4 \times \{190 - [25 - 12]\}$

$= 4 \times \{190 - 13\}$

$= 4 \times 177$

$= 708$

63. $1 + 5 \cdot 4 + 3 = 1 + 20 + 3$
$$= 24 \qquad \text{Correct answer}$$

To make the incorrect answer correct we add parentheses:
$$1 + 5 \cdot (4 + 3) = 36$$

65. $12 \div 4 + 2 \cdot 3 - 2 = 3 + 6 - 2$
$$= 7 \qquad \text{Correct answer}$$

To make the incorrect answer correct we add parentheses:
$$12 \div (4 + 2) \cdot 3 - 2 = 4$$

Chapter 2

Multiplication and Division: Fractional Notation

Exercise Set 2.1

1. We first find some factorizations:

   $18 = 1 \cdot 18 \qquad 18 = 3 \cdot 6$
   $18 = 2 \cdot 9$

   Factors: 1, 2, 3, 6, 9, 18

3. We first find some factorizations:

   $54 = 1 \cdot 54 \qquad 54 = 3 \cdot 18$
   $54 = 2 \cdot 27 \qquad 54 = 6 \cdot 9$

   Factors: 1, 2, 3, 6, 9, 18, 27, 54

5. We first find some factorizations:

   $4 = 1 \cdot 4 \qquad 4 = 2 \cdot 2$

   Factors: 1, 2, 4

7. The only factorization is $7 = 1 \cdot 7$.

   Factors: 1, 7

9. The only factorization is $1 = 1 \cdot 1$.

   Factor: 1

11. We first find some factorizations:

    $98 = 1 \cdot 98 \qquad 98 = 7 \cdot 14$
    $98 = 2 \cdot 49$

    Factors: 1, 2, 7, 14, 49, 98

13. $\begin{array}{ll} 1 \cdot 4 = 4 & 6 \cdot 4 = 24 \\ 2 \cdot 4 = 8 & 7 \cdot 4 = 28 \\ 3 \cdot 4 = 12 & 8 \cdot 4 = 32 \\ 4 \cdot 4 = 16 & 9 \cdot 4 = 36 \\ 5 \cdot 4 = 20 & 10 \cdot 4 = 40 \end{array}$

15. $\begin{array}{ll} 1 \cdot 20 = 20 & 6 \cdot 20 = 120 \\ 2 \cdot 20 = 40 & 7 \cdot 20 = 140 \\ 3 \cdot 20 = 60 & 8 \cdot 20 = 160 \\ 4 \cdot 20 = 80 & 9 \cdot 20 = 180 \\ 5 \cdot 20 = 100 & 10 \cdot 20 = 200 \end{array}$

17. $\begin{array}{ll} 1 \cdot 3 = 3 & 6 \cdot 3 = 18 \\ 2 \cdot 3 = 6 & 7 \cdot 3 = 21 \\ 3 \cdot 3 = 9 & 8 \cdot 3 = 24 \\ 4 \cdot 3 = 12 & 9 \cdot 3 = 27 \\ 5 \cdot 3 = 15 & 10 \cdot 3 = 30 \end{array}$

19. $\begin{array}{ll} 1 \cdot 12 = 12 & 6 \cdot 12 = 72 \\ 2 \cdot 12 = 24 & 7 \cdot 12 = 84 \\ 3 \cdot 12 = 36 & 8 \cdot 12 = 96 \\ 4 \cdot 12 = 48 & 9 \cdot 12 = 108 \\ 5 \cdot 12 = 60 & 10 \cdot 12 = 120 \end{array}$

21. $\begin{array}{ll} 1 \cdot 10 = 10 & 6 \cdot 10 = 60 \\ 2 \cdot 10 = 20 & 7 \cdot 10 = 70 \\ 3 \cdot 10 = 30 & 8 \cdot 10 = 80 \\ 4 \cdot 10 = 40 & 9 \cdot 10 = 90 \\ 5 \cdot 10 = 50 & 10 \cdot 10 = 100 \end{array}$

23. $\begin{array}{ll} 1 \cdot 9 = 9 & 6 \cdot 9 = 54 \\ 2 \cdot 9 = 18 & 7 \cdot 9 = 63 \\ 3 \cdot 9 = 27 & 8 \cdot 9 = 72 \\ 4 \cdot 9 = 36 & 9 \cdot 9 = 81 \\ 5 \cdot 9 = 45 & 10 \cdot 9 = 90 \end{array}$

25. We divide 26 by 6.

$$\begin{array}{r} 4 \\ 6 \overline{)2\,6} \\ 2\,4 \\ \hline 2 \end{array}$$

Since the remainder is not 0, 26 is not divisible by 6.

27. We divide 1880 by 8.

$$\begin{array}{r} 2\,3\,5 \\ 8 \overline{)1\,8\,8\,0} \\ 1\,6\,0\,0 \\ \hline 2\,8\,0 \\ 2\,4\,0 \\ \hline 4\,0 \\ 4\,0 \\ \hline 0 \end{array}$$

Since the remainder is 0, 1880 is divisible by 8.

29. We divide 256 by 16.

$$\begin{array}{r} 1\,6 \\ 16 \overline{)2\,5\,6} \\ 1\,6\,0 \\ \hline 9\,6 \\ 9\,6 \\ \hline 0 \end{array}$$

Since the remainder is 0, 256 is divisible by 16.

31. We divide 4227 by 9.

$$\begin{array}{r} 4\,6\,9 \\ 9 \overline{)4\,2\,2\,7} \\ 3\,6\,0\,0 \\ \hline 6\,2\,7 \\ 5\,4\,0 \\ \hline 8\,7 \\ 8\,1 \\ \hline 6 \end{array}$$

Since the remainder is not 0, 4227 is not divisible by 9.

33. We divide 8650 by 16.

$$\begin{array}{r} 5\,4\,0 \\ 16\,\overline{\smash{\big)}\,8\,6\,5\,0} \\ \underline{8\,0\,0\,0} \\ 6\,5\,0 \\ \underline{6\,4\,0} \\ 1\,0 \end{array}$$

Since the remainder is not 0, 8650 is not divisible by 16.

35. 1 is neither prime nor composite.

37. The number 9 has factors 1, 3, and 9.

Since 9 is not 1 and not prime, it is composite.

39. The number 11 is prime. It has only the factors 1 and 11.

41. The number 29 is prime. It has only the factors 1 and 29.

43.
$$\begin{array}{r} 2 \\ 2\,\overline{\smash{\big)}\,4} \\ 2\,\overline{\smash{\big)}\,8} \end{array}\quad \leftarrow \; 2 \text{ is prime}$$
$8 = 2 \cdot 2 \cdot 2$

45.
$$\begin{array}{r} 7 \\ 2\,\overline{\smash{\big)}\,1\,4} \end{array}\quad \leftarrow \; 7 \text{ is prime}$$
$14 = 2 \cdot 7$

47.
$$\begin{array}{r} 7 \\ 3\,\overline{\smash{\big)}\,2\,1} \\ 2\,\overline{\smash{\big)}\,4\,2} \end{array}\quad \leftarrow \; 7 \text{ is prime}$$
$42 = 2 \cdot 3 \cdot 7$

49.
$$\begin{array}{r} 5 \\ 5\,\overline{\smash{\big)}\,2\,5} \end{array}\quad \leftarrow \; 5 \text{ is prime}$$
(25 is not divisible by 2 or 3. We move to 5.)
$25 = 5 \cdot 5$

51.
$$\begin{array}{r} 5 \\ 5\,\overline{\smash{\big)}\,2\,5} \\ 2\,\overline{\smash{\big)}\,5\,0} \end{array}\quad \leftarrow \; 5 \text{ is prime}$$
(25 is not divisible by 2 or 3. We move to 5.)
$50 = 2 \cdot 5 \cdot 5$

53.
$$\begin{array}{r} 1\,3 \\ 1\,3\,\overline{\smash{\big)}\,1\,6\,9} \end{array}\quad \leftarrow \; 13 \text{ is prime}$$
(169 is not divisible by 2, 3, 5, 7 or 11. We move to 13.)
$169 = 13 \cdot 13$

55.
$$\begin{array}{r} 5 \\ 5\,\overline{\smash{\big)}\,2\,5} \\ 2\,\overline{\smash{\big)}\,5\,0} \\ 2\,\overline{\smash{\big)}\,1\,0\,0} \end{array}\quad \leftarrow \; 5 \text{ is prime}$$
(25 is not divisible by 2 or 3. We move to 5.)
$100 = 2 \cdot 2 \cdot 5 \cdot 5$

We can also use a factor tree.

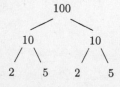

57.
$$\begin{array}{r} 7 \\ 5\,\overline{\smash{\big)}\,3\,5} \end{array}\quad \leftarrow \; 7 \text{ is prime}$$
(35 is not divisible by 2 or 3. We move to 5.)
$35 = 5 \cdot 7$

59.
$$\begin{array}{r} 3 \\ 3\,\overline{\smash{\big)}\,9} \\ 2\,\overline{\smash{\big)}\,1\,8} \\ 2\,\overline{\smash{\big)}\,3\,6} \\ 2\,\overline{\smash{\big)}\,7\,2} \end{array}\quad \leftarrow \; 3 \text{ is prime}$$
(9 is not divisible by 2. We move to 3.)
$72 = 2 \cdot 2 \cdot 2 \cdot 3 \cdot 3$

We can also use a factor tree, as shown in Example 10 in the text.

61.
$$\begin{array}{r} 1\,1 \\ 7\,\overline{\smash{\big)}\,7\,7} \end{array}\quad \leftarrow \; 11 \text{ is prime}$$
(77 is not divisible by 2, 3, or 5. We move to 7.)
$77 = 7 \cdot 11$

63.
$$\begin{array}{r} 1\,3 \\ 3\,\overline{\smash{\big)}\,3\,9} \\ 2\,\overline{\smash{\big)}\,7\,8} \\ 2\,\overline{\smash{\big)}\,1\,5\,6} \end{array}\quad \leftarrow \; 13 \text{ is prime}$$
$156 = 2 \cdot 2 \cdot 3 \cdot 13$

We can also use a factor tree.

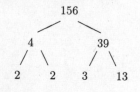

65.
$$\begin{array}{r} 5 \\ 5\,\overline{\smash{\big)}\,2\,5} \\ 3\,\overline{\smash{\big)}\,7\,5} \\ 2\,\overline{\smash{\big)}\,1\,5\,0} \\ 2\,\overline{\smash{\big)}\,3\,0\,0} \end{array}\quad \leftarrow \; 5 \text{ is prime}$$
$300 = 2 \cdot 2 \cdot 3 \cdot 5 \cdot 5$

We can also use a factor tree.

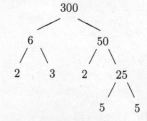

67.
$$\begin{array}{r} 1\,3 \\ \times \quad 2 \\ \hline 2\,6 \end{array}$$

69.
$$\begin{array}{r} {}^{3}\;\;\; \\ 2\,5 \\ \times \quad 1\,7 \\ \hline 1\,7\,5 \\ 2\,5\,0 \\ \hline 4\,2\,5 \end{array}$$
Multiplying by 7
Multiplying by 10
Adding

71.
$$2\,2\,\overline{\smash{\big)}\,0} \quad \begin{array}{c} 0 \\ \underline{0} \\ 0 \end{array}$$

The answer is 0.

73.
$$2\,2\,\overline{\smash{\big)}\,2\,2} \quad \begin{array}{c} 1 \\ \underline{2\,2} \\ 0 \end{array}$$

The answer is 1.

75. Since there are 2 factors a rectangular array is indicated. One factor gives the number of rows, the other the number of objects in each row. Thus, we have a rectangular array of 6 rows with 9 objects each or 9 rows with 6 objects each.

Exercise Set 2.2

1. A number is divisible by 2 if its <u>ones digit</u> is even.

4<u>6</u> is divisible by 2 because <u>6</u> is even.
22<u>4</u> is divisible by 2 because <u>4</u> is even.
1<u>9</u> is not divisible by 2 because <u>9</u> is not even.
55<u>5</u> is not divisible by 2 because <u>5</u> is not even.
30<u>0</u> is divisible by 2 because <u>0</u> is even.
3<u>6</u> is divisible by 2 because <u>6</u> is even.
45,27<u>0</u> is divisible by 2 because <u>0</u> is even.
444<u>4</u> is divisible by 2 because <u>4</u> is even.
8<u>5</u> is not divisible by 2 because <u>5</u> is not even.
71<u>1</u> is not divisible by 2 because <u>1</u> is not even.
13,25<u>1</u> is not divisible by 2 because <u>1</u> is not even.
254,76<u>5</u> is not divisible by 2 because <u>5</u> is not even.

3. A number is divisible by 4 if the <u>number</u> named by the last <u>two</u> digits is divisible by 4.

<u>46</u> is not divisible by 4 because <u>46</u> is not divisible by 4.
2<u>24</u> is divisible by 4 because <u>24</u> is divisible by 4.
<u>19</u> is not divisible by 4 because <u>19</u> is not divisible by 4.
5<u>55</u> is not divisible by 4 because <u>55</u> is not divisible by 4.
3<u>00</u> is divisible by 4 because <u>00</u> is divisible by 4.
<u>36</u> is divisible by 4 because <u>36</u> is divisible by 4.
45,2<u>70</u> is not divisible by 4 because <u>70</u> is not divisible by 4.
44<u>44</u> is divisible by 4 because <u>44</u> is divisible by 4.
<u>85</u> is not divisible by 4 because <u>85</u> is not divisible by 4.
7<u>11</u> is not divisible by 4 because <u>11</u> is not divisible by 4.
13,2<u>51</u> is not divisible by 4 because <u>51</u> is not divisible by 4.
254,7<u>65</u> is not divisible by 4 because <u>65</u> is not divisible by 4.

5. For a number to be divisible by 6, the sum of the digits must be divisible by 3 and the ones digit must be 0, 2, 4, 6 or 8 (even). It is most efficient to determine if the ones digit is even first and then, if so, to determine if the sum of the digits is divisible by 3.

46 is not divisible by 6 because 46 is not divisible by 3.

$$4 + 6 = 10$$
$$\uparrow$$
Not divisible by 3

224 is not divisible by 6 because 224 is not divisible by 3.

$$2 + 2 + 4 = 8$$
$$\uparrow$$
Not divisible by 3

19 is not divisible by 6 because 19 is not even.

19
$\uparrow$
Not even

555 is not divisible by 6 because 555 is not even.

555
$\uparrow$
Not even

300 is divisible by 6.

300          $3 + 0 + 0 = 3$
$\uparrow$ $\vert$
Even   Divisible by 3

36 is divisible by 6.

36          $3 + 6 = 9$
$\uparrow$ $\uparrow$
Even   Divisible by 3

45,270 is divisible by 6.

45,270 $4 + 5 + 2 + 7 + 0 = 18$
$\uparrow$ $\uparrow$
Even   Divisible by 3

4444 is not divisible by 6 because 4444 is not divisible by 3.

$$4 + 4 + 4 + 4 = 16$$
$$\uparrow$$

Not divisible by 3

85 is not divisible by 6 because 85 is not even.

85
$\uparrow$
Not even

711 is not divisible by 6 because 711 is not even.

711
$\uparrow$
Not even

13,251 is not divisible by 6 because 13,251 is not even.

13,251
$\uparrow$
Not even

254,765 is not divisible by 6 because 254,765 is not even.

254,765
$\uparrow$
Not even

7. A number is divisible by 9 if the sum of the digits is divisible by 9.

46 is not divisible by 9 because $4 + 6 = 10$ and 10 is not divisible by 9.

224 is not divisible by 9 because $2 + 2 + 4 = 8$ and 8 is not divisible by 9.

19 is not divisible by 9 because $1 + 9 = 10$ and 10 is not divisible by 9.

555 is not divisible by 9 because $5 + 5 + 5 = 15$ and 15 is not divisible by 9.

300 is not divisible by 9 because $3 + 0 + 0 = 3$ and 3 is not divisible by 9.

36 is divisible by 9 because $3 + 6 = 9$ and 9 is divisible by 9.

45,270 is divisible by 9 because $4 + 5 + 2 + 7 + 0 = 18$ and 18 is divisible by 9.

4444 is not divisible by 9 because $4 + 4 + 4 + 4 = 16$ and 16 is not divisible by 9.

85 is not divisible by 9 because $8 + 5 = 13$ and 13 is not divisible by 9.

711 is divisible by 9 because $7 + 1 + 1 = 9$ and 9 is divisible by 9.

13,251 is not divisible by 9 because $1 + 3 + 2 + 5 + 1 = 12$ and 12 is not divisible by 9.

254,765 is not divisible by 9 because $2+5+4+7+6+5 = 29$ and 29 is not divisible by 9.

9. A number is divisible by 3 if the sum of the digits is divisible by 3.

56 is not divisible by 3 because $5 + 6 = 11$ and 11 is not divisible by 3.

324 is divisible by 3 because $3 + 2 + 4 = 9$ and 9 is divisible by 3.

784 is not divisible by 3 because $7 + 8 + 4 = 19$ and 19 is not divisible by 3.

55,555 is not divisible by 3 because $5 + 5 + 5 + 5 + 5 = 25$ and 25 is not divisible by 3.

200 is not divisible by 3 because $2 + 0 + 0 = 2$ and 2 is not divisible by 3.

42 is divisible by 3 because $4 + 2 = 6$ and 6 is divisible by 3.

501 is divisible by 3 because $5 + 0 + 1 = 6$ and 6 is divisible by 3.

3009 is divisible by 3 because $3 + 0 + 0 + 9 = 12$ and 12 is divisible by 3.

75 is divisible by 3 because $7 + 5 = 12$ and 12 is divisible by 3.

812 is not divisible by 3 because $8 + 1 + 2 = 11$ and 11 is not divisible by 3.

2345 is not divisible by 3 because $2 + 3 + 4 + 5 = 14$ and 14 is not divisible by 3.

2001 is divisible by 3 because $2 + 0 + 0 + 1 = 3$ and 3 is divisible by 3.

11. A number is divisible by 5 if the ones digit is 0 or 5.

5<u>6</u> is not divisible by 5 because the ones digit (6) is not 0 or 5.

32<u>4</u> is not divisible by 5 because the ones digit (4) is not 0 or 5.

78<u>4</u> is not divisible by 5 because the ones digit (4) is not 0 or 5.

55,55<u>5</u> is divisible by 5 because the ones digit (5) is 5.

20<u>0</u> is divisible by 5 because the ones digit (0) is 0.

4<u>2</u> is not divisible by 5 because the ones digit (2) is not 0 or 5.

50<u>1</u> is not divisible by 5 because the ones digit (1) is not 0 or 5.

300<u>9</u> is not divisible by 5 because the ones digit (9) is not 0 or 5.

7<u>5</u> is divisible by 5 because the ones digit (5) is 5.

81<u>2</u> is not divisible by 5 because the ones digit (2) is not 0 or 5.

234<u>5</u> is divisible by 5 because the ones digit (5) is 5.

200<u>1</u> is not divisible by 5 because the ones digit (1) is not 0 or 5.

13. A number is divisible by 9 if the sum of the digits is divisible by 9.

56 is not divisible by 9 because $5 + 6 = 11$ and 11 is not divisible by 9.

324 is divisible by 9 because $3 + 2 + 4 = 9$ and 9 is divisible by 9.

784 is not divisible by 9 because $7 + 8 + 4 = 19$ and 19 is not divisible by 9.

55,555 is not divisible by 9 because $5 + 5 + 5 + 5 + 5 = 25$ and 25 is not divisible by 9.

200 is not divisible by 9 because $2 + 0 + 0 = 2$ and 2 is not divisible by 9.

42 is not divisible by 9 because $4 + 2 = 6$ and 6 is not divisible by 9.

501 is not divisible by 9 because $5 + 0 + 1 = 6$ and 6 is not divisible by 9.

3009 is not divisible by 9 because $3 + 0 + 0 + 9 = 12$ and 12 is not divisible by 9.

75 is not divisible by 9 because $7 + 5 = 12$ and 12 is not divisible by 9.

812 is not divisible by 9 because $8 + 1 + 2 = 11$ and 11 is not divisible by 9.

2345 is not divisible by 9 because $2 + 3 + 4 + 5 = 14$ and 14 is not divisible by 9.

2001 is not divisible by 9 because $2 + 0 + 0 + 1 = 3$ and 3 is not divisible by 9.

15. A number is divisible by 10 if the ones digit is 0.

Of the numbers under consideration, the only one whose ones digit is 0 is 200. Therefore, 200 is divisible by 10. None of the other numbers is divisible by 10.

17.
$$56 + x = 194$$
$$56 + x - 56 = 194 - 56 \quad \text{Subtracting 56 on both sides}$$
$$x = 138$$

The solution is 138.

19.
$$18 \cdot t = 1008$$
$$\frac{18 \cdot t}{18} = \frac{1008}{18} \quad \text{Dividing by 18 on both sides}$$
$$t = 56$$

The solution is 56.

21. *Familiarize*. This is a multistep problem. Find the total cost of the shirts and the total cost of the trousers and then find the sum of the two.

We let s = the total cost of the shirts and t = the total cost of the trousers.

Translate. We write two equations.

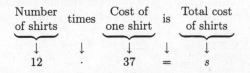

Number of shirts	times	Cost of one shirt	is	Total cost of shirts
↓	↓	↓	↓	↓
12	·	37	=	s

Number of pairs of trousers	times	Cost of one pair	is	Total cost of trousers
↓	↓	↓	↓	↓
4	·	59	=	t

Solve. We carry out the multiplication.

$$12 \cdot 37 = s$$
$$444 = s \quad \text{Doing the multiplication}$$

The total cost of the 12 shirts is $444.

$$4 \cdot 59 = t$$
$$236 = t \quad \text{Doing the multiplication}$$

The total cost of the 4 pairs of trousers is $236.

Now we find the total amount spent. We let a = this amount.

Total cost of shirts	plus	Total cost of trousers	is	Total amount spent
↓	↓	↓	↓	↓
444	+	236	=	a

To solve the equation, carry out the addition.

$$\begin{array}{r} 4\,4\,4 \\ +\,2\,3\,6 \\ \hline 6\,8\,0 \end{array}$$

Check. We can repeat the calculations. The answer checks.

State. The total cost is $680.

23. 780<u>0</u> is divisible by 2 because the ones digit (0) is even.

$7800 \div 2 = 3900$ so $7800 = 2 \cdot 3900$.

390<u>0</u> is divisible by 2 because the ones digit (0) is even.

$3900 \div 2 = 1950$ so $3900 = 2 \cdot 1950$ and $7800 = 2 \cdot 2 \cdot 1950$.

195<u>0</u> is divisible by 2 because the ones digit (0) is even.

$1950 \div 2 = 975$ so $1950 = 2 \cdot 975$ and $7800 = 2 \cdot 2 \cdot 2 \cdot 975$.

97<u>5</u> is not divisible by 2 because the ones digit (5) is not even. Move on to 3.

975 is divisible by 3 because the sum of the digits ($9 + 7 + 5 = 21$) is divisible by 3.

$975 \div 3 = 325$ so $975 = 3 \cdot 325$ and $7800 = 2 \cdot 2 \cdot 2 \cdot 3 \cdot 325$.

Since 975 is not divisible by 2, none of its factors is divisible by 2. Therefore, we no longer need to check for divisibility by 2.

325 is not divisible by 3 because the sum of the digits ($3 + 2 + 5 = 10$) is not divisible by 3. Move on to 5.

32<u>5</u> is divisible by 5 because the ones digit is 5.

$325 \div 5 = 65$ so $325 = 5 \cdot 65$ and $7800 = 2 \cdot 2 \cdot 2 \cdot 3 \cdot 5 \cdot 65$.

Since 325 is not divisible by 3, none of its factors is divisible by 3. Therefore we no longer need to check for divisibility by 3.

6<u>5</u> is divisible by 5 because the ones digit is 5.

$65 \div 5 = 13$ so $65 = 5 \cdot 13$ and $7800 = 2 \cdot 2 \cdot 2 \cdot 3 \cdot 5 \cdot 5 \cdot 13$.

13 is prime so the prime factorization of 7800 is $2 \cdot 2 \cdot 2 \cdot 3 \cdot 5 \cdot 5 \cdot 13$.

25. 277<u>2</u> is divisible by 2 because the ones digit (2) is even.

$2772 \div 2 = 1386$ so $2772 = 2 \cdot 1386$.

138<u>6</u> is divisible by 2 because the ones digit (6) is even.

$1386 \div 2 = 693$ so $1386 = 2 \cdot 693$ and $2772 = 2 \cdot 2 \cdot 693$.

69<u>3</u> is not divisible by 2 because the ones digit (3) is not even. We move to 3.

693 is divisible by 3 because the sum of the digits ($6 + 9 + 3 = 18$) is divisible by 3.

$693 \div 3 = 231$ so $693 = 3 \cdot 231$ and $2772 = 2 \cdot 2 \cdot 3 \cdot 231$.

Since 693 is not divisible by 2, none of its factors is divisible by 2. Therefore, we no longer need to check divisibility by 2.

231 is divisible by 3 because the sum of the digits ($2 + 3 + 1 = 6$) is divisible by 3.

$231 \div 3 = 77$ so $231 = 3 \cdot 77$ and $2772 = 2 \cdot 2 \cdot 3 \cdot 3 \cdot 77$.

77 is not divisible by 3 since the sum of the digits ($7 + 7 = 14$) is not divisible by 3. We move to 5.

7<u>7</u> is not divisible by 5 because the ones digit (7) is not 0 or 5. We move to 7.

We have not stated a test for divisibility by 7 so we will just try dividing by 7.

$$7 \overline{)\,7\,7\,}^{\;1\,1} \quad \leftarrow \text{11 is prime}$$

$77 \div 7 = 11$ so $77 = 7 \cdot 11$ and the prime factorization of 2772 is $2 \cdot 2 \cdot 3 \cdot 3 \cdot 7 \cdot 11$.

Exercise Set 2.3

1. The top number is the numerator, and the bottom number is the denominator.

$$\frac{3}{4} \begin{array}{l} \leftarrow \text{Numerator} \\ \leftarrow \text{Denominator} \end{array}$$

3.
$$\frac{1\,1}{2\,0} \begin{array}{l} \leftarrow \text{Numerator} \\ \leftarrow \text{Denominator} \end{array}$$

5. The dollar is divided into 4 parts of the same size, and 2 of them are shaded. This is $2 \cdot \frac{1}{4}$ or $\frac{2}{4}$. Thus, $\frac{2}{4}$ (two-fourths) of the dollar is shaded.

7. The yard is divided into 8 parts of the same size, and 1 of them is shaded. Thus, $\frac{1}{8}$ (one-eighth) of the yard is shaded.

9. We have 2 quarts, each divided into thirds. We take $\frac{1}{3}$ of each. This is $2 \cdot \frac{1}{3}$ or $\frac{2}{3}$. Thus, $\frac{2}{3}$ of a quart is shaded.

11. The triangle is divided into 4 triangles of the same size, and 3 of them are shaded. This is $3 \cdot \frac{1}{4}$ or $\frac{3}{4}$. Thus, $\frac{3}{4}$ (three-fourths) of the triangle is shaded.

13. The pie is divided into 8 parts of the same size, and 4 of them are shaded. This is $4 \cdot \frac{1}{8}$, or $\frac{4}{8}$. Thus, $\frac{4}{8}$ (four-eighths) of the pie is shaded.

15. The acre is divided into 12 parts of the same size, and 6 of them are shaded. This is $6 \cdot \frac{1}{12}$, or $\frac{6}{12}$. Thus, $\frac{6}{12}$ (six-twelfths) of the acre is shaded.

17. There are 8 circles, and 5 are shaded. Thus, $\frac{5}{8}$ of the circles are shaded.

19. There are 5 objects in the set, and 3 of the objects are shaded. Thus, $\frac{3}{5}$ of the set is shaded.

21. Remember: $\frac{0}{n} = 0$, for n that is not 0.

$$\frac{0}{8} = 0$$

Think of dividing an object into 8 parts and taking none of them. We get 0.

23. Remember: $\frac{n}{1} = n$.

$$\frac{8}{1} = 8$$

Think of taking 8 objects and dividing them into 1 part. (We do not divide them.) We have 8 objects.

25. Remember: $\frac{n}{n} = 1$, for n that is not 0.

$$\frac{20}{20} = 1$$

If we divide an object into 20 parts and take 20 of them, we get all of the object (1 whole object).

27. Remember: $\frac{n}{n} = 1$, for n that is not 0.

$$\frac{45}{45} = 1$$

If we divide an object into 45 parts and take 45 of them, we get all of the object (1 whole object).

29. Remember: $\frac{0}{n} = 0$, for n that is not 0.

$$\frac{0}{238} = 0$$

Think of dividing an object into 238 parts and taking none of them. We get 0.

31. Remember: $\frac{n}{n} = 1$, for n that is not 0.

$$\frac{238}{238} = 1$$

If we divide an object into 238 parts and take 238 of them, we get all of the object (1 whole object).

33. Remember: $\frac{n}{n} = 1$, for n that is not 0.

$$\frac{3}{3} = 1$$

If we divide an object into 3 parts and take 3 of them, we get all of the object (1 whole object).

35. Remember: $\frac{n}{n} = 1$, for n that is not 0.

$$\frac{87}{87} = 1$$

If we divide an object into 87 parts and take 87 of them, we get all of the object (1 whole object).

37. Remember: $\frac{n}{n} = 1$, for n that is not 0.

$$\frac{8}{8} = 1$$

39. Remember: $\frac{n}{1} = n$

$$\frac{8}{1} = 8$$

41. Remember: $\frac{n}{0}$ is not defined for any whole number n.

$\frac{729}{0}$ is not defined.

43. $\dfrac{5}{6 - 6} = \dfrac{5}{0}$

Remember: $\frac{n}{0}$ is not defined for any whole number n.

Thus, $\dfrac{5}{6 - 6}$ is not defined.

45. Round 3 4,5 6 $\boxed{2}$ to the nearest ten.
     $\uparrow$

The digit 6 is in the tens place. Consider the next digit to the right. Since the digit, 2, is 4 or lower, round down, meaning that 6 tens stays as 6 tens. Then change the digit to the right of the tens digit to zero.

The answer is 34,560.

47. Round 3 4, 5 6 2 to the nearest thousand.
     ↑

The digit 4 is in the thousands place. Consider the next digit to the right. Since the digit, 5, is 5 or higher, round 4 thousands up to 5 thousands. Then change all digits to the right of the thousands digit to zeros.

The answer is 35,000.

49. *Familiarize*. We visualize the situation. We let p = the number of pounds of hair cut each year. (We assume the same amount is cut each year.)

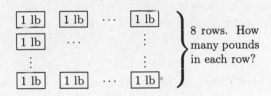

$$\left.\begin{array}{l}\end{array}\right\} \text{8 rows. How many pounds in each row?}$$

Translate and Solve. We translate to an equation and solve as follows:

$29,824 \div 8 = p$

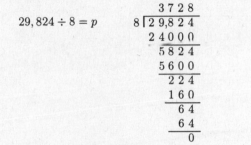

Check. We can check by multiplying the number of pounds of hair cut each year by 8:

    $8 \cdot 3728 = 29,824$

State. The pet care service cut 3728 lb of hair each year.

51. Think of dividing $2700 into 2700 parts of the same size. (Each part is $1.)

a) Think of $1200 as 1200 of the 2700 parts. Thus, we have $1200 \cdot \dfrac{1}{2700}$, or $\dfrac{1200}{2700}$.

b) Think of $540 as 540 of the 2700 parts. Thus, we have $540 \cdot \dfrac{1}{2700}$, or $\dfrac{540}{2700}$.

c) Think of $360 as 360 of the 2700 parts. Thus, we have $360 \cdot \dfrac{1}{2700}$, or $\dfrac{360}{2700}$.

d) First, find the amount of miscellaneous expenses. To do this we must find the total of the amounts spent for tuition, rent, and food and then take that amount away from the amount earned.

To find the total spent for tuition, rent, and food translate to an equation and solve.

    $1200 + 540 + 360 = x$
           $2100 = x$ Doing the addition

The total is $2100.

Now we have a "take-away" situation:

    $2700 - 2100 = m$
           $600 = m$ Doing the subtraction

Thus, $600 went for miscellaneous expenses.

Think of $600 as 600 of the 2700 parts. Thus, we have $600 \cdot \dfrac{1}{2700}$, or $\dfrac{600}{2700}$.

Exercise Set 2.4

1. $3 \cdot \dfrac{1}{5} = \dfrac{3 \cdot 1}{5} = \dfrac{3}{5}$

3. $5 \times \dfrac{1}{8} = \dfrac{5 \times 1}{8} = \dfrac{5}{8}$

5. $\dfrac{2}{11} \cdot 4 = \dfrac{2 \cdot 4}{11} = \dfrac{8}{11}$

7. $10 \cdot \dfrac{7}{9} = \dfrac{10 \cdot 7}{9} = \dfrac{70}{9}$

9. $\dfrac{2}{5} \cdot 1 = \dfrac{2 \cdot 1}{5} = \dfrac{2}{5}$

11. $\dfrac{2}{5} \cdot 3 = \dfrac{2 \cdot 3}{5} = \dfrac{6}{5}$

13. $7 \cdot \dfrac{3}{4} = \dfrac{7 \cdot 3}{4} = \dfrac{21}{4}$

15. $17 \times \dfrac{5}{6} = \dfrac{17 \times 5}{6} = \dfrac{85}{6}$

17. $\dfrac{1}{2} \cdot \dfrac{1}{3} = \dfrac{1 \cdot 1}{2 \cdot 3} = \dfrac{1}{6}$

19. $\dfrac{1}{4} \times \dfrac{1}{10} = \dfrac{1 \times 1}{4 \times 10} = \dfrac{1}{40}$

21. $\dfrac{2}{3} \times \dfrac{1}{5} = \dfrac{2 \times 1}{3 \times 5} = \dfrac{2}{15}$

23. $\dfrac{2}{5} \cdot \dfrac{2}{3} = \dfrac{2 \cdot 2}{5 \cdot 3} = \dfrac{4}{15}$

25. $\dfrac{3}{4} \cdot \dfrac{3}{4} = \dfrac{3 \cdot 3}{4 \cdot 4} = \dfrac{9}{16}$

27. $\dfrac{2}{3} \cdot \dfrac{7}{13} = \dfrac{2 \cdot 7}{3 \cdot 13} = \dfrac{14}{39}$

29. $\dfrac{1}{10} \cdot \dfrac{7}{10} = \dfrac{1 \cdot 7}{10 \cdot 10} = \dfrac{7}{100}$

31. $\dfrac{7}{8} \cdot \dfrac{7}{8} = \dfrac{7 \cdot 7}{8 \cdot 8} = \dfrac{49}{64}$

33. $\dfrac{1}{10} \cdot \dfrac{1}{100} = \dfrac{1 \cdot 1}{10 \cdot 100} = \dfrac{1}{1000}$

35. $\dfrac{14}{15} \cdot \dfrac{13}{19} = \dfrac{14 \cdot 13}{15 \cdot 19} = \dfrac{182}{285}$

37. *Familiarize*. Recall that area is length times width. We draw a picture. We will let A = the area of the table top.

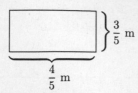

$$\left.\begin{array}{c}\end{array}\right\}\frac{3}{5}\ \text{m}$$

$$\underbrace{\qquad\qquad}_{\textstyle\frac{4}{5}\ \text{m}}$$

Translate. Then we translate.

$$\begin{array}{ccccc}\text{Area} & \text{is} & \text{length} & \text{times} & \text{width}\\ \downarrow & \downarrow & \downarrow & \downarrow & \downarrow\\ A & = & \frac{4}{5} & \times & \frac{3}{5}\end{array}$$

Solve. The sentence tells us what to do. We multiply.

$$\frac{4}{5}\times\frac{3}{5}=\frac{4\times 3}{5\times 5}=\frac{12}{25}$$

Check. We repeat the calculation. The answer checks.

State. The area is $\frac{12}{25}$ m².

39. *Familiarize*. We know that 1 of 39 high school football players plays college football. That is, $\frac{1}{39}$ of high school football players play college football. In addition, we know that 1 of 39 college players plays professional football. That is, $\frac{1}{39}$ of college football players play professional football or $\frac{1}{39}$ of the $\frac{1}{39}$ of high school players who play college football also play professionally. We let f = the fractional part of high school football players that plays professional football.

Translate. The multiplication sentence $\frac{1}{39}\cdot\frac{1}{39}$ corresponds to this situation.

Solve. We carry out the multiplication.

$$\frac{1}{39}\cdot\frac{1}{39}=\frac{1\cdot 1}{39\cdot 39}=\frac{1}{1521}$$

Check. We repeat the calculation. The answer checks.

State. The fractional part of high school players that plays professional football is $\frac{1}{1521}$.

41. *Familiarize*. We draw a picture. We let h = the amount of honey that is needed.

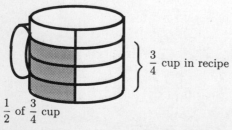

$\left.\begin{array}{c}\end{array}\right\}\frac{3}{4}$ cup in recipe

$\frac{1}{2}$ of $\frac{3}{4}$ cup

Translate. The multiplication sentence $\frac{1}{2}\cdot\frac{3}{4}=h$ corresponds to the situation.

Solve. We multiply:

$$\frac{1}{2}\cdot\frac{3}{4}=\frac{1\cdot 3}{2\cdot 4}=\frac{3}{8}$$

Check. We repeat the calculation. The answer checks.

State. $\frac{3}{8}$ cup of honey is needed.

43. *Familiarize*. Think of each container of juice as $\frac{1}{100}$ of the entire group. We let p = the fractional part of juice purchased that is orange juice.

Translate. The multiplication sentence $56\cdot\frac{1}{100}=p$ corresponds to this situation.

Solve. We carry out the multiplication.

$$56\cdot\frac{1}{100}=\frac{56\cdot 1}{100}=\frac{56}{100}$$

Check. We repeat the calculation. The answer checks.

State. $\frac{56}{100}$ of the juice purchased in grocery stores is orange juice.

45.
$$\begin{array}{r}204\\35\overline{\smash{)}7140}\\7000\\\hline 140\\140\\\hline 0\end{array}$$

The answer is 204.

47.
$$\begin{array}{r}3001\\9\overline{\smash{)}27{,}009}\\27000\\\hline 9\\9\\\hline 0\end{array}$$

The answer is 3001.

49. 4, $\boxed{6}$ 7 8, 9 5 2

The digit 6 means 6 hundred thousands.

Exercise Set 2.5

1. Since $2\cdot 5=10$, we multiply by $\frac{5}{5}$.

$$\frac{1}{2}=\frac{1}{2}\cdot\frac{5}{5}=\frac{1\cdot 5}{2\cdot 5}=\frac{5}{10}$$

3. Since $8\cdot 4=32$, we multiply by $\frac{4}{4}$.

$$\frac{5}{8}=\frac{5}{8}\cdot\frac{4}{4}=\frac{5\cdot 4}{8\cdot 4}=\frac{20}{32}$$

5. Since $10\cdot 3=30$, we multiply by $\frac{3}{3}$.

$$\frac{9}{10}=\frac{9}{10}\cdot\frac{3}{3}=\frac{9\cdot 3}{10\cdot 3}=\frac{27}{30}$$

7. Since $8\cdot 4=32$, we multiply by $\frac{4}{4}$.

$$\frac{7}{8}=\frac{7}{8}\cdot\frac{4}{4}=\frac{7\cdot 4}{8\cdot 4}=\frac{28}{32}$$

9. Since $12 \cdot 4 = 48$, we multiply by $\frac{4}{4}$.

$$\frac{5}{12} = \frac{5}{12} \cdot \frac{4}{4} = \frac{5 \cdot 4}{12 \cdot 4} = \frac{20}{48}$$

11. Since $18 \cdot 3 = 54$, we multiply by $\frac{3}{3}$.

$$\frac{17}{18} = \frac{17}{18} \cdot \frac{3}{3} = \frac{17 \cdot 3}{18 \cdot 3} = \frac{51}{54}$$

13. Since $3 \cdot 15 = 45$, we multiply by $\frac{15}{15}$.

$$\frac{5}{3} = \frac{5}{3} \cdot \frac{15}{15} = \frac{5 \cdot 15}{3 \cdot 15} = \frac{75}{45}$$

15. Since $22 \cdot 6 = 132$, we multiply by $\frac{6}{6}$.

$$\frac{7}{22} = \frac{7}{22} \cdot \frac{6}{6} = \frac{7 \cdot 6}{22 \cdot 6} = \frac{42}{132}$$

17.
$$\frac{2}{4} = \frac{1 \cdot 2}{2 \cdot 2} \quad \longleftarrow \text{ Factor the numerator}$$
$$\quad\quad\quad \longleftarrow \text{ Factor the denominator}$$
$$= \frac{1}{2} \cdot \frac{2}{2} \quad \longleftarrow \text{ Factor the fraction}$$
$$= \frac{1}{2} \cdot 1 \quad \longleftarrow \frac{2}{2} = 1$$
$$= \frac{1}{2} \quad \longleftarrow \text{ Removing a factor of 1}$$

19.
$$\frac{6}{8} = \frac{3 \cdot 2}{4 \cdot 2} \quad \longleftarrow \text{ Factor the numerator}$$
$$\quad\quad\quad \longleftarrow \text{ Factor the denominator}$$
$$= \frac{3}{4} \cdot \frac{2}{2} \quad \longleftarrow \text{ Factor the fraction}$$
$$= \frac{3}{4} \cdot 1 \quad \longleftarrow \frac{2}{2} = 1$$
$$= \frac{3}{4} \quad \longleftarrow \text{ Removing a factor of 1}$$

21.
$$\frac{2}{15} = \frac{1 \cdot 3}{5 \cdot 3} \quad \longleftarrow \text{ Factor the numerator}$$
$$\quad\quad\quad \longleftarrow \text{ Factor the denominator}$$
$$= \frac{1}{5} \cdot \frac{3}{3} \quad \longleftarrow \text{ Factor the fraction}$$
$$= \frac{1}{5} \cdot 1 \quad \longleftarrow \frac{3}{3} = 1$$
$$= \frac{1}{5} \quad \longleftarrow \text{ Removing a factor of 1}$$

23. $\dfrac{24}{8} = \dfrac{3 \cdot 8}{1 \cdot 8} = \dfrac{3}{1} \cdot \dfrac{8}{8} = \dfrac{3}{1} \cdot 1 = \dfrac{3}{1} = 3$

25. $\dfrac{18}{24} = \dfrac{3 \cdot 6}{4 \cdot 6} = \dfrac{3}{4} \cdot \dfrac{6}{6} = \dfrac{3}{4} \cdot 1 = \dfrac{3}{4}$

27. $\dfrac{14}{16} = \dfrac{7 \cdot 2}{8 \cdot 2} = \dfrac{7}{8} \cdot \dfrac{2}{2} = \dfrac{7}{8} \cdot 1 = \dfrac{7}{8}$

29. $\dfrac{12}{10} = \dfrac{6 \cdot 2}{5 \cdot 2} = \dfrac{6}{5} \cdot \dfrac{2}{2} = \dfrac{6}{5} \cdot 1 = \dfrac{6}{5}$

31. $\dfrac{16}{48} = \dfrac{1 \cdot 16}{3 \cdot 16} = \dfrac{1}{3} \cdot \dfrac{16}{16} = \dfrac{1}{3} \cdot 1 = \dfrac{1}{3}$

33. $\dfrac{150}{25} = \dfrac{6 \cdot 25}{1 \cdot 25} = \dfrac{6}{1} \cdot \dfrac{25}{25} = \dfrac{6}{1} \cdot 1 = \dfrac{6}{1} = 6$

We could also simplify $\dfrac{150}{25}$ by doing the division $150 \div 25$.
That is, $\dfrac{150}{25} = 150 \div 25 = 6$.

35. $\dfrac{17}{51} = \dfrac{1 \cdot 17}{3 \cdot 17} = \dfrac{1}{3} \cdot \dfrac{17}{17} = \dfrac{1}{3} \cdot 1 = \dfrac{1}{3}$

37. We multiply these two numbers: We multiply these two numbers:

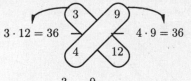

Since $36 = 36$, $\dfrac{3}{4} = \dfrac{9}{12}$.

39. We multiply these two numbers: We multiply these two numbers:

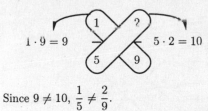

Since $9 \neq 10$, $\dfrac{1}{5} \neq \dfrac{2}{9}$.

41. We multiply these two numbers: We multiply these two numbers:

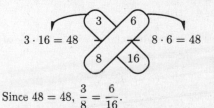

Since $48 = 48$, $\dfrac{3}{8} = \dfrac{6}{16}$.

43. We multiply these two numbers: We multiply these two numbers:

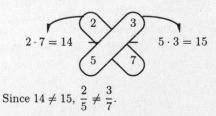

Since $14 \neq 15$, $\dfrac{2}{5} \neq \dfrac{3}{7}$.

45. We multiply these two numbers: We multiply these two numbers:

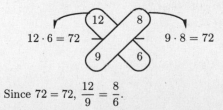

Since $72 = 72$, $\dfrac{12}{9} = \dfrac{8}{6}$.

47.

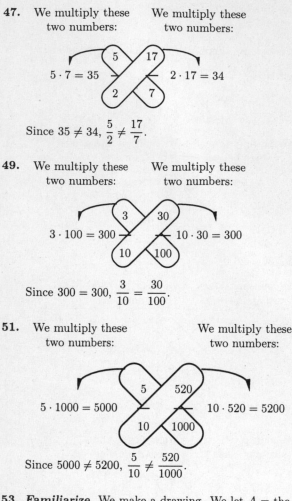

We multiply these two numbers: We multiply these two numbers:

$5 \cdot 7 = 35$ $2 \cdot 17 = 34$

Since $35 \neq 34$, $\dfrac{5}{2} \neq \dfrac{17}{7}$.

49.

We multiply these two numbers: We multiply these two numbers:

$3 \cdot 100 = 300$ $10 \cdot 30 = 300$

Since $300 = 300$, $\dfrac{3}{10} = \dfrac{30}{100}$.

51.

We multiply these two numbers: We multiply these two numbers:

$5 \cdot 1000 = 5000$ $10 \cdot 520 = 5200$

Since $5000 \neq 5200$, $\dfrac{5}{10} \neq \dfrac{520}{1000}$.

53. *Familiarize.* We make a drawing. We let A = the area.

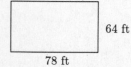

64 ft

78 ft

Translate. Using the formula for area, we have
$$A = l \cdot w = 78 \cdot 64.$$

Solve. We carry out the multiplication.

$$\begin{array}{r} 7\,8 \\ \times\ \ 6\,4 \\ \hline 3\,1\,2 \\ 4\,6\,8\,0 \\ \hline 4\,9\,9\,2 \end{array}$$

Thus, $A = 4992$.

Check. We repeat the calculation. The answer checks.

State. The area is 4992 ft^2.

55.

$$\begin{array}{r} 3\,4 \\ -\ 2\,3 \\ \hline 1\,1 \end{array}$$

57.

$$\begin{array}{r} {\scriptstyle 7\ 9\ 13} \\ \cancel{8\,0\,3} \\ -\ 6\,1\,7 \\ \hline 1\,8\,6 \end{array}$$

59. Think of each person as $\dfrac{1}{10}$. First we write fractional notation for the part of the population that is shy. The multiplication sentence $4 \cdot \dfrac{1}{10} = s$ corresponds to the situation. We multiply:
$$4 \cdot \frac{1}{10} = \frac{4 \cdot 1}{10} = \frac{4}{10}$$

Then we simplify:
$$\frac{4}{10} = \frac{2 \cdot 2}{5 \cdot 2} = \frac{2}{5} \cdot \frac{2}{2} = \frac{2}{5} \cdot 1 = \frac{2}{5}$$

Next we write fractional notation for the part of the population that is not shy. Since 4 of 10 people are shy, then $10 - 4$, or 6, of 10 people are not shy. The multiplication sentence $6 \cdot \dfrac{1}{10} = n$ corresponds to the situation. We multiply:
$$6 \cdot \frac{1}{10} = \frac{6 \cdot 1}{10} = \frac{6}{10}$$

Then we simplify:
$$\frac{6}{10} = \frac{2 \cdot 3}{2 \cdot 5} = \frac{2}{2} \cdot \frac{3}{5} = 1 \cdot \frac{3}{5} = \frac{3}{5}$$

61. Bernard Gilkey's batting average was $\dfrac{116}{384}$; Bip Robert's batting average was $\dfrac{172}{532}$. We test these fractions for equality:

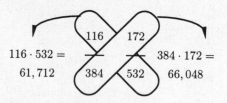

$116 \cdot 532 =$ $384 \cdot 172 =$
$61,712$ $66,048$

Since $61,712 \neq 66,048$, then $\dfrac{116}{384} \neq \dfrac{172}{532}$ and the batting averages are not the same.

Exercise Set 2.6

1. $\dfrac{2}{3} \cdot \dfrac{1}{2} = \dfrac{2 \cdot 1}{3 \cdot 2} = \dfrac{2}{2} \cdot \dfrac{1}{3} = \dfrac{1}{3}$

3. $\dfrac{7}{8} \cdot \dfrac{1}{7} = \dfrac{7 \cdot 1}{8 \cdot 7} = \dfrac{7}{7} \cdot \dfrac{1}{8} = \dfrac{1}{8}$

5. $\dfrac{1}{8} \cdot \dfrac{4}{5} = \dfrac{1 \cdot 4}{8 \cdot 5} = \dfrac{1 \cdot 4}{2 \cdot 4 \cdot 5} = \dfrac{4}{4} \cdot \dfrac{1}{2 \cdot 5} = \dfrac{1}{2 \cdot 5} = \dfrac{1}{10}$

7. $\dfrac{1}{4} \cdot \dfrac{2}{3} = \dfrac{1 \cdot 2}{4 \cdot 3} = \dfrac{1 \cdot 2}{2 \cdot 2 \cdot 3} = \dfrac{2}{2} \cdot \dfrac{1}{2 \cdot 3} = \dfrac{1}{2 \cdot 3} = \dfrac{1}{6}$

9. $\dfrac{12}{5} \cdot \dfrac{9}{8} = \dfrac{12 \cdot 9}{5 \cdot 8} = \dfrac{4 \cdot 3 \cdot 9}{5 \cdot 2 \cdot 4} = \dfrac{4}{4} \cdot \dfrac{3 \cdot 9}{5 \cdot 2} = \dfrac{3 \cdot 9}{5 \cdot 2} = \dfrac{27}{10}$

11. $\dfrac{10}{9} \cdot \dfrac{7}{5} = \dfrac{10 \cdot 7}{9 \cdot 5} = \dfrac{5 \cdot 2 \cdot 7}{9 \cdot 5} = \dfrac{5}{5} \cdot \dfrac{2 \cdot 7}{9} = \dfrac{2 \cdot 7}{9} = \dfrac{14}{9}$

13. $9 \cdot \dfrac{1}{9} = \dfrac{9 \cdot 1}{9} = \dfrac{9 \cdot 1}{9 \cdot 1} = 1$

15. $\dfrac{1}{3} \cdot 3 = \dfrac{1 \cdot 3}{3} = \dfrac{1 \cdot 3}{1 \cdot 3} = 1$

17. $\dfrac{7}{10} \cdot \dfrac{10}{7} = \dfrac{7 \cdot 10}{10 \cdot 7} = \dfrac{7 \cdot 10}{7 \cdot 10} = 1$

19. $\dfrac{7}{5} \cdot \dfrac{5}{7} = \dfrac{7 \cdot 5}{5 \cdot 7} = \dfrac{7 \cdot 5}{7 \cdot 5} = 1$

21. $\dfrac{1}{4} \cdot 8 = \dfrac{1 \cdot 8}{4} = \dfrac{8}{4} = \dfrac{4 \cdot 2}{4 \cdot 1} = \dfrac{4}{4} \cdot \dfrac{2}{1} = \dfrac{2}{1} = 2$

23. $24 \cdot \dfrac{1}{6} = \dfrac{24 \cdot 1}{6} = \dfrac{24}{6} = \dfrac{4 \cdot 6}{1 \cdot 6} = \dfrac{4}{1} \cdot \dfrac{6}{6} = \dfrac{4}{1} = 4$

25. $12 \cdot \dfrac{3}{4} = \dfrac{12 \cdot 3}{4} = \dfrac{4 \cdot 3 \cdot 3}{4 \cdot 1} = \dfrac{4}{4} \cdot \dfrac{3 \cdot 3}{1} = \dfrac{3 \cdot 3}{1} = 9$

27. $\dfrac{3}{8} \cdot 24 = \dfrac{3 \cdot 24}{8} = \dfrac{3 \cdot 3 \cdot 8}{1 \cdot 8} = \dfrac{8}{8} \cdot \dfrac{3 \cdot 3}{1} = \dfrac{3 \cdot 3}{1} = 9$

29. $13 \cdot \dfrac{2}{5} = \dfrac{13 \cdot 2}{5} = \dfrac{26}{5}$

31. $\dfrac{7}{10} \cdot 28 = \dfrac{7 \cdot 28}{10} = \dfrac{7 \cdot 2 \cdot 14}{2 \cdot 5} = \dfrac{2}{2} \cdot \dfrac{7 \cdot 14}{5} = \dfrac{7 \cdot 14}{5} = \dfrac{98}{5}$

33. $\dfrac{1}{6} \cdot 360 = \dfrac{1 \cdot 360}{6} = \dfrac{360}{6} = \dfrac{6 \cdot 60}{6 \cdot 1} = \dfrac{6}{6} \cdot \dfrac{60}{1} = \dfrac{60}{1} = 60$

35. $240 \cdot \dfrac{1}{8} = \dfrac{240 \cdot 1}{8} = \dfrac{240}{8} = \dfrac{8 \cdot 30}{8 \cdot 1} = \dfrac{8}{8} \cdot \dfrac{30}{1} = \dfrac{30}{1} = 30$

37. $\dfrac{4}{10} \cdot \dfrac{5}{10} = \dfrac{4 \cdot 5}{10 \cdot 10} = \dfrac{2 \cdot 2 \cdot 5 \cdot 1}{2 \cdot 5 \cdot 2 \cdot 5} = \dfrac{2 \cdot 2 \cdot 5}{2 \cdot 2 \cdot 5} \cdot \dfrac{1}{5} = \dfrac{1}{5}$

39. $\dfrac{8}{10} \cdot \dfrac{45}{100} = \dfrac{8 \cdot 45}{10 \cdot 100} = \dfrac{2 \cdot 2 \cdot 2 \cdot 5 \cdot 9}{2 \cdot 5 \cdot 2 \cdot 5 \cdot 2 \cdot 5}$

$= \dfrac{2 \cdot 2 \cdot 2 \cdot 5}{2 \cdot 2 \cdot 2 \cdot 5} \cdot \dfrac{9}{5 \cdot 5} = \dfrac{9}{5 \cdot 5} = \dfrac{9}{25}$

41. $\dfrac{11}{24} \cdot \dfrac{3}{5} = \dfrac{11 \cdot 3}{24 \cdot 5} = \dfrac{11 \cdot 3}{3 \cdot 8 \cdot 5} = \dfrac{3}{3} \cdot \dfrac{11}{8 \cdot 5} = \dfrac{11}{8 \cdot 5} = \dfrac{11}{40}$

43. $\dfrac{10}{21} \cdot \dfrac{3}{4} = \dfrac{10 \cdot 3}{21 \cdot 4} = \dfrac{2 \cdot 5 \cdot 3}{3 \cdot 7 \cdot 2 \cdot 2}$

$= \dfrac{2 \cdot 3}{2 \cdot 3} \cdot \dfrac{5}{7 \cdot 2} = \dfrac{5}{7 \cdot 2} = \dfrac{5}{14}$

45. *Familiarize.* We visualize the situation. We let $a =$ the amount received for working $\dfrac{3}{4}$ of a day.

1 day $36	
3/4 day a	

Translate. We write an equation.

$$\underbrace{\text{Pay for 3/4 of a day}}_{} \ \text{is} \ \dfrac{3}{4} \ \text{of} \ \$36$$

$$\begin{array}{ccccc} \downarrow & & \downarrow & \downarrow & \downarrow \\ a & = & \dfrac{3}{4} & \cdot & 36 \end{array}$$

Solve. We carry out the multiplication.

$$a = \dfrac{3}{4} \cdot 36 = \dfrac{3 \cdot 36}{4}$$
$$= \dfrac{3 \cdot 9 \cdot 4}{1 \cdot 4} = \dfrac{3 \cdot 9}{1} \cdot \dfrac{4}{4}$$
$$= 27$$

Check. We can repeat the calculation. We can also determine that the answer seems reasonable since we multiplied 36 by a number less than 1 and the result is less than 36. The answer checks.

State. \$27 is received for working $\dfrac{3}{4}$ of a day.

47. *Familiarize.* We visualize the situation. We let $n =$ the number of addresses that will be incorrect after one year.

Mailing list 2500 addresses		
1/4 of the addresses n		

Translate.

$$\underbrace{\text{Number incorrect}}_{} \ \text{is} \ \dfrac{1}{4} \ \text{of} \ \underbrace{\text{Number of addresses}}_{}$$

$$\begin{array}{ccccc} \downarrow & \downarrow & \downarrow & \downarrow & \downarrow \\ n & = & \dfrac{1}{4} & \cdot & 2500 \end{array}$$

Solve. We carry out the multiplication.

$$n = \dfrac{1}{4} \cdot 2500 = \dfrac{1 \cdot 2500}{4} = \dfrac{2500}{4}$$
$$= \dfrac{4 \cdot 625}{4 \cdot 1} = \dfrac{4}{4} \cdot \dfrac{625}{1}$$
$$= 625$$

Check. We can repeat the calculation. We can also determine that the answer seems reasonable since we multiplied 2500 by a number less than 1 and the result is less than 2500. The answer checks.

State. After one year 625 addresses will be incorrect.

49. *Familiarize.* We draw a picture.

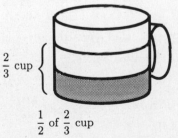

$\dfrac{2}{3}$ cup

$\dfrac{1}{2}$ of $\dfrac{2}{3}$ cup

We let $n =$ the amount of flour the chef should use.

Translate. The multiplication sentence

$$\dfrac{1}{2} \cdot \dfrac{2}{3} = n$$

corresponds to the situation.

Solve. We multiply and simplify:

$$n = \frac{1}{2} \cdot \frac{2}{3} = \frac{1 \cdot 2}{2 \cdot 3} = \frac{2}{2} \cdot \frac{1}{3} = \frac{1}{3}$$

Check. We can repeat the calculation. We can also determine that the answer seems reasonable since we multiplied $\frac{2}{3}$ by a number less than 1 and the result is less than $\frac{2}{3}$. The answer checks.

State. The chef should use $\frac{1}{3}$ cup of flour.

51. Familiarize. We visualize the situation. Let $a =$ the amount of the loan.

Tuition $2400	
2/3 of the tuition $a	

Translate. We write an equation.

$$\underbrace{\text{Amount of loan}} \quad \text{is} \quad \frac{2}{3} \quad \text{of} \quad \underbrace{\text{the tuition}}$$
$$\downarrow \qquad\qquad \downarrow \quad \downarrow \quad \downarrow \qquad \downarrow$$
$$a \qquad\qquad = \quad \frac{2}{3} \quad \cdot \quad 2400$$

Solve. We carry out the multiplication.

$$a = \frac{2}{3} \cdot 2400 = \frac{2 \cdot 2400}{3}$$
$$= \frac{2 \cdot 3 \cdot 800}{3 \cdot 1} = \frac{3}{3} \cdot \frac{2 \cdot 800}{1}$$
$$= 1600$$

Check. We can repeat the calculation. We can also determine that the answer seems reasonable since we multiplied 2400 by a number less than 1 and the result is less than 2400. The answer checks.

State. The loan was $1600.

53. Familiarize. We draw a picture.

$\frac{2}{3}$ in.

1 in.
240 miles

We let $n =$ the number of miles represented by $\frac{2}{3}$ in.

Translate. The multiplication sentence

$$n = \frac{2}{3} \cdot 240$$

corresponds to the situation.

Solve. We multiply and simplify:

$$n = \frac{2}{3} \cdot 240 = \frac{2 \cdot 240}{3} = \frac{2 \cdot 3 \cdot 80}{1 \cdot 3}$$
$$= \frac{3}{3} \cdot \frac{2 \cdot 80}{1} = \frac{2 \cdot 80}{1}$$
$$= 160$$

Check. We can repeat the calculation. We can also determine that the answer seems reasonable since we multiplied 240 by a number less than 1 and the result is less than 240.

State. $\frac{2}{3}$ in. on the map represents 160 miles.

55. Familiarize. This is a multistep problem. First we find the amount of each of the given expenses. Then we find the total of these expenses and take it away from the annual income to find how much is spent for other expenses.

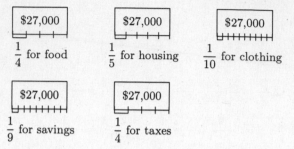

We let f, h, c, s, and t represent the amounts spent on food, housing, clothing, savings, and taxes, respectively.

Translate. The following multiplication sentences correspond to the situation.

$$\frac{1}{4} \cdot 27{,}000 = f \qquad \frac{1}{9} \cdot 27{,}000 = s$$
$$\frac{1}{5} \cdot 27{,}000 = h \qquad \frac{1}{4} \cdot 27{,}000 = t$$
$$\frac{1}{10} \cdot 27{,}000 = c$$

Solve. We multiply and simplify.

$$f = \frac{1}{4} \cdot 27{,}000 = \frac{27{,}000}{4} = \frac{4 \cdot 6750}{4 \cdot 1} = \frac{4}{4} \cdot \frac{6750}{1} =$$
$$6750$$

$$h = \frac{1}{5} \cdot 27{,}000 = \frac{27{,}000}{5} = \frac{5 \cdot 5400}{5 \cdot 1} = \frac{5}{5} \cdot \frac{5400}{1} =$$
$$5400$$

$$c = \frac{1}{10} \cdot 27{,}000 = \frac{27{,}000}{10} = \frac{10 \cdot 2700}{10 \cdot 1} = \frac{10}{10} \cdot \frac{2700}{1} =$$
$$2700$$

$$s = \frac{1}{9} \cdot 27{,}000 = \frac{27{,}000}{9} = \frac{9 \cdot 3000}{9 \cdot 1} = \frac{9}{9} \cdot \frac{3000}{1} =$$
$$3000$$

$$t = \frac{1}{4} \cdot 27{,}000 = \frac{27{,}000}{4} = \frac{4 \cdot 6750}{4 \cdot 1} = \frac{4}{4} \cdot \frac{6750}{1} =$$
$$6750$$

We add to find the total of these expenses.

$$\begin{array}{r} {}^{2}\ {}^{1}\ \\ \$\,6\ 7\ 5\ 0 \\ 5\ 4\ 0\ 0 \\ 2\ 7\ 0\ 0 \\ 3\ 0\ 0\ 0 \\ 6\ 7\ 5\ 0 \\ \hline \$\,2\,4,6\,0\,0 \end{array}$$

We let m = the amount spent on other expenses and subtract to find this amount.

Annual income	minus	Total of itemized expenses	is	Total spent on other expenses
↓	↓	↓	↓	↓
$27,000	−	$24,600	=	m
		$2400	=	m

Subtracting

Check. We repeat the calculations. The results check.

State. $6750 is spent for food, $5400 for housing, $2700 for clothing, $3000 for savings, $6750 for taxes, and $2400 for other expenses.

57. $48 \cdot t = 1680$

$$\frac{48 \cdot t}{48} = \frac{1680}{48}$$

$$t = 35$$

The solution is 35.

59. $t + 28 = 5017$

$$t = 5017 - 28$$

$$t = 4989$$

The solution is 4989.

61.
$$\begin{array}{r} {}^{8\ 9\ \overset{15}{\cancel{6}}\ 10} \\ \cancel{9\,0\,6}\ \cancel{0} \\ -\ 4\ 3\ 8\ 7 \\ \hline 4\ 6\ 7\ 3 \end{array}$$

Exercise Set 2.7

1. $\dfrac{5}{6}$ Interchange the numerator and denominator.

The reciprocal of $\dfrac{5}{6}$ is $\dfrac{6}{5}$. $\left(\dfrac{5}{6} \cdot \dfrac{6}{5} = \dfrac{30}{30} = 1\right)$

3. Think of 6 as $\dfrac{6}{1}$.

$\dfrac{6}{1}$ Interchange the numerator and denominator.

The reciprocal of $\dfrac{6}{1}$ is $\dfrac{1}{6}$. $\left(\dfrac{6}{1} \cdot \dfrac{1}{6} = \dfrac{6}{6} = 1\right)$

5. $\dfrac{1}{6}$ Interchange the numerator and denominator.

The reciprocal of $\dfrac{1}{6}$ is 6. $\left(\dfrac{6}{1} = 6;\ \dfrac{1}{6} \cdot \dfrac{6}{1} = \dfrac{6}{6} = 1\right)$

(Note that we also found that 6 and $\dfrac{1}{6}$ are reciprocals in Exercise 3.)

7. $\dfrac{10}{3}$ Interchange the numerator and denominator.

The reciprocal of $\dfrac{10}{3}$ is $\dfrac{3}{10}$. $\left(\dfrac{10}{3} \cdot \dfrac{3}{10} = \dfrac{30}{30} = 1\right)$

9. $\dfrac{3}{5} \div \dfrac{3}{4} = \dfrac{3}{5} \cdot \dfrac{4}{3}$ Multiplying the dividend $\left(\dfrac{3}{5}\right)$ by the reciprocal of the divisor $\left(\text{The reciprocal of } \dfrac{3}{4} \text{ is } \dfrac{4}{3}.\right)$

$= \dfrac{3 \cdot 4}{5 \cdot 3}$ Multiplying numerators and denominators

$= \dfrac{3}{3} \cdot \dfrac{4}{5} = \dfrac{4}{5}$ Simplifying

11. $\dfrac{3}{5} \div \dfrac{9}{4} = \dfrac{3}{5} \cdot \dfrac{4}{9}$ Multiplying the dividend $\left(\dfrac{3}{5}\right)$ by the reciprocal of the divisor $\left(\text{The reciprocal of } \dfrac{9}{4} \text{ is } \dfrac{4}{9}.\right)$

$= \dfrac{3 \cdot 4}{5 \cdot 9}$ Multiplying numerators and denominators

$= \dfrac{3 \cdot 4}{5 \cdot 3 \cdot 3}$

$= \dfrac{3}{3} \cdot \dfrac{4}{5 \cdot 3}$ Simplifying

$= \dfrac{4}{5 \cdot 3} = \dfrac{4}{15}$

13. $\dfrac{4}{3} \div \dfrac{1}{3} = \dfrac{4}{3} \cdot 3 = \dfrac{4 \cdot 3}{3} = \dfrac{3}{3} \cdot 4 = 4$

15. $\dfrac{1}{3} \div \dfrac{1}{6} = \dfrac{1}{3} \cdot 6 = \dfrac{1 \cdot 6}{3} = \dfrac{1 \cdot 2 \cdot 3}{1 \cdot 3} = \dfrac{1 \cdot 3}{1 \cdot 3} \cdot 2 = 2$

17. $\dfrac{3}{8} \div 3 = \dfrac{3}{8} \cdot \dfrac{1}{3} = \dfrac{3 \cdot 1}{8 \cdot 3} = \dfrac{3}{3} \cdot \dfrac{1}{8} = \dfrac{1}{8}$

19. $\dfrac{12}{7} \div 4 = \dfrac{12}{7} \cdot \dfrac{1}{4} = \dfrac{12 \cdot 1}{7 \cdot 4} = \dfrac{4 \cdot 3 \cdot 1}{7 \cdot 4} = \dfrac{4}{4} \cdot \dfrac{3 \cdot 1}{7} =$

$\dfrac{3 \cdot 1}{7} = \dfrac{3}{7}$

21. $12 \div \dfrac{3}{2} = 12 \cdot \dfrac{2}{3} = \dfrac{12 \cdot 2}{3} = \dfrac{3 \cdot 4 \cdot 2}{3 \cdot 1} = \dfrac{3}{3} \cdot \dfrac{4 \cdot 2}{1}$

$= \dfrac{4 \cdot 2}{1} = \dfrac{8}{1} = 8$

23. $28 \div \dfrac{4}{5} = 28 \cdot \dfrac{5}{4} = \dfrac{28 \cdot 5}{4} = \dfrac{4 \cdot 7 \cdot 5}{4 \cdot 1} = \dfrac{4}{4} \cdot \dfrac{7 \cdot 5}{1}$

$= \dfrac{7 \cdot 5}{1} = 35$

25. $\dfrac{5}{8} \div \dfrac{5}{8} = \dfrac{5}{8} \cdot \dfrac{8}{5} = \dfrac{5 \cdot 8}{8 \cdot 5} = \dfrac{5 \cdot 8}{5 \cdot 8} = 1$

27. $\dfrac{8}{15} \div \dfrac{4}{5} = \dfrac{8}{15} \cdot \dfrac{5}{4} = \dfrac{8 \cdot 5}{15 \cdot 4} = \dfrac{2 \cdot 4 \cdot 5}{3 \cdot 5 \cdot 4} = \dfrac{4 \cdot 5}{4 \cdot 5} \cdot \dfrac{2}{3} = \dfrac{2}{3}$

29. $\dfrac{9}{5} \div \dfrac{4}{5} = \dfrac{9}{5} \cdot \dfrac{5}{4} = \dfrac{9 \cdot 5}{5 \cdot 4} = \dfrac{5}{5} \cdot \dfrac{9}{4} = \dfrac{9}{4}$

31. $120 \div \dfrac{5}{6} = 120 \cdot \dfrac{6}{5} = \dfrac{120 \cdot 6}{5} = \dfrac{5 \cdot 24 \cdot 6}{5 \cdot 1} = \dfrac{5}{5} \cdot \dfrac{24 \cdot 6}{1}$

$\qquad = \dfrac{24 \cdot 6}{1} = 144$

33. $\dfrac{4}{5} \cdot x = 60$

$\qquad x = 60 \div \dfrac{4}{5} \qquad$ Dividing on both sides by $\dfrac{4}{5}$

$\qquad x = 60 \cdot \dfrac{5}{4} \qquad$ Multiplying by the reciprocal

$\qquad = \dfrac{60 \cdot 5}{4} = \dfrac{4 \cdot 15 \cdot 5}{4 \cdot 1} = \dfrac{4}{4} \cdot \dfrac{15 \cdot 5}{1} = \dfrac{15 \cdot 5}{1} = 75$

35. $\dfrac{5}{3} \cdot y = \dfrac{10}{3}$

$\qquad y = \dfrac{10}{3} \div \dfrac{5}{3} \qquad$ Dividing on both sides by $\dfrac{5}{3}$

$\qquad y = \dfrac{10}{3} \cdot \dfrac{3}{5} \qquad$ Multiplying by the reciprocal

$\qquad = \dfrac{10 \cdot 3}{3 \cdot 5} = \dfrac{2 \cdot 5 \cdot 3}{3 \cdot 5 \cdot 1} = \dfrac{5 \cdot 3}{5 \cdot 3} \cdot \dfrac{2}{1} = \dfrac{2}{1} = 2$

37. $x \cdot \dfrac{25}{36} = \dfrac{5}{12}$

$\qquad x = \dfrac{5}{12} \div \dfrac{25}{36} = \dfrac{5}{12} \cdot \dfrac{36}{25} = \dfrac{5 \cdot 36}{12 \cdot 25} = \dfrac{5 \cdot 3 \cdot 12}{12 \cdot 5 \cdot 5}$

$\qquad = \dfrac{5 \cdot 12}{5 \cdot 12} \cdot \dfrac{3}{5} = \dfrac{3}{5}$

39. $n \cdot \dfrac{8}{7} = 360$

$\qquad n = 360 \div \dfrac{8}{7} = 360 \cdot \dfrac{7}{8} = \dfrac{360 \cdot 7}{8} = \dfrac{8 \cdot 45 \cdot 7}{8 \cdot 1}$

$\qquad = \dfrac{8}{8} \cdot \dfrac{45 \cdot 7}{1} = \dfrac{45 \cdot 7}{1} = 315$

41. *Familiarize*. We draw a picture.

$\underbrace{\vdash\!+\!+\!+\!+\!+\!+\!\dashv}_{}$ 6 pieces of the same length

$\underbrace{}$
$\dfrac{3}{5}$ m

We let $n =$ the length of each piece.

***Translate*.** The multiplication that corresponds to the situation is

$\qquad 6 \cdot n = \dfrac{3}{5}.$

***Solve*.** We solve the equation by dividing on both sides by 6 and carrying out the division:

$n = \dfrac{3}{5} \div 6 = \dfrac{3}{5} \cdot \dfrac{1}{6} = \dfrac{3 \cdot 1}{5 \cdot 6} = \dfrac{3 \cdot 1}{5 \cdot 2 \cdot 3} = \dfrac{3}{3} \cdot \dfrac{1}{5 \cdot 2} =$

$\dfrac{1}{5 \cdot 2} = \dfrac{1}{10}$

***Check*.** We repeat the calculation. The answer checks.

***State*.** Each piece is $\dfrac{1}{10}$ m.

43. *Familiarize*. We draw a picture. We let $n =$ the number of pairs of soccer shorts that can be made.

$\underbrace{\boxed{\dfrac{3}{4}\ \text{yd}} \quad \boxed{\dfrac{3}{4}\ \text{yd}} \quad \cdots \quad \boxed{\dfrac{3}{4}\ \text{yd}}}_{n \text{ pairs of shorts}}$

***Translate*.** The multiplication that corresponds to the situation is

$\qquad \dfrac{3}{4} \cdot n = 24.$

***Solve*.** We solve the equation by dividing on both sides by $\dfrac{3}{4}$ and carrying out the division:

$n = 24 \div \dfrac{3}{4} = 24 \cdot \dfrac{4}{3} = \dfrac{24 \cdot 4}{3} = \dfrac{3 \cdot 8 \cdot 4}{3 \cdot 1} = \dfrac{3}{3} \cdot \dfrac{8 \cdot 4}{1}$

$\qquad = \dfrac{8 \cdot 4}{1} = 32$

***Check*.** We repeat the calculation. The answer checks.

***State*.** 32 pairs of soccer shorts can be made from 24 yd of fabric.

45. *Familiarize*. We draw a picture. We let $n =$ the number of sugar bowls that can be filled.

$\underbrace{\boxed{\dfrac{2}{3}\ \text{cup}} \quad \boxed{\dfrac{2}{3}\ \text{cup}} \quad \cdots \quad \boxed{\dfrac{2}{3}\ \text{cup}}}_{n \text{ bowls}}$

***Translate*.** We write a multiplication sentence:

$\qquad \dfrac{2}{3} \cdot n = 16$

***Solve*.** Solve the equation as follows:

$\dfrac{2}{3} \cdot n = 16$

$n = 16 \div \dfrac{2}{3} = 16 \cdot \dfrac{3}{2} = \dfrac{16 \cdot 3}{2} = \dfrac{2 \cdot 8 \cdot 3}{2 \cdot 1}$

$\qquad = \dfrac{2}{2} \cdot \dfrac{8 \cdot 3}{1} = \dfrac{8 \cdot 3}{1} = 24$

***Check*.** We repeat the calculation. The answer checks.

***State*.** 24 sugar bowls can be filled.

47. *Familiarize*. We draw a picture. We let $n =$ the amount the bucket could hold.

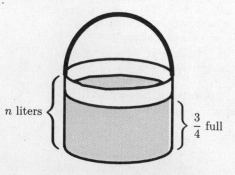

Translate. We write a multiplication sentence:

$$\frac{3}{4} \cdot n = 12$$

Solve. Solve the equation as follows:

$$\frac{3}{4} \cdot n = 12$$

$$n = 12 \div \frac{3}{4} = 12 \cdot \frac{4}{3} = \frac{12 \cdot 4}{3} = \frac{3 \cdot 4 \cdot 4}{3 \cdot 1}$$

$$= \frac{3}{3} \cdot \frac{4 \cdot 4}{1} = \frac{4 \cdot 4}{1} = 16$$

Check. We repeat the calculation. The answer checks.

State. The bucket could hold 16 L.

49. *Familiarize*. This is a multistep problem. First we find the length of the total trip. Then we find how many kilometers were left to drive. We draw a picture. We let $n =$ the length of the total trip.

$\frac{5}{8}$ of the trip

180 km

n km

Translate. We translate to an equation.

Fraction of trip completed	times	Total length of trip	is	Amount already traveled
↓	↓	↓	↓	↓
$\frac{5}{8}$	$\cdot$	n	$=$	180

Solve. We solve the equation as follows:

$$\frac{5}{8} \cdot n = 180$$

$$n = 180 \div \frac{5}{8} = 180 \cdot \frac{8}{5} = \frac{5 \cdot 36 \cdot 8}{5 \cdot 1}$$

$$= \frac{5}{5} \cdot \frac{36 \cdot 8}{1} = \frac{36 \cdot 8}{1} = 288$$

The total trip was 288 km.

Now we find how many kilometers were left to travel. Let $t =$ this number.

Length of total trip	minus	Distance traveled	is	Distance left to travel
↓	↓	↓	↓	↓
288	$-$	180	$=$	t

We carry out the subtraction:

$$288 - 180 = t$$

$$108 = t$$

Check. We repeat the calculation. The results check.

State. The total trip was 288 km. There were 108 km left to travel.

51.

```
      6 7
  4 | 2 6 8
      2 4 0
      -----
        2 8
        2 8
      -----
          0
```

The answer is 67.

53.

```
        2 8 5
  2 4 | 6 8 4 2
        4 8 0 0
        -------
        2 0 4 2
        1 9 2 0
        -------
          1 2 2
          1 2 0
          -----
              2
```

The answer is 285 R 2.

55. Let $n =$ the number.

$$\frac{1}{3} \cdot n = \frac{1}{4}$$

$$n = \frac{1}{4} \div \frac{1}{3} = \frac{1}{4} \cdot \frac{3}{1} = \frac{1 \cdot 3}{4 \cdot 1} = \frac{3}{4}$$

The number is $\frac{3}{4}$. Now we find $\frac{1}{2}$ of $\frac{3}{4}$.

$$\frac{1}{2} \cdot \frac{3}{4} = \frac{1 \cdot 3}{2 \cdot 4} = \frac{3}{8}$$

One-half of the number is $\frac{3}{8}$.

Chapter 3

Addition and Subtraction: Fractional Notation

Exercise Set 3.1

In this section we will find the LCM using the multiples method in Exercises 1 - 19 and the factorization method in Exercises 21 - 43.

1. a) 4 is a multiple of 2, so it is the LCM.

   c) The LCM = 4.

3. a) 25 is not a multiple of 10.

   b) Check multiples:

   $2 \cdot 25 = 50$ A multiple of 10

   c) The LCM = 50.

5. a) 40 is a multiple of 20, so it is the LCM.

   c) The LCM = 40.

7. a) 27 is not a multiple of 18.

   b) Check multiples:

   $2 \cdot 27 = 54$ A multiple of 18

   c) The LCM = 54.

9. a) 50 is not a multiple of 30.

   b) Check multiples:

$2 \cdot 50 = 100$	Not a multiple of 30
$3 \cdot 50 = 150$	A multiple of 30

   c) The LCM = 150.

11. a) 40 is not a multiple of 30.

    b) Check multiples:

$2 \cdot 40 = 80$	Not a multiple of 30
$3 \cdot 40 = 120$	A multiple of 30

    c) The LCM = 120.

13. a) 24 is not a multiple of 18.

    b) Check multiples:

$2 \cdot 24 = 48$	Not a multiple of 18
$3 \cdot 24 = 72$	A multiple of 18

    c) The LCM = 72.

15. a) 70 is not a multiple of 60.

    b) Check multiples:

$2 \cdot 70 = 140$	Not a multiple of 60
$3 \cdot 70 = 210$	Not a multiple of 60
$4 \cdot 70 = 280$	Not a multiple of 60
$5 \cdot 70 = 350$	Not a multiple of 60
$6 \cdot 70 = 420$	A multiple of 60

    c) The LCM = 420.

17. a) 36 is not a multiple of 16.

    b) Check multiples:

$2 \cdot 36 = 72$	Not a multiple of 16
$3 \cdot 36 = 108$	Not a multiple of 16
$4 \cdot 36 = 144$	A multiple of 16

    c) The LCM = 144.

19. a) 36 is not a multiple of 32.

    b) Check multiples:

$2 \cdot 36 = 72$	Not a multiple of 32
$3 \cdot 36 = 108$	Not a multiple of 32
$4 \cdot 36 = 144$	Not a multiple of 32
$5 \cdot 36 = 180$	Not a multiple of 32
$6 \cdot 36 = 216$	Not a multiple of 32
$7 \cdot 36 = 252$	Not a multiple of 32
$8 \cdot 36 = 288$	A multiple of 32

    c) The LCM = 288.

21. Note that each of the numbers 2, 3, and 5 is prime. They have no common prime factor. When this happens, the LCM is just the product of the numbers.

    The LCM is $2 \cdot 3 \cdot 5$, or 30.

23. Note that each of the numbers 3, 5, and 7 is prime. They have no common prime factor. When this happens, the LCM is just the product of the numbers.

    The LCM is $3 \cdot 5 \cdot 7$, or 105.

25. a) Find the prime factorization of each number.

    $$24 = 2 \cdot 2 \cdot 2 \cdot 3$$
    $$36 = 2 \cdot 2 \cdot 3 \cdot 3$$
    $$12 = 2 \cdot 2 \cdot 3$$

    b) Create a product by writing factors, using each the greatest number of times it occurs in any one factorization.

    Consider the factor 2. The greatest number of times 2 occurs in any one factorization is three. We write 2 as a factor three times.

    $$2 \cdot 2 \cdot 2 \cdot ?$$

    Consider the factor 3. The greatest number of times 3 occurs in any one factorization is two. We write 3 as a factor two times.

    $$2 \cdot 2 \cdot 2 \cdot 3 \cdot 3 \cdot ?$$

    Since there are no other prime factors in any of the factorizations, the LCM is $2 \cdot 2 \cdot 2 \cdot 3 \cdot 3$, or 72.

27. a) Find the prime factorization of each number.

$$5 = 5 \qquad \text{(5 is prime.)}$$
$$12 = 2 \cdot 2 \cdot 3$$
$$15 = 3 \cdot 5$$

b) Create a product by writing each factor the greatest number of times it occurs in any one factorization.

The greatest number of times 2 occurs in any one factorization is two times.

The greatest number of times 3 occurs in any one factorization is one time.

The greatest number of times 5 occurs in any one factorization is one time.

Since there are no other prime factors in any of the factorizations, the LCM is $2 \cdot 2 \cdot 3 \cdot 5$, or 60.

29. a) Find the prime factorization of each number.

$$9 = 3 \cdot 3$$
$$12 = 2 \cdot 2 \cdot 3$$
$$6 = 2 \cdot 3$$

b) Create a product by writing each factor the greatest number of times it occurs in any one factorization.

The greatest number of times 2 occurs in any one factorization is two times.

The greatest number of times 3 occurs in any one factorization is two times.

Since there are no other prime factors in any of the factorizations, the LCM is $2 \cdot 2 \cdot 3 \cdot 3$, or 36.

31. a) Find the prime factorization of each number.

$$3 = 3 \qquad \text{(3 is prime.)}$$
$$6 = 2 \cdot 3$$
$$8 = 2 \cdot 2 \cdot 2$$

b) Create a product by writing each factor the greatest number of times it occurs in any one factorization.

The greatest number of times 2 occurs in any one factorization is three times.

The greatest number of times 3 occurs in any one factorization is one time.

Since there are no other prime factors in any of the factorizations, the LCM is $2 \cdot 2 \cdot 2 \cdot 3$, or 24.

33. Note that 8 is a factor of 48. If one number is a factor of another, the LCM is the greater number.

The LCM is 48.

The factorization method will also work here if you do not recognize at the outset that 8 is a factor of 48.

35. Note that 5 is a factor of 50. If one number is a factor of another, the LCM is the greater number.

The LCM is 50.

37. Note that 11 and 13 are prime. They have no common prime factor. When this happens, the LCM is just the product of the numbers.

The LCM is $11 \cdot 13$, or 143.

39. a) Find the prime factorization of each number.

$$12 = 2 \cdot 2 \cdot 3$$
$$35 = 5 \cdot 7$$

b) Note that the two numbers have no common prime factor. When this happens, the LCM is just the product of the numbers.

The LCM is $12 \cdot 35$, or 420.

41. a) Find the prime factorization of each number.

$$54 = 3 \cdot 3 \cdot 3 \cdot 2$$
$$63 = 3 \cdot 3 \cdot 7$$

b) Create a product by writing each factor the greatest number of times it occurs in any one factorization.

The greatest number of times 2 occurs in any one factorization is one time.

The greatest number of times 3 occurs in any one factorization is three times.

The greatest number of times 7 occurs in any one factorization is one time.

Since there are no other prime factors in any of the factorizations, the LCM is $2 \cdot 3 \cdot 3 \cdot 3 \cdot 7$, or 378.

43. a) Find the prime factorization of each number.

$$81 = 3 \cdot 3 \cdot 3 \cdot 3$$
$$90 = 2 \cdot 3 \cdot 3 \cdot 5$$

b) Create a product by writing each factor the greatest number of times it occurs in any one factorization.

The greatest number of times 2 occurs in any one factorization is one time.

The greatest number of times 3 occurs in any one factorization is four times.

The greatest number of times 5 occurs in any one factorization is one time.

Since there are no other prime factors in any of the factorizations, the LCM is $2 \cdot 3 \cdot 3 \cdot 3 \cdot 3 \cdot 5$, or 810.

45. *Familiarize*. We draw a picture. Repeated addition applies here.

$3250				
$13	$13	$\cdots$		$13

Translate. We must determine how many 13's there are in 3250. This number is the number of seats in the auditorium. We let $x =$ the number of seats in the auditorium.

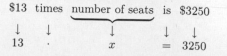

Solve. To solve the equation, we divide on both sides by 13.

$$13 \cdot x = 3250$$
$$\frac{13 \cdot x}{13} = \frac{3250}{13}$$
$$x = 250$$

Check. If 250 seats are sold at \$13 each, the total receipts are $250 \cdot \$13$, or \$3250. The result checks.

State. The auditorium contains 250 seats.

47. $\dfrac{4}{5} \cdot \dfrac{10}{12} = \dfrac{4 \cdot 10}{5 \cdot 12} = \dfrac{40}{60} = \dfrac{2 \cdot 20}{3 \cdot 20} = \dfrac{2}{3} \cdot \dfrac{20}{20} = \dfrac{2}{3} \cdot 1 = \dfrac{2}{3}$

49. The length of the carton must be a multiple of both 6 and 8. The shortest length carton will be the least common multiple of 6 and 8.

$$6 = 2 \cdot 3$$
$$8 = 2 \cdot 2 \cdot 2$$

LCM is $2 \cdot 2 \cdot 2 \cdot 3$, or 24.

The shortest carton is 24 in. long.

51. a) 324 is not a multiple of 288.

b) Check multiples, using a calculator.

$2 \cdot 324 = 648$ Not a multiple of 288
$3 \cdot 324 = 972$ Not a multiple of 288
$4 \cdot 324 = 1296$ Not a multiple of 288
$5 \cdot 324 = 1620$ Not a multiple of 288
$6 \cdot 324 = 1944$ Not a multiple of 288
$7 \cdot 324 = 2268$ Not a multiple of 288
$8 \cdot 324 = 2592$ A multiple of 288

c) The LCM = 2592.

Exercise Set 3.2

1. $\dfrac{7}{8} + \dfrac{1}{8} = \dfrac{7+1}{8} = \dfrac{8}{8} = 1$

3. $\dfrac{1}{8} + \dfrac{5}{8} = \dfrac{1+5}{8} = \dfrac{6}{8} = \dfrac{3 \cdot 2}{4 \cdot 2} = \dfrac{3}{4} \cdot \dfrac{2}{2} = \dfrac{3}{4} \cdot 1 = \dfrac{3}{4}$

5. $\dfrac{2}{3} + \dfrac{5}{6}$ 3 is a factor of 6, so the LCD is 6.

$= \dfrac{2}{3} \cdot \dfrac{2}{2} + \dfrac{5}{6}$ ← This fraction already has the LCD as denominator.

Think: $3 \times \square = 6$. The answer is 2, so we multiply by 1, using $\dfrac{2}{2}$.

$= \dfrac{4}{6} + \dfrac{5}{6} = \dfrac{9}{6}$

$= \dfrac{3}{2}$ Simplifying

7. $\dfrac{1}{8} + \dfrac{1}{6}$ $8 = 2 \cdot 2 \cdot 2$ and $6 = 2 \cdot 3$, so the LCD is $2 \cdot 2 \cdot 2 \cdot 3$, or 24

$= \underbrace{\dfrac{1}{8} \cdot \dfrac{3}{3}} + \underbrace{\dfrac{1}{6} \cdot \dfrac{4}{4}}$

Think: $6 \times \square = 24$. The answer is 4, so we multiply by 1, using $\dfrac{4}{4}$.

Think: $8 \times \square = 24$. The answer is 3, so we multiply by 1, using $\dfrac{3}{3}$.

$= \dfrac{3}{24} + \dfrac{4}{24}$

$= \dfrac{7}{24}$

9. $\dfrac{4}{5} + \dfrac{7}{10}$ 5 is a factor of 10, so the LCD is 10.

$= \underbrace{\dfrac{4}{5} \cdot \dfrac{2}{2}} + \dfrac{7}{10}$ ← This fraction already has the LCD as denominator.

Think: $5 \times \square = 10$. The answer is 2, so we multiply by 1, using $\dfrac{2}{2}$.

$= \dfrac{8}{10} + \dfrac{7}{10} = \dfrac{15}{10}$

$= \dfrac{3}{2}$ Simplifying

11. $\dfrac{5}{12} + \dfrac{3}{8}$ $12 = 2 \cdot 2 \cdot 3$ and $8 = 2 \cdot 2 \cdot 2$, so the LCD is $2 \cdot 2 \cdot 2 \cdot 3$, or 24.

$= \underbrace{\dfrac{5}{12} \cdot \dfrac{2}{2}} + \underbrace{\dfrac{3}{8} \cdot \dfrac{3}{3}}$

Think: $8 \times \square = 24$. The answer is 3, so we multiply by 1, using $\dfrac{3}{3}$.

Think: $12 \times \square = 24$. The answer is 2, so we multiply by 1, using $\dfrac{2}{2}$.

$= \dfrac{10}{24} + \dfrac{9}{24} = \dfrac{19}{24}$

13. $\dfrac{3}{20} + \dfrac{3}{4}$ 4 is a factor of 20, so the LCD is 20.

$= \dfrac{3}{20} + \dfrac{3}{4} \cdot \dfrac{5}{5}$ Multiplying by 1

$= \dfrac{3}{20} + \dfrac{15}{20} = \dfrac{18}{20} = \dfrac{9}{10}$

15. $\dfrac{5}{6} + \dfrac{7}{9}$ $6 = 2 \cdot 3$ and $9 = 3 \cdot 3$, so the LCD is $2 \cdot 3 \cdot 3$, or 18.

$= \dfrac{5}{6} \cdot \dfrac{3}{3} + \dfrac{7}{9} \cdot \dfrac{2}{2}$ Multiplying by 1

$= \dfrac{15}{18} + \dfrac{14}{18} = \dfrac{29}{18}$

17. $\dfrac{3}{10} + \dfrac{1}{100}$ 10 is a factor of 100, so the LCD is 100.

$$= \dfrac{3}{10} \cdot \dfrac{10}{10} + \dfrac{1}{100}$$

$$= \dfrac{30}{100} + \dfrac{1}{100} = \dfrac{31}{100}$$

19. $\dfrac{5}{12} + \dfrac{4}{15}$ $12 = 2 \cdot 2 \cdot 3$ and $15 = 3 \cdot 5$, so the LCD is $2 \cdot 2 \cdot 3 \cdot 5$, or 60.

$$= \dfrac{5}{12} \cdot \dfrac{5}{5} + \dfrac{4}{15} \cdot \dfrac{4}{4}$$

$$= \dfrac{25}{60} + \dfrac{16}{60} = \dfrac{41}{60}$$

21. $\dfrac{9}{10} + \dfrac{99}{100}$ 10 is a factor of 100, so the LCD is 100.

$$= \dfrac{9}{10} \cdot \dfrac{10}{10} + \dfrac{99}{100}$$

$$= \dfrac{90}{100} + \dfrac{99}{100} = \dfrac{189}{100}$$

23. $\dfrac{7}{8} + \dfrac{0}{1}$ 1 is a factor of 8, so the LCD is 8.

$$= \dfrac{7}{8} + \dfrac{0}{1} \cdot \dfrac{8}{8}$$

$$= \dfrac{7}{8} + \dfrac{0}{8} = \dfrac{7}{8}$$

Note that if we had observed at the outset that $\dfrac{0}{1} = 0$, the computation becomes $\dfrac{7}{8} + 0 = \dfrac{7}{8}$.

25. $\dfrac{3}{8} + \dfrac{1}{6}$ $8 = 2 \cdot 2 \cdot 2$ and $6 = 2 \cdot 3$, so the LCD is $2 \cdot 2 \cdot 2 \cdot 3$, or 24.

$$= \dfrac{3}{8} \cdot \dfrac{3}{3} + \dfrac{1}{6} \cdot \dfrac{4}{4}$$

$$= \dfrac{9}{24} + \dfrac{4}{24} = \dfrac{13}{24}$$

27. $\dfrac{5}{12} + \dfrac{7}{24}$ 12 is a factor of 24, so the LCD is 24.

$$= \dfrac{5}{12} \cdot \dfrac{2}{2} + \dfrac{7}{24}$$

$$= \dfrac{10}{24} + \dfrac{7}{24} = \dfrac{17}{24}$$

29. $\dfrac{3}{16} + \dfrac{5}{16} + \dfrac{4}{16} = \dfrac{3+5+4}{16} = \dfrac{12}{16} = \dfrac{3}{4}$

31. $\dfrac{8}{10} + \dfrac{7}{100} + \dfrac{4}{1000}$ 10 and 100 are factors of 1000, so the LCD is 1000.

$$= \dfrac{8}{10} \cdot \dfrac{100}{100} + \dfrac{7}{100} \cdot \dfrac{10}{10} + \dfrac{4}{1000}$$

$$= \dfrac{800}{1000} + \dfrac{70}{1000} + \dfrac{4}{1000} = \dfrac{874}{1000}$$

$$= \dfrac{437}{500}$$

33. $\dfrac{3}{8} + \dfrac{5}{12} + \dfrac{8}{15}$

$$= \dfrac{3}{2 \cdot 2 \cdot 2} + \dfrac{5}{2 \cdot 2 \cdot 3} + \dfrac{8}{3 \cdot 5} \quad \text{Factoring the denominators}$$

The LCM is $2 \cdot 2 \cdot 2 \cdot 3 \cdot 5$, or 120.

$$= \dfrac{3}{2 \cdot 2 \cdot 2} \cdot \dfrac{3 \cdot 5}{3 \cdot 5} + \dfrac{5}{2 \cdot 2 \cdot 3} \cdot \dfrac{2 \cdot 5}{2 \cdot 5} + \dfrac{8}{3 \cdot 5} \cdot \dfrac{2 \cdot 2 \cdot 2}{2 \cdot 2 \cdot 2}$$

In each case we multiply by 1 to obtain the LCD in the denominator.

$$= \dfrac{3 \cdot 3 \cdot 5}{2 \cdot 2 \cdot 2 \cdot 3 \cdot 5} + \dfrac{5 \cdot 2 \cdot 5}{2 \cdot 2 \cdot 3 \cdot 2 \cdot 5} + \dfrac{8 \cdot 2 \cdot 2 \cdot 2}{3 \cdot 5 \cdot 2 \cdot 2 \cdot 2}$$

$$= \dfrac{45}{120} + \dfrac{50}{120} + \dfrac{64}{120}$$

$$= \dfrac{159}{120} = \dfrac{53}{40}$$

35. $\dfrac{15}{24} + \dfrac{7}{36} + \dfrac{91}{48}$

$$= \dfrac{15}{2 \cdot 2 \cdot 2 \cdot 3} + \dfrac{7}{2 \cdot 2 \cdot 3 \cdot 3} + \dfrac{91}{2 \cdot 2 \cdot 2 \cdot 2 \cdot 3}$$

Factoring the denominators.

The LCM is $2 \cdot 2 \cdot 2 \cdot 2 \cdot 3 \cdot 3$, or 144.

$$= \dfrac{15}{2 \cdot 2 \cdot 2 \cdot 3} \cdot \dfrac{2 \cdot 3}{2 \cdot 3} + \dfrac{7}{2 \cdot 2 \cdot 3 \cdot 3} \cdot \dfrac{2 \cdot 2}{2 \cdot 2} +$$

$$\dfrac{91}{2 \cdot 2 \cdot 2 \cdot 3} \cdot \dfrac{3}{3}$$

In each case we multiply by 1 to obtain the LCD in the denominator.

$$= \dfrac{15 \cdot 2 \cdot 3}{2 \cdot 2 \cdot 2 \cdot 3 \cdot 2 \cdot 3} + \dfrac{7 \cdot 2 \cdot 2}{2 \cdot 2 \cdot 3 \cdot 3 \cdot 2 \cdot 2} + \dfrac{91 \cdot 3}{2 \cdot 2 \cdot 2 \cdot 2 \cdot 3 \cdot 3}$$

$$= \dfrac{90}{144} + \dfrac{28}{144} + \dfrac{273}{144} = \dfrac{391}{144}$$

37. Familiarize. We draw a picture. We let $p =$ the number of pounds of tea the shopper bought.

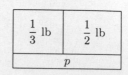

$\frac{1}{3}$ lb	$\frac{1}{2}$ lb
p	

Translate. An addition sentence corresponds to this situation.

Pounds of orange pekoe plus Pounds of English cinnamon is Total pounds of tea

$$\frac{1}{3} + \frac{1}{2} = p$$

Solve. We carry out the addition. Since 3 and 2 are both prime numbers, the LCM of the denominators is their product $3 \cdot 2$, or 6.

$$\frac{1}{3} \cdot \frac{2}{2} + \frac{1}{2} \cdot \frac{3}{3} = p$$
$$\frac{2}{6} + \frac{3}{6} = p$$
$$\frac{5}{6} = p$$

Check. We check by repeating the calculation. We also note that the sum is larger than either of the individual weights, so the answer seems reasonable.

State. The shopper bought $\frac{5}{6}$ lb of tea.

39. *Familiarize*. We draw a picture. We let D = the total distance walked.

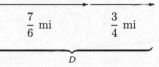

$\frac{7}{6}$ mi $\frac{3}{4}$ mi

D

Translate. An addition sentence corresponds to this situation.

Distance to friend's house plus Distance to class is Total distance

$$\frac{7}{6} + \frac{3}{4} = D$$

Solve. To solve the equation, carry out the addition. Since $6 = 2 \cdot 3$ and $4 = 2 \cdot 2$, the LCM of the denominators is $2 \cdot 2 \cdot 3$, or 12.

$$\frac{7}{6} \cdot \frac{2}{2} + \frac{3}{4} \cdot \frac{3}{3} = D$$
$$\frac{14}{12} + \frac{9}{12} = D$$
$$\frac{23}{12} = D$$

Check. We repeat the calculation. We also note that the sum is larger than either of the original distances, so the answer seems reasonable.

State. The student walked $\frac{23}{12}$ mi.

41. *Familiarize*. This is a multistep problem. First we find the total weight of the cubic meter of concrete mix. We visualize the situation, letting w = the total weight.

420 kg	150 kg	120 kg
w		

Translate. An addition sentence corresponds to this situation.

Weight of cement plus Weight of stone plus Weight of sand is Total weight

$$420 + 150 + 120 = w$$

Solve. We carry out the addition.

$$420 + 150 + 120 = w$$
$$690 = w$$

Since the mix contains 420 kg of cement, the part that is cement is $\frac{420}{690} = \frac{14 \cdot 30}{23 \cdot 30} = \frac{14}{23} \cdot \frac{30}{30} = \frac{14}{23}$.

Since the mix contains 150 kg of stone, the part that is stone is $\frac{150}{690} = \frac{5 \cdot 30}{23 \cdot 30} = \frac{5}{23} \cdot \frac{30}{30} = \frac{5}{23}$.

Since the mix contains 120 kg of sand, the part that is sand is $\frac{120}{690} = \frac{4 \cdot 30}{23 \cdot 30} = \frac{4}{23} \cdot \frac{30}{30} = \frac{4}{23}$.

We add these amounts: $\frac{14}{23} + \frac{5}{23} + \frac{4}{23} = \frac{14 + 5 + 4}{23} = \frac{23}{23} = 1$.

Check. We repeat the calculations. We also note that since the total of the fractional parts is 1, the answer is probably correct.

State. The total weight of the cubic meter of concrete mix is 690 kg. Of this, $\frac{14}{23}$ is cement, $\frac{5}{23}$ is stone, and $\frac{4}{23}$ is sand. The result when we add these amounts is 1.

43. *Familiarize*. We draw a picture. We let t = the total thickness.

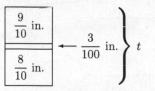

$\frac{9}{10}$ in.

$\frac{8}{10}$ in.

$\frac{3}{100}$ in. t

Translate. An addition sentence corresponds to this situation.

Thickness of one board plus Thickness of glue plus

$$\frac{9}{10} + \frac{3}{100} +$$

Thickness of second board is Total thickness

$$\frac{8}{10} = t$$

Solve. To solve the equation, carry out the addition.

The LCD is 100 since 10 is a factor of 100.

$$\frac{9}{10} \cdot \frac{10}{10} + \frac{3}{100} + \frac{8}{10} \cdot \frac{10}{10} = t$$

$$\frac{90}{100} + \frac{3}{100} + \frac{80}{100} = t$$

$$\frac{173}{100} = t$$

Check. We repeat the calculation. We also note that the sum is larger than any of the individual thicknesses, so the answer seems reasonable.

State. The result is $\frac{173}{100}$ in. thick.

45.
$$\begin{array}{r} {\scriptstyle 2} \\ {\scriptstyle 1\ 4} \\ 5\ 1\ 6 \\ \times \quad 4\ 0\ 8 \\ \hline 4\ 1\ 2\ 8 \\ 2\ 0\ 6\ 4\ 0\ 0 \\ \hline 2\ 1\ 0,5\ 2\ 8 \end{array}$$
Multiplying by 8
Multiplying by 400
Adding

47.
$$\begin{array}{r} {\scriptstyle 3} \\ {\scriptstyle 1} \\ {\scriptstyle 2} \\ 8\ 0\ 0\ 9 \\ \times \quad 4\ 2\ 3 \\ \hline 2\ 4\ 0\ 2\ 7 \\ 1\ 6\ 0\ 1\ 8\ 0 \\ 3\ 2\ 0\ 3\ 6\ 0\ 0 \\ \hline 3,3\ 8\ 7,8\ 0\ 7 \end{array}$$
Multiplying by 3
Multiplying by 20
Multiplying by 400
Adding

49. Familiarize. This is a multistep problem. We must find the total cost of the 5 sweaters and the total cost of the 6 shirts. Repeated addition works well in finding each total. The total cost is the sum of these two totals. We let $x =$ the total cost of the 5 sweaters and $y =$ the total cost of the 6 shirts.

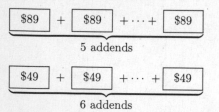
5 addends

$$\boxed{\$49} + \boxed{\$49} + \cdots + \boxed{\$49}$$
6 addends

Translate. We translate to two equations.

Number of sweaters	times	Cost per sweater	is	Total cost of sweaters
↓	↓	↓	↓	↓
5	·	89	=	x

Number of shirts	times	Cost per shirt	is	Total cost of shirts
↓	↓	↓	↓	↓
6	·	49	=	y

Solve. We carry out the multiplication.

$$\begin{array}{r} 8\ 9 \\ \times \quad 5 \\ \hline 4\ 4\ 5 \end{array}$$
Thus, $x = \$445$.

$$\begin{array}{r} 4\ 9 \\ \times \quad 6 \\ \hline 2\ 9\ 4 \end{array}$$
Thus, $y = \$294$.

We let $c =$ the total cost of the sweaters and shirts.

Total cost of sweaters	plus	Total cost of shirts	is	Total cost
↓	↓	↓	↓	↓
$445	+	$294	=	c

To solve the equation we carry out the addition.

$$\begin{array}{r} \$\ 4\ 4\ 5 \\ +\ \ 2\ 9\ 4 \\ \hline \$\ 7\ 3\ 9 \end{array}$$

Check. We repeat the calculations. The answer checks.

State. The total cost of the sweaters and shirts is $739.

Exercise Set 3.3

1. When denominators are the same, subtract the numerators and keep the denominator.

$$\frac{5}{6} - \frac{1}{6} = \frac{5-1}{6} = \frac{4}{6} = \frac{2 \cdot 2}{2 \cdot 3} = \frac{2}{2} \cdot \frac{2}{3} = \frac{2}{3}$$

3. When denominators are the same, subtract the numerators and keep the denominator.

$$\frac{11}{12} - \frac{2}{12} = \frac{11-2}{12} = \frac{9}{12} = \frac{3 \cdot 3}{3 \cdot 4} = \frac{3}{3} \cdot \frac{3}{4} = \frac{3}{4}$$

5. The LCM of 4 and 8 is 8.

$$\frac{3}{4} - \frac{1}{8} = \frac{3}{4} \cdot \frac{2}{2} - \frac{1}{8} \quad \longleftarrow \text{This fraction already has the LCM as the denominator.}$$

Think: $4 \times \boxed{} = 8$. The answer is 2, so we multiply by 1, using $\frac{2}{2}$.

$$= \frac{6}{8} - \frac{1}{8} = \frac{5}{8}$$

7. The LCM of 8 and 12 is 24.

$$\frac{1}{8} - \frac{1}{12} = \frac{1}{8} \cdot \frac{3}{3} - \frac{1}{12} \cdot \frac{2}{2}$$

Think: $12 \times \boxed{} = 24$. The answer is 2, so we multiply by 1, using $\frac{2}{2}$.

Think: $8 \times \boxed{} = 24$. The answer is 3, so we multiply by 1, using $\frac{3}{3}$.

$$= \frac{3}{24} - \frac{2}{24} = \frac{1}{24}$$

9. The LCM of 3 and 6 is 6.

$$\frac{4}{3} - \frac{5}{6} = \frac{4}{3} \cdot \frac{2}{2} - \frac{5}{6}$$

$$= \frac{8}{6} - \frac{5}{6} = \frac{3}{6}$$

$$= \frac{1 \cdot 3}{2 \cdot 3} = \frac{1}{2} \cdot \frac{3}{3}$$

$$= \frac{1}{2}$$

11. The LCM of 4 and 28 is 28.

$$\frac{3}{4} - \frac{3}{28} = \frac{3}{4} \cdot \frac{7}{7} - \frac{3}{28}$$

$$= \frac{21}{28} - \frac{3}{28}$$

$$= \frac{18}{28} = \frac{9 \cdot 2}{14 \cdot 2}$$

$$= \frac{9}{14} \cdot \frac{2}{2} = \frac{9}{14}$$

13. The LCM of 4 and 20 is 20.

$$\frac{3}{4} - \frac{3}{20} = \frac{3}{4} \cdot \frac{5}{5} - \frac{3}{20}$$

$$= \frac{15}{20} - \frac{3}{20} = \frac{12}{20}$$

$$= \frac{3 \cdot 4}{5 \cdot 4} = \frac{3}{5} \cdot \frac{4}{4}$$

$$= \frac{3}{5}$$

15. The LCM of 4 and 20 is 20.

$$\frac{3}{4} - \frac{1}{20} = \frac{3}{4} \cdot \frac{5}{5} - \frac{1}{20}$$

$$= \frac{15}{20} - \frac{1}{20} = \frac{14}{20}$$

$$= \frac{2 \cdot 7}{2 \cdot 10} = \frac{2}{2} \cdot \frac{7}{10}$$

$$= \frac{7}{10}$$

17. The LCM of 12 and 15 is 60.

$$\frac{5}{12} - \frac{2}{15} = \frac{5}{12} \cdot \frac{5}{5} - \frac{2}{15} \cdot \frac{4}{4}$$

$$= \frac{25}{60} - \frac{8}{60} = \frac{17}{60}$$

19. The LCM of 10 and 100 is 100.

$$\frac{6}{10} - \frac{7}{100} = \frac{6}{10} \cdot \frac{10}{10} - \frac{7}{100}$$

$$= \frac{60}{100} - \frac{7}{100} = \frac{53}{100}$$

21. The LCM of 15 and 25 is 75.

$$\frac{7}{15} - \frac{3}{25} = \frac{7}{15} \cdot \frac{5}{5} - \frac{3}{25} \cdot \frac{3}{3}$$

$$= \frac{35}{75} - \frac{9}{75} = \frac{26}{75}$$

23. The LCM of 10 and 100 is 100.

$$\frac{99}{100} - \frac{9}{10} = \frac{99}{100} - \frac{9}{10} \cdot \frac{10}{10}$$

$$= \frac{99}{100} - \frac{90}{100} = \frac{9}{100}$$

25. The LCM of 3 and 8 is 24.

$$\frac{2}{3} - \frac{1}{8} = \frac{2}{3} \cdot \frac{8}{8} - \frac{1}{8} \cdot \frac{3}{3}$$

$$= \frac{16}{24} - \frac{3}{24}$$

$$= \frac{13}{24}$$

27. The LCM of 5 and 2 is 10.

$$\frac{3}{5} - \frac{1}{2} = \frac{3}{5} \cdot \frac{2}{2} - \frac{1}{2} \cdot \frac{5}{5}$$

$$= \frac{6}{10} - \frac{5}{10}$$

$$= \frac{1}{10}$$

29. The LCM of 12 and 8 is 24.

$$\frac{5}{12} - \frac{3}{8} = \frac{5}{12} \cdot \frac{2}{2} - \frac{3}{8} \cdot \frac{3}{3}$$

$$= \frac{10}{24} - \frac{9}{24}$$

$$= \frac{1}{24}$$

31. The LCM of 8 and 16 is 16.

$$\frac{7}{8} - \frac{1}{16} = \frac{7}{8} \cdot \frac{2}{2} - \frac{1}{16}$$

$$= \frac{14}{16} - \frac{1}{16}$$

$$= \frac{13}{16}$$

33. The LCM of 25 and 15 is 75.

$$\frac{17}{25} - \frac{4}{15} = \frac{17}{25} \cdot \frac{3}{3} - \frac{4}{15} \cdot \frac{5}{5}$$

$$= \frac{51}{75} - \frac{20}{75}$$

$$= \frac{31}{75}$$

35. The LCM of 25 and 150 is 150.

$$\frac{23}{25} - \frac{112}{150} = \frac{23}{25} \cdot \frac{6}{6} - \frac{112}{150}$$

$$= \frac{138}{150} - \frac{112}{150} = \frac{26}{150}$$

$$= \frac{2 \cdot 13}{2 \cdot 75} = \frac{2}{2} \cdot \frac{13}{75}$$

$$= \frac{13}{75}$$

37. Since there is a common denominator, compare the numerators.

$$5 < 6, \text{ so } \frac{5}{8} < \frac{6}{8}.$$

39. The LCD is 12.

$$\frac{1}{3} \cdot \frac{4}{4} = \frac{4}{12} \quad \text{We multiply by 1 to get the LCD.}$$

$$\frac{1}{4} \cdot \frac{3}{3} = \frac{3}{12} \quad \text{We multiply by 1 to get the LCD.}$$

Since $4 > 3$, it follows that $\frac{4}{12} > \frac{3}{12}$, so $\frac{1}{3} > \frac{1}{4}$.

41. The LCD is 21.

$$\frac{2}{3} \cdot \frac{7}{7} = \frac{14}{21} \quad \text{We multiply by 1 to get the LCD.}$$

$$\frac{5}{7} \cdot \frac{3}{3} = \frac{15}{21} \quad \text{We multiply by 1 to get the LCD.}$$

Since $14 < 15$, it follows that $\frac{14}{21} < \frac{15}{21}$, so $\frac{2}{3} < \frac{5}{7}$.

43. The LCD is 30.

$$\frac{4}{5} \cdot \frac{6}{6} = \frac{24}{30}$$

$$\frac{5}{6} \cdot \frac{5}{5} = \frac{25}{30}$$

Since $24 < 25$, it follows that $\frac{24}{30} < \frac{25}{30}$, so $\frac{4}{5} < \frac{5}{6}$.

45. The LCD is 20.

The denominator of $\frac{19}{20}$ is the LCD.

$$\frac{4}{5} \cdot \frac{4}{4} = \frac{16}{20}$$

Since $19 > 16$, it follows that $\frac{19}{20} > \frac{16}{20}$, so $\frac{19}{20} > \frac{4}{5}$.

47. The LCD is 20.

The denominator of $\frac{19}{20}$ is the LCD.

$$\frac{9}{10} \cdot \frac{2}{2} = \frac{18}{20}$$

Since $19 > 18$, it follows that $\frac{19}{20} > \frac{18}{20}$, so $\frac{19}{20} > \frac{9}{10}$.

49. The LCD is $21 \cdot 13$, or 273.

$$\frac{31}{21} \cdot \frac{13}{13} = \frac{403}{273}$$

$$\frac{41}{13} \cdot \frac{21}{21} = \frac{861}{273}$$

Since $403 < 861$, it follows that $\frac{403}{273} < \frac{861}{273}$, so $\frac{31}{21} < \frac{41}{13}$.

51.
$$x + \frac{1}{30} = \frac{1}{10}$$

$$x + \frac{1}{30} - \frac{1}{30} = \frac{1}{10} - \frac{1}{30} \quad \text{Subtracting } \frac{1}{30} \text{ on both sides}$$

$$x + 0 = \frac{1}{10} \cdot \frac{3}{3} - \frac{1}{30} \quad \begin{array}{l}\text{The LCD is 30. We} \\ \text{multiply by 1 to get} \\ \text{the LCD.}\end{array}$$

$$x = \frac{3}{30} - \frac{1}{30} = \frac{2}{30}$$

$$x = \frac{1 \cdot 2}{2 \cdot 15} = \frac{1}{15} \cdot \frac{2}{2}$$

$$x = \frac{1}{15}$$

The solution is $\frac{1}{15}$.

53.
$$\frac{2}{3} + t = \frac{4}{5}$$

$$\frac{2}{3} + t - \frac{2}{3} = \frac{4}{5} - \frac{2}{3} \quad \text{Subtracting } \frac{2}{3} \text{ on both sides.}$$

$$t + 0 = \frac{4}{5} \cdot \frac{3}{3} - \frac{2}{3} \cdot \frac{5}{5} \quad \begin{array}{l}\text{The LCD is 15. We} \\ \text{multiply by 1 to get} \\ \text{the LCD.}\end{array}$$

$$t = \frac{12}{15} - \frac{10}{15} = \frac{2}{15}$$

The solution is $\frac{2}{15}$.

55.
$$m + \frac{5}{6} = \frac{9}{10}$$

$$m + \frac{5}{6} - \frac{5}{6} = \frac{9}{10} - \frac{5}{6}$$

$$m + 0 = \frac{9}{10} \cdot \frac{3}{3} - \frac{5}{6} \cdot \frac{5}{5}$$

$$m = \frac{27}{30} - \frac{25}{30} = \frac{2}{30}$$

$$m = \frac{1 \cdot 2}{2 \cdot 15} = \frac{1}{15} \cdot \frac{2}{2}$$

$$m = \frac{1}{15}$$

The solution is $\frac{1}{15}$.

57. _Familiarize_. We draw a picture. We let x = the portion of the estate the fourth child received.

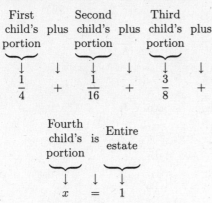

Translate. We have a "how much more" situation.

First
child's plus
portion

Second
child's plus
portion

Third
child's plus
portion

$$\frac{1}{4} \quad + \quad \frac{1}{16} \quad + \quad \frac{3}{8} \quad +$$

Fourth
child's is
portion

Entire
estate

$$x \quad = \quad 1$$

Solve. To solve the equation, we begin by adding the fractions on the left side.

$$\frac{1}{4} + \frac{1}{16} + \frac{3}{8} + x = 1$$

$$\frac{1}{4} \cdot \frac{4}{4} + \frac{1}{16} + \frac{3}{8} \cdot \frac{2}{2} + x = 1 \quad \text{The LCD is 16.}$$

$$\frac{4}{16} + \frac{1}{16} + \frac{6}{16} + x = 1$$

$$\frac{11}{16} + x = 1$$

$$x = 1 - \frac{11}{16}$$

$$x = 1 \cdot \frac{16}{16} - \frac{11}{16} = \frac{16}{16} - \frac{11}{16}$$

$$\text{The LCD is 16.}$$

$$x = \frac{5}{16}$$

Check. We return to the original problem and add:

$$\frac{1}{4} + \frac{1}{16} + \frac{3}{8} + \frac{5}{16} = \frac{1}{4} \cdot \frac{4}{4} + \frac{1}{16} + \frac{3}{8} \cdot \frac{2}{2} + \frac{5}{16} =$$

$$\frac{4}{16} + \frac{1}{16} + \frac{6}{16} + \frac{5}{16} = \frac{16}{16} = 1. \text{ The answer checks.}$$

State. The fourth child received $\frac{5}{16}$ of the estate.

59. $\frac{9}{10} \div \frac{3}{5} = \frac{9}{10} \cdot \frac{5}{3} = \frac{9 \cdot 5}{10 \cdot 3} = \frac{3 \cdot 3 \cdot 5}{2 \cdot 5 \cdot 3} = \frac{3 \cdot 5}{3 \cdot 5} \cdot \frac{3}{2} = \frac{3}{2}$

61. The batting average is a fraction with the number of hits as its numerator and the number of at bats as its denominator.

Bernard Gilkey's batting average was $\frac{116}{384}$.

Bip Roberts' batting average was $\frac{172}{532}$.

Using a calculator, we multiply by 1 to make the denominators the same.

$$\frac{116}{384} \cdot \frac{532}{532} = \frac{61,712}{204,288}$$

$$\frac{172}{532} \cdot \frac{384}{384} = \frac{66,048}{204,288}$$

Since $66,048 > 61,712$, it follows that $\frac{66,048}{204,288} > \frac{61,712}{204,288}$, so $\frac{172}{532} > \frac{116}{384}$.

Thus, Bip Roberts had the higher batting average.

63. $\frac{7}{8} - \frac{1}{10} \times \frac{5}{6} = \frac{7}{8} - \frac{1 \times 5}{10 \times 6} = \frac{7}{8} - \frac{1 \times 5}{2 \times 5 \times 6}$

$$= \frac{7}{8} - \frac{5}{5} \times \frac{1}{2 \times 6} = \frac{7}{8} - \frac{1}{2 \times 6} = \frac{7}{8} - \frac{1}{12}$$

$$= \frac{7}{8} \cdot \frac{3}{3} - \frac{1}{12} \cdot \frac{2}{2} = \frac{21}{24} - \frac{2}{24} = \frac{19}{24}$$

65. $\left(\frac{2}{3}\right)^2 + \left(\frac{3}{4}\right)^2 = \frac{4}{9} + \frac{9}{16} = \frac{4}{9} \cdot \frac{16}{16} + \frac{9}{16} \cdot \frac{9}{9} =$

$$\frac{64}{144} + \frac{81}{144} = \frac{145}{144}$$

Exercise Set 3.4

1. [b]   [a] Multiply: $5 \cdot 3 = 15$.

$5\frac{2}{3} = \frac{17}{3}$ [b] Add: $15 + 2 = 17$.

[a]   [c] Keep the denominator.

3. [b]   [a] Multiply: $3 \cdot 4 = 12$.

$3\frac{1}{4} = \frac{13}{4}$ [b] Add: $12 + 1 = 13$.

[a]   [c] Keep the denominator.

5. $10\frac{1}{8} = \frac{81}{8}$ $(10 \cdot 8 = 80, 80 + 1 = 81)$

7. $5\frac{1}{10} = \frac{51}{10}$ $(5 \cdot 10 = 50, 50 + 1 = 51)$

9. $20\frac{3}{5} = \frac{103}{5}$ $(20 \cdot 5 = 100, 100 + 3 = 103)$

11. $9\frac{5}{6} = \frac{59}{6}$ $(9 \cdot 6 = 54, 54 + 5 = 59)$

13. $7\frac{3}{10} = \frac{73}{10}$ $(7 \cdot 10 = 70, \ 70 + 3 = 73)$

15. $1\frac{5}{8} = \frac{13}{8}$ $(1 \cdot 8 = 8, \ 8 + 5 = 13)$

17. $12\frac{3}{4} = \frac{51}{4}$ $(12 \cdot 4 = 48, \ 48 + 3 = 51)$

19. $4\frac{3}{10} = \frac{43}{10}$ $(4 \cdot 10 = 40, \ 40 + 3 = 43)$

21. $2\frac{3}{100} = \frac{203}{100}$ $(2 \cdot 100 = 200, \ 200 + 3 = 203)$

23. $66\frac{2}{3} = \frac{200}{3}$ $(66 \cdot 3 = 198, \ 198 + 2 = 200)$

25. $5\frac{29}{50} = \frac{279}{50}$ $(5 \cdot 50 = 250, \ 250 + 29 = 279)$

27. To convert $\frac{18}{5}$ to a mixed numeral, we divide.

$$
\begin{array}{r}
3 \\
5\overline{)1\,8} \\
1\,5 \\
\hline
3
\end{array}
\qquad \frac{18}{5} = 3\frac{3}{5}
$$

29. To convert $\frac{14}{3}$ to a mixed numeral, we divide.

$$
\begin{array}{r}
4 \\
3\overline{)1\,4} \\
1\,2 \\
\hline
2
\end{array}
\qquad \frac{14}{3} = 4\frac{2}{3}
$$

31.

$$
\begin{array}{r}
4 \\
6\overline{)2\,7} \\
2\,4 \\
\hline
3
\end{array}
\qquad \frac{27}{6} = 4\frac{3}{6} = 4\frac{1}{2}
$$

33.

$$
\begin{array}{r}
5 \\
1\,0\overline{)5\,7} \\
5\,0 \\
\hline
7
\end{array}
\qquad \frac{57}{10} = 5\frac{7}{10}
$$

35.

$$
\begin{array}{r}
7 \\
7\overline{)5\,3} \\
4\,9 \\
\hline
4
\end{array}
\qquad \frac{53}{7} = 7\frac{4}{7}
$$

37.

$$
\begin{array}{r}
7 \\
6\overline{)4\,5} \\
4\,2 \\
\hline
3
\end{array}
\qquad \frac{45}{6} = 7\frac{3}{6} = 7\frac{1}{2}
$$

39.

$$
\begin{array}{r}
1\,1 \\
4\overline{)4\,6} \\
4\,0 \\
\hline
6 \\
4 \\
\hline
2
\end{array}
\qquad \frac{46}{4} = 11\frac{2}{4} = 11\frac{1}{2}
$$

41.

$$
\begin{array}{r}
1 \\
8\overline{)1\,2} \\
8 \\
\hline
4
\end{array}
\qquad \frac{12}{8} = 1\frac{4}{8} = 1\frac{1}{2}
$$

43.

$$
\begin{array}{r}
7 \\
1\,0\,0\overline{)7\,5\,7} \\
7\,0\,0 \\
\hline
5\,7
\end{array}
\qquad \frac{757}{100} = 7\frac{57}{100}
$$

45.

$$
\begin{array}{r}
4\,3 \\
8\overline{)3\,4\,5} \\
3\,2\,0 \\
\hline
2\,5 \\
2\,4 \\
\hline
1
\end{array}
\qquad \frac{345}{8} = 43\frac{1}{8}
$$

47. We first divide as usual.

$$
\begin{array}{r}
1\,0\,8 \\
8\overline{)8\,6\,9} \\
8\,0\,0 \\
\hline
6\,9 \\
6\,4 \\
\hline
5
\end{array}
$$

The answer is 108 R 5. We write a mixed numeral for the quotient as follows: $108\frac{5}{8}$.

49. We first divide as usual.

$$
\begin{array}{r}
6\,1\,8 \\
5\overline{)3\,0\,9\,1} \\
3\,0\,0\,0 \\
\hline
9\,1 \\
5\,0 \\
\hline
4\,1 \\
4\,0 \\
\hline
1
\end{array}
$$

The answer is 618 R 1. We write a mixed numeral for the quotient as follows: $618\frac{1}{5}$.

51.

$$
\begin{array}{r}
4\,0 \\
2\,1\overline{)8\,5\,2} \\
8\,4\,0 \\
\hline
1\,2
\end{array}
$$

We get $40\frac{12}{21}$. This simplifies as $40\frac{4}{7}$.

53.

$$
\begin{array}{r}
5\,5 \\
1\,0\,2\overline{)5\,6\,1\,2} \\
5\,1\,0\,0 \\
\hline
5\,1\,2 \\
5\,1\,0 \\
\hline
2
\end{array}
$$

We get $55\frac{2}{102}$. This simplifies as $55\frac{1}{51}$.

55. $\frac{6}{5} \cdot 15 = \frac{6 \cdot 15}{5} = \frac{6 \cdot 3 \cdot 5}{5 \cdot 1} = \frac{5}{5} \cdot \frac{6 \cdot 3}{1} = \frac{6 \cdot 3}{1} = \frac{18}{1} = 18$

57. $\dfrac{7}{10} \cdot \dfrac{5}{14} = \dfrac{7 \cdot 5}{10 \cdot 14} = \dfrac{7 \cdot 5 \cdot 1}{2 \cdot 5 \cdot 2 \cdot 7} = \dfrac{7 \cdot 5}{7 \cdot 5} \cdot \dfrac{1}{2 \cdot 2} = \dfrac{1}{2 \cdot 2} = \dfrac{1}{4}$

59. $\dfrac{2}{3} \div \dfrac{1}{36} = \dfrac{2}{3} \cdot \dfrac{36}{1} = \dfrac{2 \cdot 36}{3 \cdot 1} = \dfrac{2 \cdot 3 \cdot 12}{3 \cdot 1} = \dfrac{3}{3} \cdot \dfrac{2 \cdot 12}{1} =$

$\dfrac{2 \cdot 12}{1} = \dfrac{24}{1} = 24$

61. $200 \div \dfrac{15}{64} = 200 \cdot \dfrac{64}{15} = \dfrac{200 \cdot 64}{15} = \dfrac{5 \cdot 40 \cdot 64}{5 \cdot 3} =$

$\dfrac{5}{5} \cdot \dfrac{40 \cdot 64}{3} = \dfrac{40 \cdot 64}{3} = \dfrac{2560}{3}$

63. $\dfrac{56}{7} + \dfrac{2}{3} = 8 + \dfrac{2}{3}$ $\qquad (56 \div 7 = 8)$

$\qquad\qquad = 8\dfrac{2}{3}$

65.
$$7 \overline{)\,3\,6\,6\,}\;\;\begin{array}{r} 5\,2 \\ \hline \end{array}$$
$$\begin{array}{r} 3\,5\,0 \\ \hline 1\,6 \\ 1\,4 \\ \hline 2 \end{array}$$
$\qquad\qquad \dfrac{366}{7} = 52\dfrac{2}{7}$

Exercise Set 3.5

1.
$$\begin{array}{r} 2\dfrac{7}{8} \\[4pt] +3\dfrac{5}{8} \\ \hline 5\dfrac{12}{8} = 5 + \dfrac{12}{8} \\[4pt] = 5 + 1\dfrac{1}{2} \\[4pt] = 6\dfrac{1}{2} \end{array}$$

To find a mixed numeral for $\dfrac{12}{8}$ we divide:

$$8\overline{)\,1\,2\,}\;\begin{array}{r}1\\\hline\end{array} \qquad \dfrac{12}{8} = 1\dfrac{4}{8} = 1\dfrac{1}{2}$$
$$\begin{array}{r}8\\\hline 4\end{array}$$

3. The LCD is 12.

$$\begin{array}{r} 1\;\boxed{\dfrac{1}{4} \cdot \dfrac{3}{3}} = 1\dfrac{3}{12} \\[6pt] +1\;\boxed{\dfrac{2}{3} \cdot \dfrac{4}{4}} = +1\dfrac{8}{12} \\ \hline 2\dfrac{11}{12} \end{array}$$

5. The LCD is 12.

$$\begin{array}{r} 8\;\boxed{\dfrac{3}{4} \cdot \dfrac{3}{3}} = 8\dfrac{9}{12} \\[6pt] +5\;\boxed{\dfrac{5}{6} \cdot \dfrac{2}{2}} = +5\dfrac{10}{12} \\ \hline 13\dfrac{19}{12} = 13 + \dfrac{19}{12} \\[4pt] = 13 + 1\dfrac{7}{12} \\[4pt] = 14\dfrac{7}{12} \end{array}$$

7. The LCD is 10.

$$\begin{array}{r} 3\;\boxed{\dfrac{2}{5} \cdot \dfrac{2}{2}} = 3\dfrac{4}{10} \\[6pt] +8\;\dfrac{7}{10} = +8\dfrac{7}{10} \\ \hline 11\dfrac{11}{10} = 11 + \dfrac{11}{10} \\[4pt] = 11 + 1\dfrac{1}{10} \\[4pt] = 12\dfrac{1}{10} \end{array}$$

9. The LCD is 24.

$$\begin{array}{r} 5\;\boxed{\dfrac{3}{8} \cdot \dfrac{3}{3}} = 5\dfrac{9}{24} \\[6pt] +10\;\boxed{\dfrac{5}{6} \cdot \dfrac{4}{4}} = +10\dfrac{20}{24} \\ \hline 15\dfrac{29}{24} = 15 + \dfrac{29}{24} \\[4pt] = 15 + 1\dfrac{5}{24} \\[4pt] = 16\dfrac{5}{24} \end{array}$$

11. The LCD is 10.

$$\begin{array}{r} 12\;\boxed{\dfrac{4}{5} \cdot \dfrac{2}{2}} = 12\dfrac{8}{10} \\[6pt] +8\;\dfrac{7}{10} = +8\dfrac{7}{10} \\ \hline 20\dfrac{15}{10} = 20 + \dfrac{15}{10} \\[4pt] = 20 + 1\dfrac{5}{10} \\[4pt] = 21\dfrac{5}{10} \\[4pt] = 21\dfrac{1}{2} \end{array}$$

13. The LCD is 8.

$$\begin{array}{r} 14\dfrac{5}{8} = 14\dfrac{5}{8} \\[6pt] +13\;\boxed{\dfrac{1}{4} \cdot \dfrac{2}{2}} = +13\dfrac{2}{8} \\ \hline 27\dfrac{7}{8} \end{array}$$

15. The LCD is 24.

$$
\begin{array}{r}
7\,\boxed{\dfrac{1}{8}\cdot\dfrac{3}{3}} = 7\,\dfrac{3}{24} \\[2mm]
9\,\boxed{\dfrac{2}{3}\cdot\dfrac{8}{8}} = 9\,\dfrac{16}{24} \\[2mm]
+10\,\boxed{\dfrac{3}{4}\cdot\dfrac{6}{6}} = +10\,\dfrac{18}{24} \\[1mm]
\hline
\end{array}
$$

$$
26\,\dfrac{37}{24} = 26 + \dfrac{37}{24}
$$
$$
= 26 + 1\dfrac{13}{24}
$$
$$
= 27\dfrac{13}{24}
$$

17.
$$
\begin{array}{r}
4\dfrac{1}{5} = 3\dfrac{6}{5} \\[2mm]
-2\dfrac{3}{5} = -2\dfrac{3}{5} \\[1mm]
\hline
1\dfrac{3}{5}
\end{array}
$$

> Since $\dfrac{1}{5}$ is smaller than $\dfrac{3}{5}$, we cannot subtract until we borrow:
>
> $$4\dfrac{1}{5} = 3+\dfrac{5}{5}+\dfrac{1}{5} = 3+\dfrac{6}{5} = 3\dfrac{6}{5}$$

19. The LCD is 10.

$$
\begin{array}{r}
6\,\boxed{\dfrac{3}{5}\cdot\dfrac{2}{2}} = 6\,\dfrac{6}{10} \\[2mm]
-2\,\boxed{\dfrac{1}{2}\cdot\dfrac{5}{5}} = -2\,\dfrac{5}{10} \\[1mm]
\hline
4\,\dfrac{1}{10}
\end{array}
$$

21. The LCD is 24.

$$
\begin{array}{r}
34\,\boxed{\dfrac{1}{3}\cdot\dfrac{8}{8}} = 34\,\dfrac{8}{24} = 33\,\dfrac{32}{24} \\[2mm]
-12\,\boxed{\dfrac{5}{8}\cdot\dfrac{3}{3}} = -12\,\dfrac{15}{24} = -12\,\dfrac{15}{24} \\[1mm]
\hline
21\,\dfrac{17}{24}
\end{array}
$$

$\Big($Since $\dfrac{8}{24}$ is smaller than $\dfrac{15}{24}$, we cannot subtract until we borrow: $34\dfrac{8}{24} = 33 + \dfrac{24}{24} + \dfrac{8}{24} = 33 + \dfrac{32}{24} = 33\dfrac{32}{24}.\Big)$

23.
$$
\begin{array}{r}
21 = 20\dfrac{4}{4} \\[2mm]
- 8\dfrac{3}{4} = - 8\dfrac{3}{4} \\[1mm]
\hline
12\dfrac{1}{4}
\end{array}
$$
$\Big(21 = 20 + 1 = 20 + \dfrac{4}{4} = 20\dfrac{4}{4}\Big)$

25.
$$
\begin{array}{r}
34 = 33\dfrac{8}{8} \\[2mm]
- 18\dfrac{5}{8} = - 18\dfrac{5}{8} \\[1mm]
\hline
15\dfrac{3}{8}
\end{array}
$$
$\Big(34 = 33 + 1 = 33 + \dfrac{8}{8} = 33\dfrac{8}{8}\Big)$

27. The LCD is 12.

$$
\begin{array}{r}
21\,\boxed{\dfrac{1}{6}\cdot\dfrac{2}{2}} = 21\,\dfrac{2}{12} = 20\,\dfrac{14}{12} \\[2mm]
-13\,\boxed{\dfrac{3}{4}\cdot\dfrac{3}{3}} = -13\,\dfrac{9}{12} = -13\,\dfrac{9}{12} \\[1mm]
\hline
7\,\dfrac{5}{12}
\end{array}
$$

$\Big($Since $\dfrac{2}{12}$ is smaller than $\dfrac{9}{12}$, we cannot subtract until we borrow: $21\dfrac{2}{12} = 20 + \dfrac{12}{12} + \dfrac{2}{12} = 20 + \dfrac{14}{12} = 20\dfrac{14}{12}.\Big)$

29. The LCD is 8.

$$
\begin{array}{r}
14\,\dfrac{1}{8} = 14\,\dfrac{1}{8} = 13\,\dfrac{9}{8} \\[2mm]
- \boxed{\dfrac{3}{4}\cdot\dfrac{2}{2}} = - \dfrac{6}{8} = - \dfrac{6}{8} \\[1mm]
\hline
13\,\dfrac{3}{8}
\end{array}
$$

$\Big($Since $\dfrac{1}{8}$ is smaller than $\dfrac{6}{8}$, we cannot subtract until we borrow: $14\dfrac{1}{8} = 13 + \dfrac{8}{8} + \dfrac{1}{8} = 13 + \dfrac{9}{8} = 13\dfrac{9}{8}.\Big)$

31. The LCD is 18.

$$
\begin{array}{r}
25\,\boxed{\dfrac{1}{9}\cdot\dfrac{2}{2}} = 25\,\dfrac{2}{18} = 24\,\dfrac{20}{18} \\[2mm]
-13\,\boxed{\dfrac{5}{6}\cdot\dfrac{3}{3}} = -13\,\dfrac{15}{18} = -13\,\dfrac{15}{18} \\[1mm]
\hline
11\,\dfrac{5}{18}
\end{array}
$$

$\Big($Since $\dfrac{2}{18}$ is smaller than $\dfrac{15}{18}$, we cannot subtract until we borrow: $25\dfrac{2}{18} = 24 + \dfrac{18}{18} + \dfrac{2}{18} = 24 + \dfrac{20}{18} = 24\dfrac{20}{18}.\Big)$

33. **Familiarize.** We let $w =$ the total weight of the meat.

Translate. We write an equation.

Weight of one package	plus	Weight of second package	is	Total weight
↓	↓	↓	↓	↓
$1\dfrac{2}{3}$	$+$	$5\dfrac{3}{4}$	$=$	w

Solve. We carry out the addition. The LCD is 12.

$$1\;\boxed{\dfrac{2}{3}\cdot\dfrac{4}{4}}=\;1\dfrac{8}{12}$$

$$+5\;\boxed{\dfrac{3}{4}\cdot\dfrac{3}{3}}=+5\dfrac{9}{12}$$

$$6\dfrac{17}{12}=6+\dfrac{17}{12}$$

$$=6+1\dfrac{5}{12}$$

$$=7\dfrac{5}{12}$$

Check. We repeat the calculation. We also note that the answer is larger than either of the individual weights, so the answer seems reasonable.

State. The total weight of the meat was $7\dfrac{5}{12}$ lb.

35. *Familiarize*. We let h = the woman's excess weight.

Translate. We have a "how much more" situation.

Height of son	plus	How much more height	is	Height of woman
↓	↓	↓	↓	↓
$59\dfrac{7}{12}$	$+$	h	$=$	66

Solve. We solve the equation as follows:

$$h=66-59\dfrac{7}{12}$$

$$66\;\;=\;\;65\dfrac{12}{12}$$

$$-\,59\dfrac{7}{12}=-\,59\dfrac{7}{12}$$

$$6\dfrac{5}{12}$$

Check. We add the woman's excess height to her son's height:

$$6\dfrac{5}{12}+59\dfrac{7}{12}=65\dfrac{12}{12}=66$$

The answer checks.

State. The woman is $6\dfrac{5}{12}$ in. taller.

37. *Familiarize*. We draw a picture, letting x = the amount of pipe that was used.

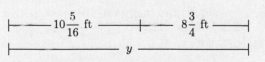

Translate. We write an addition sentence.

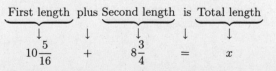

First length	plus	Second length	is	Total length
↓	↓	↓	↓	↓
$10\dfrac{5}{16}$	$+$	$8\dfrac{3}{4}$	$=$	x

Solve. We carry out the addition. The LCD is 16.

$$10\dfrac{5}{16}\;\;=\;\;10\dfrac{5}{16}$$

$$+\;8\;\boxed{\dfrac{3}{4}\cdot\dfrac{4}{4}}=+\;8\dfrac{12}{16}$$

$$18\dfrac{17}{16}=18+\dfrac{17}{16}$$

$$=18+1\dfrac{1}{16}$$

$$=19\dfrac{1}{16}$$

Check. We repeat the calculation. We also note that the total length is larger than either of the individual lengths, so the answer seems reasonable.

State. The plumber used $19\dfrac{1}{16}$ ft of pipe.

39. *Familiarize*. We draw a picture. We let D = the distance from Los Angeles at the end of the second day.

$$180\dfrac{7}{10}\text{ km}$$

L. A. ├───────────────────────────────────┤

├ ─ ─ ─ ─ ─ ─ ─ ─ ┼───────────┤

$$D\qquad\qquad 85\dfrac{1}{2}\text{ km}$$

Translate. We write an equation.

Distance away from Los Angeles	minus	Distance toward Los Angeles	is	Distance from Los Angeles
↓	↓	↓	↓	↓
$180\dfrac{7}{10}$	$-$	$85\dfrac{1}{2}$	$=$	D

Solve. To solve the equation we carry out the subtraction. The LCD is 10.

$$180\dfrac{7}{10}\;\;=\;\;180\dfrac{7}{10}$$

$$-\;85\;\boxed{\dfrac{1}{2}\cdot\dfrac{5}{5}}=-\;85\dfrac{5}{10}$$

$$95\dfrac{2}{10}=95\dfrac{1}{5}$$

Check. We add the distance from Los Angeles to the distance the person drove toward Los Angeles:

$$95\dfrac{1}{5}+85\dfrac{1}{2}=95\dfrac{2}{10}+85\dfrac{5}{10}=180\dfrac{7}{10}$$

This checks.

State. The person was $95\dfrac{1}{5}$ mi from Los Angeles.

41. Familiarize. We draw a picture.

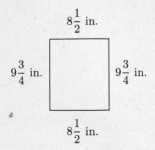

$8\frac{1}{2}$ in.

$9\frac{3}{4}$ in. $9\frac{3}{4}$ in.

$8\frac{1}{2}$ in.

Translate. We let $D =$ the distance around the paper.

Top distance	plus	Right-side distance	plus	Bottom distance	plus	Left-side distance	is	Total distance
$\downarrow$	$\downarrow$	$\downarrow$	$\downarrow$	$\downarrow$	$\downarrow$	$\downarrow$	$\downarrow$	$\downarrow$
$8\frac{1}{2}$	$+$	$9\frac{3}{4}$	$+$	$8\frac{1}{2}$	$+$	$9\frac{3}{4}$	$=$	D

Solve. To solve we carry out the addition. The LCD is 4.

$$8\,\boxed{\frac{1}{2}\cdot\frac{2}{2}} = \quad 8\,\frac{2}{4}$$
$$9\,\frac{3}{4} = \quad 9\,\frac{3}{4}$$
$$8\,\boxed{\frac{1}{2}\cdot\frac{2}{2}} = \quad 8\,\frac{2}{4}$$
$$+9\,\frac{3}{4} = + \;9\,\frac{3}{4}$$
$$\overline{\qquad\qquad\quad 34\,\frac{10}{4} = 36\frac{2}{4} = 36\frac{1}{2}}$$

Check. We repeat the calculation.

State. The distance around the book is $36\frac{1}{2}$ in.

43. Familiarize. We let $c =$ the closing price.

Translate. We write an equation.

Amount at opening	$-$	Amount dropped	$=$	Amount at closing
$\downarrow$	$\downarrow$	$\downarrow$	$\downarrow$	$\downarrow$
$104\frac{5}{8}$	$-$	$1\frac{1}{4}$	$=$	c

Solve. To solve we carry out the subtraction. The LCD is 8.

$$104\,\frac{5}{8} = \quad 104\,\frac{5}{8}$$
$$-\;\;1\,\boxed{\frac{1}{4}\cdot\frac{2}{2}} = - \;\;1\,\frac{2}{8}$$
$$\overline{\qquad\qquad\qquad 103\,\frac{3}{8}}$$

Check. We add the amount the stock dropped to the closing price:

$$1\frac{1}{4} + 103\frac{3}{8} = 1\frac{2}{8} + 103\frac{3}{8} = 104\frac{5}{8}$$

This checks.

State. The closing price was $\$103\frac{3}{8}$.

45. Familiarize. We let $t =$ the total amount of paint used.

Translate. We write an equation.

Family room paint	$+$	Bedroom paint	$=$	Total paint
$\downarrow$	$\downarrow$	$\downarrow$	$\downarrow$	$\downarrow$
$1\frac{2}{3}$	$+$	$1\frac{1}{2}$	$=$	t

Solve. To solve we carry out the addition. The LCD is 6.

$$1\,\boxed{\frac{2}{3}\cdot\frac{2}{2}} = \;\; 1\,\frac{4}{6}$$
$$+1\,\boxed{\frac{1}{2}\cdot\frac{3}{3}} = +1\,\frac{3}{6}$$
$$\overline{\qquad\qquad\quad 2\,\frac{7}{6} = 3\frac{1}{6}}$$

Check. We repeat the calculation. We also note that the total is larger than either of the individual amounts, so the answer seems reasonable.

State. In all, $3\frac{1}{6}$ gal of paint were used.

47. Familiarize. We let $h =$ the man's height.

Translate. We write an equation.

Son's height	$+$	Additional height of man	$=$	Man's height
$\downarrow$	$\downarrow$	$\downarrow$	$\downarrow$	$\downarrow$
$72\frac{5}{6}$	$+$	$5\frac{1}{4}$	$=$	h

Solve. To solve we carry out the addition. The LCD is 12.

$$72\,\boxed{\frac{5}{6}\cdot\frac{2}{2}} = \;\; 72\,\frac{10}{12}$$
$$+\;5\,\boxed{\frac{1}{4}\cdot\frac{3}{3}} = + \;\;5\,\frac{3}{12}$$
$$\overline{\qquad\qquad\quad 77\,\frac{13}{12} = 78\frac{1}{12}}$$

Check. We repeat the calculation. We also note that the man's height is larger than his son's height, so the answer seems reasonable.

State. The man is $78\frac{1}{12}$ in. tall.

49. Familiarize. We make a drawing. We let $t =$ the number of hours the designer worked on the third day.

$$\vdash\!\!-\; 2\tfrac{1}{2}\ \text{hr}\ -\!\!\!-\!\!\vdash\!\!-\; 4\tfrac{1}{5}\ \text{hr}\ -\!\!\!-\!\!\vdash\!\!- t -\!\!\dashv$$
$$\vdash\!\!-\!\!\!-\!\!\!-\!\!\!-\; 10\tfrac{1}{2}\ \text{hr}\ -\!\!\!-\!\!\!-\!\!\!-\!\!\dashv$$

Translate. We write an addition sentence.

$$2\tfrac{1}{2} + 4\tfrac{1}{5} + t = 10\tfrac{1}{2}$$

Solve. This is a two-step problem.

a) First we add $2\tfrac{1}{2} + 4\tfrac{1}{5}$ to find the time worked on the first two days. The LCD is 10.

$$2\boxed{\tfrac{1}{2}\cdot\tfrac{5}{5}} = \quad 2\tfrac{5}{10}$$
$$+\,4\boxed{\tfrac{1}{5}\cdot\tfrac{2}{2}} = +\,4\tfrac{2}{10}$$
$$\rule{3cm}{0.4pt}$$
$$6\tfrac{7}{10}$$

b) Then we subtract $6\tfrac{7}{10}$ from $10\tfrac{1}{2}$ to find the time worked on the third day. The LCD is 10.

$$6\tfrac{7}{10} + t = 10\tfrac{1}{2}$$
$$t = 10\tfrac{1}{2} - 6\tfrac{7}{10}$$

$$10\boxed{\tfrac{1}{2}\cdot\tfrac{5}{5}} = \quad 10\tfrac{5}{10} = \quad 9\tfrac{15}{10}$$
$$-\,6\tfrac{7}{10} = -\,6\tfrac{7}{10} = -\,6\tfrac{7}{10}$$
$$\rule{3cm}{0.4pt}$$
$$3\tfrac{8}{10} = 3\tfrac{4}{5}$$

Check. We repeat the calculations.

State. The designer worked $3\tfrac{4}{5}$ hr the third day.

51. The length of each of the five sides is $5\tfrac{3}{4}$ yd. We add to find the distance around the figure.

$$5\tfrac{3}{4} + 5\tfrac{3}{4} + 5\tfrac{3}{4} + 5\tfrac{3}{4} + 5\tfrac{3}{4} = 25\tfrac{15}{4} = 25 + 3\tfrac{3}{4} = 28\tfrac{3}{4}$$

The distance is $28\tfrac{3}{4}$ yd.

53. We see that d and the two smallest distances combined are the same as the largest distance. We translate and solve.

$$2\tfrac{3}{4} + d + 2\tfrac{3}{4} = 12\tfrac{7}{8}$$
$$d = 12\tfrac{7}{8} - 2\tfrac{3}{4} - 2\tfrac{3}{4}$$
$$= 10\tfrac{1}{8} - 2\tfrac{3}{4} \quad \text{Subtracting } 2\tfrac{3}{4} \text{ from } 12\tfrac{7}{8}$$
$$= 7\tfrac{3}{8} \qquad \text{Subtracting } 2\tfrac{3}{4} \text{ from } 10\tfrac{1}{8}$$

The length of d is $7\tfrac{3}{8}$ ft.

55. Familiarize. We let $b =$ the length of the bolt.

Translate. From the drawing we see that the length of the small bolt is the sum of the diameters of the two tubes and the thicknesses of the two washers and the nut. Thus, we have

$$b = \tfrac{1}{2} + \tfrac{1}{16} + \tfrac{3}{4} + \tfrac{1}{16} + \tfrac{3}{16}.$$

Solve. We carry out the addition. The LCD is 16.

$$b = \tfrac{1}{2} + \tfrac{1}{16} + \tfrac{3}{4} + \tfrac{1}{16} + \tfrac{3}{16} =$$
$$\tfrac{1}{2}\cdot\tfrac{8}{8} + \tfrac{1}{16} + \tfrac{3}{4}\cdot\tfrac{4}{4} + \tfrac{1}{16} + \tfrac{3}{16} =$$
$$\tfrac{8}{16} + \tfrac{1}{16} + \tfrac{12}{16} + \tfrac{1}{16} + \tfrac{3}{16} = \tfrac{25}{16} = 1\tfrac{9}{16}$$

Check. We repeat the calculation.

State. The smallest bolt is $1\tfrac{9}{16}$ in. long.

57. $\dfrac{12}{25} \div \dfrac{24}{5} = \dfrac{12}{25} \cdot \dfrac{5}{24} = \dfrac{12\cdot 5}{25\cdot 24} = \dfrac{12\cdot 5\cdot 1}{5\cdot 5\cdot 12\cdot 2} =$

$\dfrac{12\cdot 5}{12\cdot 5}\cdot\dfrac{1}{5\cdot 2} = \dfrac{1}{10}$

59.
$$47\tfrac{2}{3} + n = 56\tfrac{1}{4}$$
$$47\tfrac{2}{3} + n - 47\tfrac{2}{3} = 56\tfrac{1}{4} - 47\tfrac{2}{3} \qquad \text{Subtracting } 47\tfrac{2}{3}$$
$$n + 0 = 56\tfrac{3}{12} - 47\tfrac{8}{12} \qquad \text{The LCD is 12.}$$
$$n = 55\tfrac{15}{12} - 47\tfrac{8}{12} \qquad \text{Borrowing}$$
$$n = 8\tfrac{7}{12}$$

Exercise Set 3.6

1. $8 \cdot 2\tfrac{5}{6}$

$= \dfrac{8}{1} \cdot \dfrac{17}{6} \qquad \text{Writing fractional notation}$

$= \dfrac{8\cdot 17}{1\cdot 6} = \dfrac{2\cdot 4\cdot 17}{1\cdot 2\cdot 3} = \dfrac{2}{2}\cdot\dfrac{4\cdot 17}{1\cdot 3} = \dfrac{68}{3} = 22\tfrac{2}{3}$

3. $3\tfrac{5}{8} \cdot \tfrac{2}{3}$

$= \dfrac{29}{8}\cdot\dfrac{2}{3} \qquad \text{Writing fractional notation}$

$= \dfrac{29\cdot 2}{8\cdot 3} = \dfrac{29\cdot 2}{2\cdot 4\cdot 3} = \dfrac{2}{2}\cdot\dfrac{29}{4\cdot 3} = \dfrac{29}{12} = 2\tfrac{5}{12}$

5. $3\tfrac{1}{2} \cdot 2\tfrac{1}{3} = \tfrac{7}{2}\cdot\tfrac{7}{3} = \tfrac{49}{6} = 8\tfrac{1}{6}$

7. $3\tfrac{2}{5}\cdot 2\tfrac{7}{8} = \tfrac{17}{5}\cdot\tfrac{23}{8} = \tfrac{391}{40} = 9\tfrac{31}{40}$

9. $4\frac{7}{10} \cdot 5\frac{3}{10} = \frac{47}{10} \cdot \frac{53}{10} = \frac{2491}{100} = 24\frac{91}{100}$

11. $20\frac{1}{2} \cdot 10\frac{1}{5} = \frac{41}{2} \cdot \frac{51}{5} = \frac{2091}{10} = 209\frac{1}{10}$

13.
$$20 \div 3\frac{1}{5}$$
$$= 20 \div \frac{16}{5} \quad \text{Writing fractional notation}$$
$$= 20 \cdot \frac{5}{16} \quad \text{Multiplying by the reciprocal}$$
$$= \frac{20 \cdot 5}{16} = \frac{4 \cdot 5 \cdot 5}{4 \cdot 4} = \frac{4}{4} \cdot \frac{5 \cdot 5}{4} = \frac{25}{4} = 6\frac{1}{4}$$

15. $8\frac{2}{5} \div 7$
$$= \frac{42}{5} \div 7 \quad \text{Writing fractional notation}$$
$$= \frac{42}{5} \cdot \frac{1}{7} \quad \text{Multiplying by the reciprocal}$$
$$= \frac{42 \cdot 1}{5 \cdot 7} = \frac{6 \cdot 7}{5 \cdot 7} = \frac{7}{7} \cdot \frac{6}{5} = \frac{6}{5} = 1\frac{1}{5}$$

17. $4\frac{3}{4} \div 1\frac{1}{3} = \frac{19}{4} \div \frac{4}{3} = \frac{19}{4} \cdot \frac{3}{4} = \frac{19 \cdot 3}{4 \cdot 4} = \frac{57}{16} = 3\frac{9}{16}$

19. $1\frac{7}{8} \div 1\frac{2}{3} = \frac{15}{8} \div \frac{5}{3} = \frac{15}{8} \cdot \frac{3}{5} = \frac{15 \cdot 3}{8 \cdot 5} = \frac{5 \cdot 3 \cdot 3}{8 \cdot 5}$
$$= \frac{5}{5} \cdot \frac{3 \cdot 3}{8} = \frac{3 \cdot 3}{8} = \frac{9}{8} = 1\frac{1}{8}$$

21. $5\frac{1}{10} \div 4\frac{3}{10} = \frac{51}{10} \div \frac{43}{10} = \frac{51}{10} \cdot \frac{10}{43} = \frac{51 \cdot 10}{10 \cdot 43}$
$$= \frac{10}{10} \cdot \frac{51}{43} = \frac{51}{43} = 1\frac{8}{43}$$

23. $20\frac{1}{4} \div 90 = \frac{81}{4} \div 90 = \frac{81}{4} \cdot \frac{1}{90} = \frac{81 \cdot 1}{4 \cdot 90} = \frac{9 \cdot 9 \cdot 1}{4 \cdot 9 \cdot 10}$
$$= \frac{9}{9} \cdot \frac{9 \cdot 1}{4 \cdot 10} = \frac{9}{40}$$

25. **Familiarize.** Visualize this situation as a rectangular array with $16\frac{3}{4}$ revolutions in each row and 56 rows. We let n = the number of revolutions.

Translate. We write an equation.

Revolutions per minute	$\cdot$	Number of minutes	$=$	Total number of revolutions
$\downarrow$	$\downarrow$	$\downarrow$	$\downarrow$	$\downarrow$
$16\frac{3}{4}$	$\cdot$	56	$=$	n

Solve. We carry out the multiplication.

$$n = 16\frac{3}{4} \cdot 56 = \frac{67}{4} \cdot \frac{56}{1} = \frac{67 \cdot 4 \cdot 14}{4 \cdot 1} = \frac{4}{4} \cdot \frac{67 \cdot 14}{1} = 938$$

Check. We repeat the calculation. We can also do a partial check by estimating: $16\frac{3}{4} \cdot 56 \approx 17 \cdot 60 = 1020$. Thus, our answer seems reasonable.

State. The water wheel makes 938 revolutions in 56 min.

27. **Familiarize.** We let n = the number of ounces of meat required.

Translate. We write an equation.

Ounces per serving	$\cdot$	Number of servings	$=$	Total number of ounces
$\downarrow$	$\downarrow$	$\downarrow$	$\downarrow$	$\downarrow$
$3\frac{1}{2}$	$\cdot$	2	$=$	n

Solve. To solve the equation we carry out the multiplication.

$$n = 3\frac{1}{2} \cdot 2$$
$$= \frac{7}{2} \cdot \frac{2}{1} = \frac{7 \cdot 2}{2 \cdot 1} = \frac{2}{2} \cdot \frac{7}{1} = 7$$

Check. We repeat the calculation.

State. 7 ounces of meat are needed.

29. **Familiarize.** We let w = the weight of $5\frac{1}{2}$ cubic feet of water.

Translate. We write an equation.

Weight per cubic foot	$\cdot$	Number of cubic feet	$=$	Total weight
$\downarrow$	$\downarrow$	$\downarrow$	$\downarrow$	$\downarrow$
$62\frac{1}{2}$	$\cdot$	$5\frac{1}{2}$	$=$	w

Solve. To solve the equation we carry out the multiplication.

$$w = 62\frac{1}{2} \cdot 5\frac{1}{2}$$
$$= \frac{125}{2} \cdot \frac{11}{2} = \frac{125 \cdot 11}{2 \cdot 2}$$
$$= \frac{1375}{4} = 343\frac{3}{4}$$

Check. We repeat the calculation. We also note that $62\frac{1}{2} \approx 60$ and $5\frac{1}{2} \approx 5$. Then the product is about 300. Our answer seems reasonable.

State. The weight of $5\frac{1}{2}$ cubic feet of water is $343\frac{3}{4}$ lb.

31. **Familiarize, Translate, and Solve.** To find the ingredients for $\frac{1}{2}$ recipe, we multiply each ingredient by $\frac{1}{2}$.

$$2 \cdot \frac{1}{2} = \frac{2 \cdot 1}{2} = \frac{2}{2} = 1$$
$$1\frac{1}{2} \cdot \frac{1}{2} = \frac{3}{2} \cdot \frac{1}{2} = \frac{3 \cdot 1}{2 \cdot 2} = \frac{3}{4}$$
$$3 \cdot \frac{1}{2} = \frac{3 \cdot 1}{2} = \frac{3}{2} = 1\frac{1}{2}$$
$$2\frac{1}{2} \cdot \frac{1}{2} = \frac{5}{2} \cdot \frac{1}{2} = \frac{5 \cdot 1}{2 \cdot 2} = \frac{5}{4} = 1\frac{1}{4}$$

$$1 \cdot \frac{1}{2} = \frac{1 \cdot 1}{2} = \frac{1}{2}$$

$$4 \cdot \frac{1}{2} = \frac{4 \cdot 1}{2} = \frac{4}{2} = 2$$

$$\frac{1}{3} \cdot \frac{1}{2} = \frac{1 \cdot 1}{3 \cdot 2} = \frac{1}{6}$$

$$\frac{1}{4} \cdot \frac{1}{2} = \frac{1 \cdot 1}{4 \cdot 2} = \frac{1}{8}$$

Check. We repeat the calculation.

State. The ingredients for $\frac{1}{2}$ recipe are 1 chicken bouillon cube, $\frac{3}{4}$ cup hot water, $1\frac{1}{2}$ tablespoons margarine, $1\frac{1}{2}$ tablespoons flour, $1\frac{1}{4}$ cups diced cooked chicken, $\frac{1}{2}$ cup cooked peas, 2 oz sliced mushrooms (drained), $\frac{1}{6}$ cup sliced cooked carrots, $\frac{1}{8}$ cup chopped onions, 1 tablespoon chopped pimiento, $\frac{1}{2}$ teaspoon salt.

Familiarize, Translate, and Solve. To find the ingredients for 3 recipes, we multiply each ingredient by 3.

$$2 \cdot 3 = 6$$

$$1\frac{1}{2} \cdot 3 = \frac{3}{2} \cdot \frac{3}{1} = \frac{3 \cdot 3}{2 \cdot 1} = \frac{9}{2} = 4\frac{1}{2}$$

$$3 \cdot 3 = 9$$

$$2\frac{1}{2} \cdot 3 = \frac{5}{2} \cdot \frac{3}{1} = \frac{5 \cdot 3}{2 \cdot 1} = \frac{15}{2} = 7\frac{1}{2}$$

$$1 \cdot 3 = 3$$

$$\frac{1}{3} \cdot 3 = \frac{1 \cdot 3}{3} = \frac{3}{3} = 1$$

$$\frac{1}{4} \cdot 3 = \frac{1 \cdot 3}{4} = \frac{3}{4}$$

Check. We repeat the calculation.

State. The ingredients for 3 recipes are 6 chicken bouillon cubes, $4\frac{1}{2}$ cups hot water, 9 tablespoons margarine, 9 tablespoons flour, $7\frac{1}{2}$ cups diced cooked chicken, 3 cups cooked peas, 3 4-oz cans sliced mushrooms (drained), 1 cup sliced cooked carrots, $\frac{3}{4}$ cup chopped onions, 6 tablespoons chopped pimiento, 3 teaspoons salt.

33. Familiarize. We let t = the Fahrenheit temperature.

Translate.

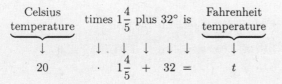

Solve. We multiply and then add, according to the rules for order of operations.

$$t = 20 \cdot 1\frac{4}{5} + 32 = \frac{20}{1} \cdot \frac{9}{5} + 32 = \frac{20 \cdot 9}{1 \cdot 5} + 32 =$$

$$\frac{4 \cdot 5 \cdot 9}{1 \cdot 5} + 32 = \frac{5}{5} \cdot \frac{4 \cdot 9}{1} + 32 = 36 + 32 = 68$$

Check. We repeat the calculation.

State. 68° Fahrenheit corresponds to 20° Celsius.

35. Familiarize. Visualize the situation as a rectangular array containing 469 revolutions with $16\frac{3}{4}$ revolutions in each row. We must determine how many rows the array has. (The last row may be incomplete.) We let t = the time the wheel rotates.

Translate. The division that corresponds to the situation is

$$469 \div 16\frac{3}{4} = t.$$

Solve. We carry out the division.

$$t = 469 \div 16\frac{3}{4} = 469 \div \frac{67}{4} = 469 \cdot \frac{4}{67} =$$

$$\frac{67 \cdot 7 \cdot 4}{67 \cdot 1} = \frac{67}{67} \cdot \frac{7 \cdot 4}{1} = 28$$

Check. We check by multiplying the time by the number of revolutions per minute.

$$16\frac{3}{4} \cdot 28 = \frac{67}{4} \cdot 28 = \frac{67 \cdot 7 \cdot 4}{4 \cdot 1} = \frac{4}{4} \cdot \frac{67 \cdot 7}{1} = 469$$

The answer checks.

State. The water wheel rotated for 28 min.

37. Familiarize. We let m = the number of miles per gallon the car got.

Translate. We write an equation.

Total number of miles traveled	÷	Number of gallons of gas used	=	Miles per gallon
↓	↓	↓	↓	↓
213	÷	$14\frac{2}{10}$	=	m

Solve. To solve the equation we carry out the division.

$$m = 213 \div 14\frac{2}{10} = 213 \div \frac{142}{10}$$

$$= 213 \cdot \frac{10}{142} = \frac{3 \cdot 71 \cdot 2 \cdot 5}{2 \cdot 71 \cdot 1}$$

$$= \frac{2 \cdot 71}{2 \cdot 71} \cdot \frac{3 \cdot 5}{1} = 15$$

Check. We repeat the calculation.

State. The car got 15 miles per gallon of gas.

39. Familiarize. We let n = the number of cubic feet occupied by 250 lb of water.

Translate. We write an equation.

$$\underbrace{\text{Total weight}}_{\downarrow} \div \underbrace{\genfrac{}{}{0pt}{}{\text{Weight per}}{\text{cubic foot}}}_{\downarrow} = \underbrace{\genfrac{}{}{0pt}{}{\text{Number of}}{\text{cubic feet}}}_{\downarrow}$$

$$250 \qquad \div \qquad 62\frac{1}{2} \qquad = \qquad n$$

Solve. To solve the equation we carry out the division.

$$n = 250 \div 62\frac{1}{2} = 250 \div \frac{125}{2}$$

$$= 250 \cdot \frac{2}{125} = \frac{2 \cdot 125 \cdot 2}{125 \cdot 1}$$

$$= \frac{125}{125} \cdot \frac{2 \cdot 2}{1} = 4$$

Check. We repeat the calculation.

State. 4 cubic feet would be occupied.

41. *Familiarize*. Visualize the situation as a rectangular array containing 24 hours with $1\frac{1}{2}$ hours in each row. We must determine how many rows the array has. (The last row may be incomplete.) We let $n =$ the number of orbits made every 24 hours.

Translate. The division that corresponds to this situation is

$$24 \div 1\frac{1}{2} = n.$$

Solve. We carry out the division.

$$n = 24 \div 1\frac{1}{2} = 24 \div \frac{3}{2} = 24 \cdot \frac{2}{3} = \frac{24 \cdot 2}{3} = \frac{3 \cdot 8 \cdot 2}{3 \cdot 1} =$$

$$\frac{3}{3} \cdot \frac{8 \cdot 2}{1} = 16$$

Check. We check by multiplying the number of orbits by the time it takes to make one orbit.

$$16 \cdot 1\frac{1}{2} = 16 \cdot \frac{3}{2} = \frac{16 \cdot 3}{2} = \frac{2 \cdot 8 \cdot 3}{2 \cdot 1} = \frac{2}{2} \cdot \frac{8 \cdot 3}{1} = 24$$

The answer checks.

State. The shuttle makes 16 orbits every 24 hr.

43. *Familiarize*. The figure is composed of two rectangles. One has dimensions s by $\frac{1}{2} \cdot s$, or $6\frac{7}{8}$ in. by $\frac{1}{2} \cdot 6\frac{7}{8}$ in. The other has dimensions $\frac{1}{2} \cdot s$ by $\frac{1}{2} \cdot s$, or $\frac{1}{2} \cdot 6\frac{7}{8}$ in. by $\frac{1}{2} \cdot 6\frac{7}{8}$ in. The total area is the sum of the areas of these two rectangles. We let $A =$ the total area.

Translate. We write an equation.

$$A = \left(6\frac{7}{8}\right) \cdot \left(\frac{1}{2} \cdot 6\frac{7}{8}\right) + \left(\frac{1}{2} \cdot 6\frac{7}{8}\right) \cdot \left(\frac{1}{2} \cdot 6\frac{7}{8}\right)$$

Solve. We carry out each multiplication and then add.

$$A = \left(6\frac{7}{8}\right) \cdot \left(\frac{1}{2} \cdot 6\frac{7}{8}\right) + \left(\frac{1}{2} \cdot 6\frac{7}{8}\right) \cdot \left(\frac{1}{2} \cdot 6\frac{7}{8}\right)$$

$$= \frac{55}{8} \cdot \left(\frac{1}{2} \cdot \frac{55}{8}\right) + \left(\frac{1}{2} \cdot \frac{55}{8}\right) \cdot \left(\frac{1}{2} \cdot \frac{55}{8}\right)$$

$$= \frac{55}{8} \cdot \frac{55}{16} + \frac{55}{16} \cdot \frac{55}{16}$$

$$= \frac{3025}{128} + \frac{3025}{256} = \frac{3025}{128} \cdot \frac{2}{2} + \frac{3025}{256}$$

$$= \frac{6050}{256} + \frac{3025}{256} = \frac{9075}{256}$$

$$= 35\frac{115}{256}$$

Check. We repeat the calculation.

State. The area is $35\frac{115}{256}$ sq in.

45. *Familiarize*. We make a drawing.

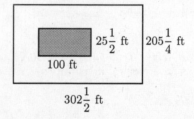

Translate. We let $A =$ the area of the lot not covered by the building.

$$\underbrace{\genfrac{}{}{0pt}{}{\text{Area}}{\text{left over}}}_{\downarrow} \underset{\downarrow}{\text{is}} \underbrace{\genfrac{}{}{0pt}{}{\text{Area}}{\text{of lot}}}_{\downarrow} \underset{\downarrow}{\text{minus}} \underbrace{\genfrac{}{}{0pt}{}{\text{Area of}}{\text{building}}}_{\downarrow}$$

$$A \qquad = \left(302\frac{1}{2}\right) \cdot \left(205\frac{1}{4}\right) \quad - \quad (100) \cdot \left(25\frac{1}{2}\right)$$

Solve. We do each multiplication and then find the difference.

$$A = \left(302\frac{1}{2}\right) \cdot \left(205\frac{1}{4}\right) - (100) \cdot \left(25\frac{1}{2}\right)$$

$$= \frac{605}{2} \cdot \frac{821}{4} - \frac{100}{1} \cdot \frac{51}{2}$$

$$= \frac{605 \cdot 821}{2 \cdot 4} - \frac{100 \cdot 51}{1 \cdot 2}$$

$$= \frac{605 \cdot 821}{2 \cdot 4} - \frac{2 \cdot 50 \cdot 51}{1 \cdot 2} = \frac{605 \cdot 821}{2 \cdot 4} - \frac{2}{2} \cdot \frac{50 \cdot 51}{1}$$

$$= \frac{496,705}{8} - 2550 = 62,088\frac{1}{8} - 2550$$

$$= 59,538\frac{1}{8}$$

Check. We repeat the calculation.

State. The area left over is $59,538\frac{1}{8}$ sq ft.

47.
$$
\begin{array}{r}
6\ 7\ 0\ 9 \\
\times\quad 2\ 1\ 3 \\
\hline
2\ 0\ 1\ 2\ 7 \\
6\ 7\ 0\ 9\ 0 \\
1\ 3\ 4\ 1\ 8\ 0\ 0 \\
\hline
1,4\ 2\ 9,0\ 1\ 7
\end{array}
$$

49. $\dfrac{5}{7} \cdot t = 420$

$$t = 420 \div \frac{5}{7} = 420 \cdot \frac{7}{5} = \frac{5 \cdot 84 \cdot 7}{5 \cdot 1} = \frac{5}{5} \cdot \frac{84 \cdot 7}{1}$$

$$= 588$$

51. $\dfrac{3}{8} \cdot \dfrac{4}{9} = \dfrac{3 \cdot 4}{8 \cdot 9} = \dfrac{3 \cdot 4 \cdot 1}{2 \cdot 4 \cdot 3 \cdot 3} = \dfrac{3 \cdot 4}{3 \cdot 4} \cdot \dfrac{1}{2 \cdot 3} = \dfrac{1}{6}$

53. $8 \div \dfrac{1}{2} + \dfrac{3}{4} + \left(5 - \dfrac{5}{8}\right)^2$

$$= 8 \div \frac{1}{2} + \frac{3}{4} + \left(\frac{40}{8} - \frac{5}{8}\right)^2$$

$$= 8 \div \frac{1}{2} + \frac{3}{4} + \left(\frac{35}{8}\right)^2$$

$$= 8 \div \frac{1}{2} + \frac{3}{4} + \frac{1225}{64}$$

$$= 8 \cdot 2 + \frac{3}{4} + \frac{1225}{64}$$

$$= 16 + \frac{3}{4} + \frac{1225}{64}$$

$$= \frac{1024}{64} + \frac{48}{64} + \frac{1225}{64}$$

$$= \frac{2297}{64} = 35\frac{57}{64}$$

55. $\dfrac{1}{3} \div \left(\dfrac{1}{2} - \dfrac{1}{5}\right) \times \dfrac{1}{4} + \dfrac{1}{6}$

$$= \frac{1}{3} \div \left(\frac{5}{10} - \frac{2}{10}\right) \times \frac{1}{4} + \frac{1}{6}$$

$$= \frac{1}{3} \div \frac{3}{10} \times \frac{1}{4} + \frac{1}{6}$$

$$= \frac{1}{3} \times \frac{10}{3} \times \frac{1}{4} + \frac{1}{6}$$

$$= \frac{10}{9} \times \frac{1}{4} + \frac{1}{6}$$

$$= \frac{2 \times 5 \times 1}{9 \times 2 \times 2} + \frac{1}{6} = \frac{2}{2} \times \frac{5 \times 1}{9 \times 2} + \frac{1}{6}$$

$$= \frac{5}{18} + \frac{1}{6} = \frac{5}{18} + \frac{3}{18} = \frac{8}{18} = \frac{4}{9}$$

57. $4\dfrac{1}{2} \div 2\dfrac{1}{2} + 8 - 4 \div \dfrac{1}{2}$

$$= \frac{9}{2} \div \frac{5}{2} + 8 - 4 \div \frac{1}{2}$$

$$= \frac{9}{2} \cdot \frac{2}{5} + 8 - 4 \div \frac{1}{2}$$

$$= \frac{2}{2} \cdot \frac{9}{5} + 8 - 4 \div \frac{1}{2}$$

$$= \frac{9}{5} + 8 - 4 \div \frac{1}{2}$$

$$= \frac{9}{5} + 8 - 4 \cdot 2$$

$$= \frac{9}{5} + 8 - 8$$

$$= \frac{9}{5} + \frac{40}{5} - \frac{40}{5} = \frac{9}{5}, \text{ or } 1\frac{4}{5}$$

Exercise Set 3.7

1. $\dfrac{1}{2} \cdot \dfrac{1}{3} \cdot \dfrac{1}{4}$

$= \dfrac{1}{6} \cdot \dfrac{1}{4}$ Doing the multiplications in

$= \dfrac{1}{24}$ order from left to right

3. $6 \div 3 \div 5$

$= 2 \div 5$ Doing the divisions in

$= \dfrac{2}{5}$ order from left to right

5. $\dfrac{2}{3} \div \dfrac{4}{3} \div \dfrac{7}{8}$

$= \dfrac{2}{3} \cdot \dfrac{3}{4} \div \dfrac{7}{8}$ Doing the first division; multiplying by the reciprocal of $\dfrac{4}{3}$

$= \dfrac{2 \cdot 3 \cdot 1}{3 \cdot 2 \cdot 2} \div \dfrac{7}{8}$

$= \dfrac{1}{2} \div \dfrac{7}{8}$ Removing a factor of 1

$= \dfrac{1}{2} \cdot \dfrac{8}{7}$ Dividing; multiplying by the reciprocal of $\dfrac{7}{8}$

$= \dfrac{1 \cdot 2 \cdot 4}{2 \cdot 7}$

$= \dfrac{4}{7}$ Removing a factor of 1

7. $\dfrac{5}{8} \div \dfrac{1}{4} - \dfrac{2}{3} \cdot \dfrac{4}{5}$

$= \dfrac{5}{8} \cdot \dfrac{4}{1} - \dfrac{2}{3} \cdot \dfrac{4}{5}$ Dividing

$= \dfrac{5 \cdot 4}{2 \cdot 4 \cdot 1} - \dfrac{2}{3} \cdot \dfrac{4}{5}$

$= \dfrac{5}{2} - \dfrac{2}{3} \cdot \dfrac{4}{5}$ Removing a factor of 1

$= \dfrac{5}{2} - \dfrac{2 \cdot 4}{3 \cdot 5}$ Multiplying

$= \dfrac{5}{2} - \dfrac{8}{15}$

$= \dfrac{5}{2} \cdot \dfrac{15}{15} - \dfrac{8}{15} \cdot \dfrac{2}{2}$ Multiplying by 1 to obtain the LCD

$= \dfrac{75}{30} - \dfrac{16}{30}$

$= \dfrac{59}{30}$, or $1\dfrac{29}{30}$ Subtracting

9. $\dfrac{1}{2} \cdot \left(\dfrac{3}{4} - \dfrac{2}{3} \right)$

$= \dfrac{1}{2} \cdot \left(\dfrac{3}{4} \cdot \dfrac{3}{3} - \dfrac{2}{3} \cdot \dfrac{4}{4} \right)$ Carrying out the operation in parentheses

$= \dfrac{1}{2} \cdot \left(\dfrac{9}{12} - \dfrac{8}{12} \right)$

$= \dfrac{1}{2} \cdot \dfrac{1}{12}$ Subtracting

$= \dfrac{1 \cdot 1}{2 \cdot 12}$ Multiplying

$= \dfrac{1}{24}$

11. $28\dfrac{1}{8} - 5\dfrac{1}{4} + 3\dfrac{1}{2}$

$= 28\dfrac{1}{8} - 5\dfrac{2}{8} + 3\dfrac{1}{2}$ Doing the additions and

$= 27\dfrac{9}{8} - 5\dfrac{2}{8} + 3\dfrac{1}{2}$ subtractions in order

$= 22\dfrac{7}{8} + 3\dfrac{1}{2}$ from left to right

$= 22\dfrac{7}{8} + 3\dfrac{4}{8}$

$= 25\dfrac{11}{8}$

$= 26\dfrac{3}{8}$, or $\dfrac{211}{8}$

13. $\dfrac{7}{8} \div \dfrac{1}{2} \cdot \dfrac{1}{4}$

$= \dfrac{7}{8} \cdot \dfrac{2}{1} \cdot \dfrac{1}{4}$ Dividing

$= \dfrac{7 \cdot 2}{2 \cdot 4 \cdot 1} \cdot \dfrac{1}{4}$

$= \dfrac{7}{4} \cdot \dfrac{1}{4}$ Removing a factor of 1

$= \dfrac{7}{16}$ Multiplying

15. $\left(\dfrac{2}{3} \right)^2 - \dfrac{1}{3} \cdot 1\dfrac{1}{4}$

$= \dfrac{4}{9} - \dfrac{1}{3} \cdot 1\dfrac{1}{4}$ Evaluating the exponental expression

$= \dfrac{4}{9} - \dfrac{1}{3} \cdot \dfrac{5}{4}$

$= \dfrac{4}{9} - \dfrac{5}{12}$ Multiplying

$= \dfrac{4}{9} \cdot \dfrac{4}{4} - \dfrac{5}{12} \cdot \dfrac{3}{3}$

$= \dfrac{16}{36} - \dfrac{15}{36}$

$= \dfrac{1}{36}$ Subtracting

17. $\dfrac{1}{2} - \left(\dfrac{1}{2} \right)^2 + \left(\dfrac{1}{2} \right)^3$

$= \dfrac{1}{2} - \dfrac{1}{4} + \dfrac{1}{8}$ Evaluating the exponential expressions

$= \dfrac{2}{4} - \dfrac{1}{4} + \dfrac{1}{8}$ Doing the additions and

$= \dfrac{1}{4} + \dfrac{1}{8}$ subtractions in order

$= \dfrac{2}{8} + \dfrac{1}{8}$ from left to right

$= \dfrac{3}{8}$

19. Add the numbers and divide by the number of addends.

$\left(\dfrac{2}{3} + \dfrac{7}{8} \right) \div 2$

$= \left(\dfrac{16}{24} + \dfrac{21}{24} \right) \div 2$ Doing the operations inside parentheses first; finding the LCD

$= \dfrac{37}{24} \div 2$ Adding

$= \dfrac{37}{24} \cdot \dfrac{1}{2}$ Dividing

$= \dfrac{37}{48}$

21. Add the numbers and divide by the number of addends.

$$\left(\frac{1}{6} + \frac{1}{8} + \frac{3}{4}\right) \div 3$$

$$= \left(\frac{4}{24} + \frac{3}{24} + \frac{18}{24}\right) \div 3$$

$$= \frac{25}{24} \div 3$$

$$= \frac{25}{24} \cdot \frac{1}{3}$$

$$= \frac{25}{72}$$

23. Add the numbers and divide by the number of addends.

$$\left(3\frac{1}{2} + 9\frac{3}{8}\right) \div 2$$

$$= \left(3\frac{4}{8} + 9\frac{3}{8}\right) \div 2$$

$$= 12\frac{7}{8} \div 2$$

$$= \frac{103}{8} \div 2$$

$$= \frac{103}{8} \cdot \frac{1}{2}$$

$$= \frac{103}{16}, \text{ or } 6\frac{7}{16}$$

25. $\left(\frac{2}{3} + \frac{3}{4}\right) \div \left(\frac{5}{6} - \frac{1}{3}\right)$

$$= \left(\frac{8}{12} + \frac{9}{12}\right) \div \left(\frac{5}{6} - \frac{2}{6}\right) \quad \begin{array}{l}\text{Doing the operations}\\\text{inside parentheses}\end{array}$$

$$= \frac{17}{12} \div \frac{3}{6}$$

$$= \frac{17}{12} \cdot \frac{6}{3}$$

$$= \frac{17 \cdot \cancel{6}}{2 \cdot \cancel{6} \cdot 3}$$

$$= \frac{17}{6}, \text{ or } 2\frac{5}{6}$$

27. $\left(\frac{1}{2} + \frac{1}{3}\right)^2 \cdot 144 - \frac{5}{8} \div 10\frac{1}{2}$

$$= \left(\frac{3}{6} + \frac{2}{6}\right)^2 \cdot 144 - \frac{5}{8} \div 10\frac{1}{2}$$

$$= \left(\frac{5}{6}\right)^2 \cdot 144 - \frac{5}{8} \div 10\frac{1}{2}$$

$$= \frac{25}{36} \cdot 144 - \frac{5}{8} \div 10\frac{1}{2}$$

$$= \frac{25 \cdot \cancel{36} \cdot 4}{\cancel{36} \cdot 1} - \frac{5}{8} \div 10\frac{1}{2}$$

$$= 100 - \frac{5}{8} \div 10\frac{1}{2}$$

$$= 100 - \frac{5}{8} \div \frac{21}{2}$$

$$= 100 - \frac{5}{8} \cdot \frac{2}{21}$$

$$= 100 - \frac{5 \cdot \cancel{2}}{\cancel{2} \cdot 4 \cdot 21}$$

$$= 100 - \frac{5}{84}$$

$$= 99\frac{79}{84}, \text{ or } \frac{8395}{84}$$

29.

$$\begin{array}{r} 1\,2\,6 \\ \times \quad 2\,7 \\ \hline 8\,8\,2 \\ 2\,5\,2\,0 \\ \hline 3\,4\,0\,2 \end{array}$$

31.

$$\begin{array}{r} 5\,9 \\ 1\,3\,2\,\overline{)\,7\,8\,6\,5} \\ 6\,6\,0\,0 \\ \hline 1\,2\,6\,5 \\ 1\,1\,8\,8 \\ \hline 7\,7 \end{array}$$

The answer is 59 R 77, or $59\frac{77}{132}$.

33. $\frac{4}{5} \div \frac{3}{10} = \frac{4}{5} \cdot \frac{10}{3} = \frac{4 \cdot 10}{5 \cdot 3} = \frac{4 \cdot 5 \cdot 2}{5 \cdot 3} = \frac{5}{5} \cdot \frac{4 \cdot 2}{3} = \frac{8}{3}, \text{ or } 2\frac{2}{3}$

35. a) The area of the larger rectangle is $13 \cdot 9\frac{1}{4}$, and the area of the smaller rectangle is $8\frac{1}{4} \cdot 7\frac{1}{4}$. We express the sum of their areas as $13 \cdot 9\frac{1}{4} + 8\frac{1}{4} \cdot 7\frac{1}{4}$.

b) $13 \cdot 9\frac{1}{4} + 8\frac{1}{4} \cdot 7\frac{1}{4}$

$$= 13 \cdot \frac{37}{4} + \frac{33}{4} \cdot \frac{29}{4}$$

$$= \frac{481}{4} + \frac{957}{16}$$

$$= \frac{1924}{16} + \frac{957}{16}$$

$$= \frac{2881}{16}, \text{ or } 180\frac{1}{16} \text{in}^2$$

c) In order to simplify the expression, we must multiply before adding.

Chapter 4

Addition and Subtraction: Decimal Notation

Exercise Set 4.1

1. a) Write a word name for
   the whole number.   [Twenty-three]

   b) Write "and" for the   Twenty-three
   decimal point   [and]

   c) Write a word name for
   the number to the right   Twenty-three
   of the decimal point,   and
   followed by the place   [nine-tenths]
   value of the last digit.

   A word name for 23.9 is twenty-three and nine-tenths.

3. One hundred seventeen—
   and———
   sixty-five hundredths———

   $$\underbrace{117}\quad.\quad\underbrace{65}$$

5.
   Thirty-four—
   and———
   eight hundred ninety-one thousandths———

   $$\underbrace{34}\quad.\quad\underbrace{891}$$

7. Write "and 48 cents" as "and $\frac{48}{100}$ dollars." A word name for $326.48 is three hundred twenty-six and $\frac{48}{100}$ dollars.

9. Write "and 72 cents" as "and $\frac{72}{100}$ dollars." A word name for $36.72 is thirty-six and $\frac{72}{100}$ dollars.

11.   6.$\underline{8}$ 6.8. $\frac{68}{10}$

   1 place   Move 1 place.   1 zero

   $6.8 = \frac{68}{10}$

13.   0.$\underline{17}$ 0.17. $\frac{17}{100}$

   2 places   Move 2 places.   2 zeros

   $0.17 = \frac{17}{100}$

15.   8.$\underline{21}$ 8.21. $\frac{821}{100}$

   2 places   Move 2 places.   2 zeros

   $8.21 = \frac{821}{100}$

17.   204.$\underline{6}$ 204.6. $\frac{2046}{10}$

   1 place   Move 1 place.   1 zero

   $204.6 = \frac{2046}{10}$

19.   1.$\underline{509}$ 1.509. $\frac{1509}{1000}$

   3 places   Move 3 places.   3 zeros

   $1.509 = \frac{1509}{1000}$

21.   46.$\underline{03}$ 46.03. $\frac{4603}{100}$

   2 places   Move 2 places.   2 zeros

   $46.03 = \frac{4603}{100}$

23.   0.$\underline{00013}$ 0.00013. $\frac{13}{100,000}$

   5 places   Move 5 places.   5 zeros

   $0.00013 = \frac{13}{100,000}$

25.   20.$\underline{003}$ 20.003. $\frac{20,003}{1000}$

   3 places   Move 3 places.   3 zeros

   $20.003 = \frac{20,003}{1000}$

27.   1.$\underline{0008}$ 1.0008. $\frac{10,008}{10,000}$

   4 places   Move 4 places.   4 zeros

   $1.0008 = \frac{10,008}{10,000}$

29.   0.$\underline{5321}$ 0.5321. $\frac{5321}{10,000}$

   4 places   Move 4 places.   4 zeros

   $0.5321 = \frac{5321}{10,000}$

31. $\dfrac{8}{10}$ 0.8.

1 zero   Move 1 place.

$\dfrac{8}{10} = 0.8$

33. $\dfrac{92}{100}$ 0.92.

2 zeros   Move 2 places.

$\dfrac{92}{100} = 0.92$

35. $\dfrac{93}{10}$ 9.3.

1 zero   Move 1 place.

$\dfrac{93}{10} = 9.3$

37. $\dfrac{889}{100}$ 8.89.

2 zeros   Move 2 places.

$\dfrac{889}{100} = 8.89$

39. $\dfrac{2508}{10}$ 250.8.

1 zero   Move 1 place.

$\dfrac{2508}{10} = 250.8$

41. $\dfrac{3798}{1000}$ 3.798.

3 zeros   Move 3 places.

$\dfrac{3798}{1000} = 3.798$

43. $\dfrac{78}{10,000}$ 0.0078.

4 zeros   Move 4 places.

$\dfrac{78}{10,000} = 0.0078$

45. $\dfrac{56,788}{100,000}$ 0.56788.

5 zeros   Move 5 places.

$\dfrac{56,788}{100,000} = 0.56788$

47. $\dfrac{2173}{100}$ 21.73.

2 zeros   Move 2 places.

$\dfrac{2173}{100} = 21.73$

49. $\dfrac{66}{100}$ 0.66.

2 zeros   Move 2 places.

$\dfrac{66}{100} = 0.66$

51. $\dfrac{853}{1000}$ 0.853.

3 zeros   Move 3 places.

$\dfrac{853}{1000} = 0.853$

53. $\dfrac{376,193}{1,000,000}$ 0.376193.

6 zeros   Move 6 places.

$\dfrac{376,193}{1,000,000} = 0.376193$

55. Round 617 $\boxed{2}$ to the nearest ten.

The digit 7 is in the tens place. Since the next digit to the right, 2, is 4 or lower, round down, meaning that 7 tens stays as 7 tens. Then change the digit to the right of the tens digit to zero.

The answer is 6170.

57. Round 6 $\boxed{1}$ 72 to the nearest thousand.

The digit 6 is in the thousands place. Since the next digit to the right, 1, is 4 or lower, round down, meaning that 6 thousands stays as 6 thousands. Then change all digits to the right of the thousands digit to zeros.

The answer is 6000.

59.
$$\begin{array}{r} {\scriptstyle 2\ 2}\\ 1\ 3\ 8\\ 6\ 7\\ 9\ 5\\ +\ \ 7\ 9\\ \hline 3\ 7\ 9 \end{array}$$

61. $4\dfrac{909}{1000} = 4 + \dfrac{909}{1000}$

Write decimal notation for $\dfrac{909}{1000}$.

$\dfrac{909}{1000}$ 0.909.

3 zeros   Move 3 places.

$4\dfrac{909}{1000} = 4.909$

Exercise Set 4.2

1. To compare two numbers in decimal notation, start at the left and compare corresponding digits moving from left to right. When two digits differ, the number with the larger digit is the larger of the two numbers.

0.06

↑   Different; 5 is larger than 0.

0.58

Thus, 0.58 is larger.

3. 0.905

↑   Starting at the left, these digits are the first to differ; 1 is larger than 0.

0.91

Thus, 0.91 is larger.

5. 0.0009

↑   Starting at the left, these digits are the first to differ, and 1 is larger than 0.

0.001

Thus, 0.001 is larger.

7. 234.07

↑   Starting at the left, these digits are the first to differ, and 5 is larger than 4.

235.07

Thus, 235.07 is larger.

9. 0.4545

↑   These digits differ, and 4 is larger than 0.

0.05454

Thus, 0.4545 is larger.

11. $\dfrac{4}{100} = 0.04$ so we compare 0.004 and 0.04.

0.004

↑   Starting at the left, these digits are the first to differ, and 4 is larger than 0.

0.04

Thus, 0.04 or $\dfrac{4}{100}$ is larger.

13. 0.54

↑   These digits differ, and 7 is larger than 5.

0.78

Thus, 0.78 is larger.

15. 0.8437

↑   Starting at the left, these digits are the first to differ, and 8 is larger than 7.

0.84384

Thus, 0.84384 is larger.

17. 0.19

↑   These digits differ, and 1 is larger than 0.

1.9

Thus, 1.9 is larger.

19. 0.1|1| Hundredths digit is 4 or lower.
↓   Round down.
0.1

21. 0.1|6| Hundredths digit is 5 or higher.
↓   Round up.
0.2

23. 0.5|7|94   Hundredths digit is 5 or higher.
↓   Round up.
0.6

25. 2.7|4|49   Hundredths digit is 4 or lower.
↓   Round down.
2.7

27. 13.4|1| Hundredths digit is 4 or lower.
↓   Round down.
13.4

29. 123.6|5| Hundredths digit is 5 or higher.
↓   Round up.
123.7

31. 0.89|3| Thousandths digit is 4 or lower.
↓   Round down.
0.89

33. 0.66|6| Thousandths digit is 5 or higher.
↓   Round up.
0.67

35. 0.42|4|6   Thousandths digit is 4 or lower.
↓   Round down.
0.42

37. 1.43|5| Thousandths digit is 5 or higher.
↓   Round up.
1.44

39. 3.58|1| Thousandths digit is 4 or lower.
↓   Round down.
3.58

41. 0.00|7| Thousandths digit is 5 or higher.
↓   Round up.
0.01

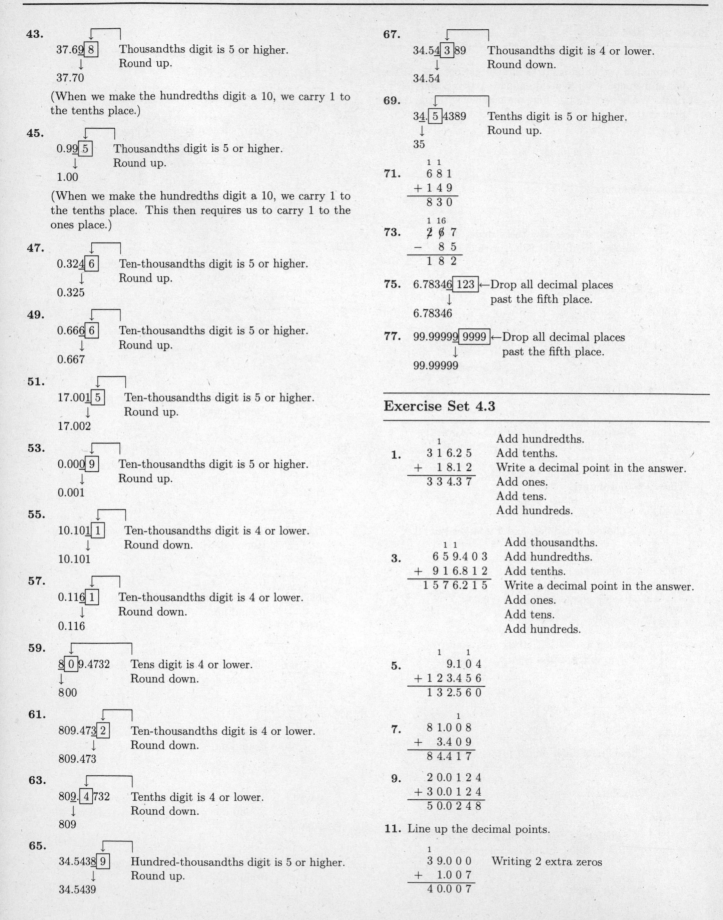

43.

37.69⟨8⟩   Thousandths digit is 5 or higher.
↓   Round up.
37.70

(When we make the hundredths digit a 10, we carry 1 to the tenths place.)

45.

0.99⟨5⟩   Thousandths digit is 5 or higher.
↓   Round up.
1.00

(When we make the hundredths digit a 10, we carry 1 to the tenths place. This then requires us to carry 1 to the ones place.)

47.

0.324⟨6⟩   Ten-thousandths digit is 5 or higher.
↓   Round up.
0.325

49.

0.666⟨6⟩   Ten-thousandths digit is 5 or higher.
↓   Round up.
0.667

51.

17.001⟨5⟩   Ten-thousandths digit is 5 or higher.
↓   Round up.
17.002

53.

0.000⟨9⟩   Ten-thousandths digit is 5 or higher.
↓   Round up.
0.001

55.

10.101⟨1⟩   Ten-thousandths digit is 4 or lower.
↓   Round down.
10.101

57.

0.116⟨1⟩   Ten-thousandths digit is 4 or lower.
↓   Round down.
0.116

59.

8⟨0⟩9.4732   Tens digit is 4 or lower.
↓   Round down.
800

61.

809.473⟨2⟩   Ten-thousandths digit is 4 or lower.
↓   Round down.
809.473

63.

809.⟨4⟩732   Tenths digit is 4 or lower.
↓   Round down.
809

65.

34.5438⟨9⟩   Hundred-thousandths digit is 5 or higher.
↓   Round up.
34.5439

67.

34.54⟨3⟩89   Thousandths digit is 4 or lower.
↓   Round down.
34.54

69.

34.⟨5⟩4389   Tenths digit is 5 or higher.
↓   Round up.
35

71.
$$\begin{array}{r} {}^{1}\ {}^{1}\ \\ 6\ 8\ 1 \\ +\ 1\ 4\ 9 \\ \hline 8\ 3\ 0 \end{array}$$

73.
$$\begin{array}{r} {}^{1}\ {}^{16} \\ \cancel{2}\ \cancel{0}\ 7 \\ -\ \ \ 8\ 5 \\ \hline 1\ 8\ 2 \end{array}$$

75. 6.78346⟨123⟩ ←Drop all decimal places
↓   past the fifth place.
6.78346

77. 99.99999⟨9999⟩ ←Drop all decimal places
↓   past the fifth place.
99.99999

Exercise Set 4.3

1.
$$\begin{array}{r} {}^{1}\quad\quad \\ 3\ 1\ 6.2\ 5 \\ +\ \ \ 1\ 8.1\ 2 \\ \hline 3\ 3\ 4.3\ 7 \end{array}$$
Add hundredths.
Add tenths.
Write a decimal point in the answer.
Add ones.
Add tens.
Add hundreds.

3.
$$\begin{array}{r} {}^{1}\ {}^{1}\quad\quad \\ 6\ 5\ 9.4\ 0\ 3 \\ +\ 9\ 1\ 6.8\ 1\ 2 \\ \hline 1\ 5\ 7\ 6.2\ 1\ 5 \end{array}$$
Add thousandths.
Add hundredths.
Add tenths.
Write a decimal point in the answer.
Add ones.
Add tens.
Add hundreds.

5.
$$\begin{array}{r} {}^{1}\quad\ {}^{1}\quad \\ 9.1\ 0\ 4 \\ +\ 1\ 2\ 3.4\ 5\ 6 \\ \hline 1\ 3\ 2.5\ 6\ 0 \end{array}$$

7.
$$\begin{array}{r} {}^{1}\quad \\ 8\ 1.0\ 0\ 8 \\ +\ \ \ 3.4\ 0\ 9 \\ \hline 8\ 4.4\ 1\ 7 \end{array}$$

9.
$$\begin{array}{r} 2\ 0.0\ 1\ 2\ 4 \\ +\ 3\ 0.0\ 1\ 2\ 4 \\ \hline 5\ 0.0\ 2\ 4\ 8 \end{array}$$

11. Line up the decimal points.

$$\begin{array}{r} {}^{1}\quad\quad \\ 3\ 9.0\ 0\ 0 \\ +\ \ \ 1.0\ 0\ 7 \\ \hline 4\ 0.0\ 0\ 7 \end{array}$$
Writing 2 extra zeros

13. Line up the decimal points.

```
    1
    0.3 4 0    Writing an extra zero
    3.5 0 0    Writing 2 extra zeros
    0.1 2 7
+ 7 6 8.0 0 0  Writing in the decimal point
                and 3 extra zeros
  7 7 1.9 6 7  Adding
```

15.
```
  1   1 1
  1 7.0 0 0 0   Writing in the decimal point.
  3.2 4 0 0     You may find it helpful to
  0.2 5 6 0     write extra zeros.
+   0.3 6 8 9
  2 0.8 6 4 9
```

17.
```
  1 2 1     1
      2.7 0 3 0
    7 8.3 3 0 0
    2 8.0 0 0 9
+ 1 1 8.4 3 4 1
  2 2 7.4 6 8 0
```

19.
```
  1 2 1    1
       9 9.6 0 0 1
   7 2 8 5.1 8 0 0
   5 0 0 0.4 2 0
+   8 7 0 0.0 0 0 0
   8 7 5 4.8 2 2 1
```

21.
```
  4 12
  5.2    Borrow ones to subtract tenths.
- 3.9    Subtract tenths.
  1.3    Write a decimal point in the answer.
         Subtract ones.
```

23.
```
  4 11 2 11   Borrow tenths to subtract hundredths.
  5 1.3 1     Subtract hundredths.
-   2.2 9     Subtract tenths.
  4 9.0 2     Write a decimal point in the answer.
              Borrow tens to subtract ones.
              Subtract ones.
              Subtract tens.
```

25.
```
  4 8.7 6
-   3.1 5
  4 5.6 1
```

27.
```
      11
  8 7 13
  9 2.3 4 1
-   6.4 2
  8 5.9 2 1
```

29.
```
  4 9 9 10
  2.5 0 0 0   Writing 3 extra zeros
- 0.0 0 2 5
  2.4 9 7 5
```

31.
```
  3 9 10
  3.4 0 0   Writing 2 extra zeros
- 0.0 0 3
  3.3 9 7
```

33. Line up the decimal points. Write an extra zero if desired.
```
  17 11
  1 7 7 10
  2 8.2 0
-1 9.3 5
    8.8 5
```

35.
```
  3 10
  3 4.0 7
- 3 0.7
    3.3 7
```

37.
```
     4 10
  8.4 5 0
- 7.4 0 5
  1.0 4 5
```

39.
```
  5 10
  6.0 0 3
- 2.3
  3.7 0 3
```

41.
```
  9 9 9 10
  1 0 0 0 0   Writing in the decimal point
- 0.0 0 9 8   and 4 extra zeros
  0.9 9 0 2   Subtracting
```

43.
```
  9 9 9 10
  1 0 0.0 0
-     0.3 4
   9 9.6 6
```

45.
```
  6 14
  7.4 8
- 2.6
  4.8 8
```

47.
```
  2 9 9 10
  3.0 0 0
-2.0 0 6
  0.9 9 4
```

49.
```
  8 9 9 10
  1 9.0 0 0
-  1.1 9 8
  1 7.8 0 2
```

51.
```
  4 9 10
  6 5.0 0
-1 3.8 7
  5 1.1 3
```

53.
```
  8 10
  3.9 0 7
-1.4 1 6
  2.4 9 1
```

55.
```
      8 17
  3 2.7 9 7 8
-   0.0 5 9 2
  3 2.7 3 8 6
```

57.
```
  2 9 10 6 14
  3.0 0 7 4
-1.3 4 0 8
  1.6 6 6 6
```

59.
```
           18
       4 8 9 17
  2 3 4 5.9 0 7 8 6
-      0.9 9 9
  2 3 4 4.9 0 8 8 6
```

61.
$$x + 17.5 = 29.15$$
$$x + 17.5 - 17.5 = 29.15 - 17.5 \qquad \text{Subtracting 17.5}$$
$$\text{on both sides}$$
$$x = 11.65$$

$$\begin{array}{r} \overset{8}{}\overset{11}{} \\ 2\,\cancel{9}.\,\cancel{1}\,5 \\ -\ 1\,7.\,5 \\ \hline 1\,1.\,6\,5 \end{array}$$

63.
$$3.205 + m = 22.456$$
$$3.205 + m - 3.205 = 22.456 - 3.205 \qquad$$
$$\text{Subtracting 3.205}$$
$$\text{on both sides}$$
$$m = 19.251$$

$$\begin{array}{r} \overset{1}{}\overset{12}{} \\ \cancel{2}\,\cancel{2}.\,4\,5\,6 \\ -\ \ 3.\,2\,0\,5 \\ \hline 1\,9.\,2\,5\,1 \end{array}$$

65.
$$17.95 + p = 402.63$$
$$17.95 + p - 17.95 = 402.63 - 17.95$$
$$\text{Subtracting 17.95}$$
$$\text{on both sides}$$
$$p = 384.68$$

$$\begin{array}{r} \overset{11}{}\overset{15}{} \\ \overset{3}{}\,\overset{9}{}\,\cancel{1}\,\cancel{5}\,{}^{13} \\ \cancel{4}\,\cancel{0}\,\cancel{2}.\,\cancel{6}\,\cancel{3} \\ -\ \ \ 1\,7.\,9\,5 \\ \hline 3\,8\,4.\,6\,8 \end{array}$$

67.
$$13{,}083.3 = x + 12{,}500.33$$
$$13{,}083.3 - 12{,}500.33 = x + 12{,}500.33 - 12{,}500.33$$
$$\text{Subtracting 12,500.33}$$
$$\text{on both sides}$$
$$582.97 = x$$

$$\begin{array}{r} \overset{2}{}\,\overset{10}{}\ \ \overset{2}{}\,\overset{2}{}\,\overset{10}{} \\ 1\,\cancel{3},\,\cancel{0}\,8\,\cancel{3}.\,\cancel{3}\,\cancel{0} \\ -\ 1\,2,\,5\,0\,0.\,3\,3 \\ \hline 5\,8\,2.\,9\,7 \end{array}$$

69.
$$x + 2349 = 17{,}684.3$$
$$x + 2349 - 2349 = 17{,}684.3 - 2349$$
$$\text{Subtracting 2349}$$
$$\text{on both sides}$$
$$x = 15{,}335.3$$

$$\begin{array}{r} \overset{7}{}\,\overset{14}{} \\ 1\,7,\,6\,\cancel{8}\,\cancel{4}.\,3 \\ -\ \ \ 2\,3\,4\,9.\,0 \\ \hline 1\,5,\,3\,3\,5.\,3 \end{array}$$

71.

$$34,\boxed{5}\,67 \qquad \text{Hundreds digit is 5 or higher.}$$
$$\downarrow \qquad\qquad \text{Round up.}$$
$$35{,}000$$

73.
$$\frac{13}{24} - \frac{3}{8} = \frac{13}{24} - \frac{3}{8} \cdot \frac{3}{3}$$
$$= \frac{13}{24} - \frac{9}{24}$$
$$= \frac{13 - 9}{24} = \frac{4}{24}$$
$$= \frac{4 \cdot 1}{4 \cdot 6} = \frac{4}{4} \cdot \frac{1}{6}$$
$$= \frac{1}{6}$$

75.
$$\begin{array}{r} \overset{7}{}\,\overset{9}{}\,\overset{15}{} \\ 8\,\cancel{8}\,\cancel{0}\,\cancel{5} \\ -\ 2\,6\,3\,9 \\ \hline 6\,1\,6\,6 \end{array}$$

77. First, "undo" the incorrect addition by subtracting 235.7 from the incorrect answer:

$$\begin{array}{r} 8\,1\,7.\,2 \\ -\ 2\,3\,5.\,7 \\ \hline 5\,8\,1.\,5 \end{array}$$

The original minuend was 581.5. Now subtract 235.7 from this as the student originally intended:

$$\begin{array}{r} 5\,8\,1.\,5 \\ -\ 2\,3\,5.\,7 \\ \hline 3\,4\,5.\,8 \end{array}$$

The correct answer is 345.8.

Exercise Set 4.4

1. *Familiarize.* We visualize the situation. We let $c =$ the amount of change.

$20	
$16.99	c

Translate. This is a "take-away" situation.

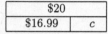

$$\begin{array}{ccccc} \text{Amount} & \text{minus} & \text{Amount of} & \text{is} & \text{Amount of} \\ \text{paid} & & \text{purchase} & & \text{change} \\ \downarrow & \downarrow & \downarrow & \downarrow & \downarrow \\ \$20 & - & \$16.99 & = & c \end{array}$$

Solve. To solve the equation we carry out the subtraction.

$$\begin{array}{r} \overset{1}{}\,\overset{9}{}\,\overset{9}{}\,\overset{10}{} \\ \cancel{2}\,\cancel{0}.\,\cancel{0}\,\cancel{0} \\ -\ 1\,6.\,9\,9 \\ \hline 3.\,0\,1 \end{array}$$

Thus, $c = \$3.01$.

Check. We check by adding 3.01 to 16.99 to get 20. This checks.

State. The change was $3.01.

3. *Familiarize.* We visualize the situation. We let $g =$ the number of gallons purchased.

23.6 gal	17.7 gal	20.8 gal	17.2 gal	25.4 gal	13.8 gal
g					

Translate. Amounts are being combined. We translate to an equation.

$$\begin{array}{ccccccccc} \text{First} & \text{plus} & \text{second} & \text{plus} & \text{third} & \text{plus} & \text{fourth} & \text{plus} \\ \downarrow & \downarrow & \downarrow & \downarrow & \downarrow & \downarrow & \downarrow & \downarrow \\ 23.6 & + & 17.7 & + & 20.8 & + & 17.2 & + \end{array}$$

$$\begin{array}{ccccc} \text{fifth} & \text{plus} & \text{sixth} & \text{is} & \text{total.} \\ \downarrow & \downarrow & \downarrow & \downarrow & \downarrow \\ 25.4 & + & 13.8 & = & g \end{array}$$

Solve. To solve the equation we carry out the addition.

$$
\begin{array}{r}
{\scriptstyle 2\;\;3} \\
2\,3.\,6 \\
1\,7.\,7 \\
2\,0.\,8 \\
1\,7.\,2 \\
2\,5.\,4 \\
+\;1\,3.\,8 \\
\hline
1\,1\,8.\,5
\end{array}
$$

Thus, $g = 118.5$.

Check. We can check by repeating the addition. We also note that the answer seems reasonable since it is larger than any of the numbers being added. We can also check by rounding:

$23.6 + 17.7 + 20.8 + 17.2 + 25.4 + 13.8 \approx 24 + 18 + 21 + 17 + 25 + 14 = 119 \approx 118.5$

State. 118.5 gallons of gasoline were purchased.

5. *Familiarize*. We visualize the situation. We let $n =$ the new temperature.

98.6°	4.2°
n	

Translate. We are combining amounts.

Normal body temperature	plus	Degrees temperature rises	is	New temperature
↓	↓	↓	↓	↓
98.6	+	4.2	=	n

Solve. To solve the equation we carry out the addition.

$$
\begin{array}{r}
{\scriptstyle 1} \\
9\,8.\,6 \\
+\;\;\;4.\,2 \\
\hline
1\,0\,2.\,8
\end{array}
$$

Thus, $n = 102.8$.

Check. We can check by repeating the addition. We can also check by rounding:

$98.6 + 4.2 \approx 99 + 4 = 103 \approx 102.8$

State. The new temperature was 102.8°F.

7. *Familiarize*. We visualize the situation. We let $m =$ the odometer reading at the end of the trip.

22,456.8 mi	234.7 mi
m	

Translate. We are combining amounts.

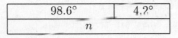

Reading before trip	plus	Miles driven	is	Reading at end of trip
↓	↓	↓	↓	↓
22,456.8	+	234.7	=	m

Solve. To solve the equation we carry out the addition.

$$
\begin{array}{r}
{\scriptstyle 1\;\;\;\;1} \\
2\,2{,}4\,5\,6.\,8 \\
+\;\;\;\;2\,3\,4.\,7 \\
\hline
2\,2{,}6\,9\,1.\,5
\end{array}
$$

Thus, $m = 22,691.5$.

Check. We can check by repeating the addition. We can also check by rounding:

$22{,}456.8 + 234.7 \approx 22{,}460 + 230 = 22{,}690 \approx 22{,}691.5$

State. The odometer reading at the end of the trip was 22,691.5.

9. *Familiarize*. We visualize the situation. We let $n =$ the number by which hamburgers exceed hot dogs, in billions.

24.8 billion	
15.9 billion	n

Translate. This is a "how-much-more" situation.

Number of hot dogs eaten	plus	Additional number of hamburgers	is	Number of hamburgers eaten
↓	↓	↓	↓	↓
15.9	+	n	=	24.8

Solve. We subtract 15.9 on both sides.

$$
n = 24.8 - 15.9
$$
$$
n = 8.9
$$

$$
\begin{array}{r}
{\scriptstyle 13} \\
{\scriptstyle 1\;\;\cancel{8}\;\;18} \\
\cancel{2}\,\cancel{4}.\,\cancel{8} \\
-\;1\,5.\,9 \\
\hline
8.\,9
\end{array}
$$

Check. We check by adding 8.9 to 15.9 to get 24.8. This checks.

State. Americans eat 8.9 billion more hamburgers than hot dogs.

11. *Familiarize*. We visualize the situation. We let $t =$ the total amount spent.

$176.20	$141.87	$38.82
t		

Translate. We are combining amounts.

Cost of greens fees	plus	Cost of cart rentals	plus	Cost of golf balls	is	Total amount spent
↓	↓	↓	↓	↓	↓	↓
$176.20	+	$141.87	+	$38.82	=	t

Solve. We carry out the addition.

$$
\begin{array}{r}
{\scriptstyle 1\;\;1\;\;1} \\
1\,7\,6.\,2\,0 \\
1\,4\,1.\,8\,7 \\
+\;\;\;3\,8.\,8\,2 \\
\hline
3\,5\,6.\,8\,9
\end{array}
$$

Thus, $t = \$356.89$.

Check. We can repeat the addition. We can also check by rounding:

$$176.20 + 141.87 + 38.82 \approx 180 + 140 + 40 = 360 \approx 356.89$$

State. The total amount spent is $356.89.

13. *Familiarize*. We visualize the situation. We let $s =$ the extra length of the Suez Canal.

175.5 km	
81.6 km	s

Translate. This is a "how-much-more" situation.

Length of Panama Canal	plus	Extra length of Suez Canal	is	Length of Suez Canal
↓	↓	↓	↓	↓
81.6	+	s	=	175.5

Solve. We subtract 81.6 on both sides.

$s = 175.5 - 81.6$
$s = 93.9$

$$\begin{array}{r} \overset{17\ \ 4\ 15}{1\ 7\ 5.5} \\ -\ \ \ 8\ 1.6 \\ \hline 9\ 3.9 \end{array}$$

Check. We add 93.9 to 81.6 to get 175.5. This checks.

State. The Suez Canal is 93.9 km longer than the Panama Canal.

15. *Familiarize*. We visualize the situation. We let $t =$ the combined time.

4.25 min	4.86 min	3.98 min	5.0 min
t			

Translate. We are combining amounts.

First plus second plus third plus

↓	↓	↓	↓	↓	↓
4.25	+	4.86	+	3.98	+

fourth is combined time.

↓	↓	↓
5.0	=	t

Solve. To solve the equation we carry out the addition.

$$\begin{array}{r} \overset{2\ 1}{}\ \ \\ 4.2\ 5 \\ 4.8\ 6 \\ 3.9\ 8 \\ +\ 5.0 \\ \hline 1\ 8.0\ 9 \end{array}$$

Thus, $t = 18.09$ min.

Check. We can check by repeating the addition. We can also check by rounding:

$$4.25 + 4.86 + 3.98 + 5.0 \approx 4 + 5 + 4 + 5 = 18 \approx 18.09$$

State. The combined time was 18.09 min.

17. *Familiarize*. We visualize the situation. We let $p =$ the amount of precipitation in inches and $c =$ the amount of precipitation in centimeters.

34.55 in., or 87.757 cm	2.3 in., or 5.842 cm
p in., or c cm	

Translate and Solve. We are combining amounts.

Amount of rain	plus	Amount of snow	is	Total precipitation

a) Taking the amounts in inches, we translate to an equation:

$$34.55 + 2.3 = p$$

To solve we carry out the addition.

$$\begin{array}{r} 3\ 4.5\ 5 \\ +\ \ \ 2.3 \\ \hline 3\ 6.8\ 5 \end{array}$$

Thus, $p = 36.85$ in.

b) Taking the amounts in centimeters, we translate to an equation:

$$87.757 + 5.842 = c$$

To solve we carry out the addition.

$$\begin{array}{r} \overset{1\ \ 1}{8\ 7.7\ 5\ 7} \\ +\ \ \ 5.8\ 4\ 2 \\ \hline 9\ 3.5\ 9\ 9 \end{array}$$

Thus, $c = 93.599$ cm.

Check. We can repeat the addition. We can also check by rounding:

$$34.55 + 2.3 \approx 35 + 2 = 37 \approx 36.85;$$

$$87.757 + 5.842 \approx 88 + 6 = 94 \approx 93.599.$$

State. a) The total average annual precipitation in Dallas, Texas, is 36.85 in.

b) The total average annual precipitation in Dallas, Texas, is 93.599 cm.

19. *Familiarize*. We visualize the situation. We let $m =$ the number of miles driven.

28,576.8	m
28,802.6	

Translate. This is a "how-much-more" situation.

First odometer reading	plus	Miles driven	is	Second odometer reading
↓	↓	↓	↓	↓
28,576.8	+	m	=	28,802.6

Solve. To solve the equation we subtract 28,576.8 on both sides.

$m = 28,802.6 - 28,576.8$
$m = 225.8$

$$\begin{array}{r} \overset{\ 7\ \ 9\ \ \overset{11}{1}\ 16}{2\ 8,\cancel{8}\ \cancel{0}\ \cancel{2}.\cancel{6}} \\ -\ 2\ 8,5\ 7\ 6.8 \\ \hline 2\ 2\ 5.8 \end{array}$$

Check. We check by adding 225.8 to 28,576.8 to get 28,802.6. This checks.

State. 225.8 miles were driven.

21. Familiarize. This is a multistep problem. We will first find the total amount of the checks. Then we will find how much is left in the account after the checks are written. Finally, we will use this amount and the amount of the deposit to find the balance in the account after all the changes. We will let c = the total amount of the checks.

Translate and Solve. We are combining amounts.

First check	plus	Second check	plus	Third check	is	Total amount of checks
↓	↓	↓	↓	↓	↓	↓
23.82	+	507.88	+	98.32	=	c

To solve the equation we carry out the addition.

```
  1 2 2 1
    2 3.8 2
    5 0 7.8 8
  +   9 8.3 2
    6 3 0.0 2
```

Thus, $c = 630.02$.

Now we let a = the amount in the account after the checks are written.

Original amount	less	Check amount	is	New amount
↓	↓	↓	↓	↓
1123.56	−	630.02	=	a

To solve the equation we carry out the subtraction.

```
        10
      ⁰ 12
    1 1 2 3.5 6
  −   6 3 0.0 2
      4 9 3.5 4
```

Thus, $a = 493.54$.

Finally, we let f = the amount in the account after the paycheck is deposited.

Amount after checks	plus	Amount of deposit	is	Final amount
↓	↓	↓	↓	↓
493.54	+	678.20	=	f

We carry out the addition.

```
  1 1
    4 9 3.5 4
  + 6 7 8.2 0
    1 1 7 1.7 4
```

Thus, $f = 1171.74$.

Check. We repeat the calculations.

State. There is $1171.74 in the account after the changes.

23. Familiarize. We visualize the situation. We let p = the number by which the O'Hare passengers exceed the San Francisco passengers in millions.

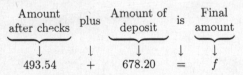

Translate. We have a "how much more" situation.

San Francisco passengers	plus	Additional O'Hare passengers	is	O'Hare passengers
↓	↓	↓	↓	↓
31.1	+	p	=	59.9

Solve. We subtract 31.1 on both sides of the equation:

$p = 59.9 - 31.1$

$p = 28.8$

```
    5 9.9
  − 3 1.1
    2 8.8
```

Check. We add 28.8 to 31.1 to get 59.9. This checks.

State. O'Hare handles 28.8 million more passengers per year than San Francisco.

25. Familiarize. We visualize the situation. We let t = the number of passengers handled by Dallas/Ft. Worth and San Francicsco together in millions.

48.5 million	31.1 million
t	

Translate. We are combining amounts.

Dallas/ Ft. Worth passengers	plus	San Francisco passengers	is	Number handled together
↓	↓	↓	↓	↓
48.5	+	31.1	=	t

Solve. We carry out the addition.

```
    4 8.5
  + 3 1.1
    7 9.6
```

Thus $t = 79.6$.

Check. We can repeat the addition. We can also check by rounding: $48.5 + 31.1 \approx 50 + 30 = 80 \approx 79.6$.

State. Dallas/Ft. Worth and San Francisco handle 79.6 million passengers per year together.

27. Familiarize. We visualize the situation. We let a = the total amount the student paid.

$39.95	$39.95	$4.80
a		

Translate. We are combining amounts.

First pair of slacks	plus	Second pair of slacks	plus	Tax	is	Total Cost
↓	↓	↓	↓	↓	↓	↓
$39.95	+	$39.95	+	$4.80	=	a

Solve. To solve we carry out the addition.

$$\begin{array}{r} {\scriptstyle 2\ 2\ 1} \\ 3\ 9.9\ 5 \\ 3\ 9.9\ 5 \\ +\ \ \ 4.8\ 0 \\ \hline 8\ 4.7\ 0 \end{array}$$

Thus $a = \$84.70$.

Check. To check we can repeat the addition. We can also check by rounding:

$$39.95 + 39.95 + 4.80 \approx 40 + 40 + 5 = 85 \approx 84.70$$

State. The student paid $84.70.

29. *Familiarize*. We let d = the distance around the figure.

Translate. We are combining lengths.

The sum of the lengths of the 5 sides	is	the distance around the figure.
↓	↓	↓
$8.9 + 23.8 + 4.7 + 22.1 + 18.6$	$=$	d

Solve. To solve we carry out the addition.

$$\begin{array}{r} {\scriptstyle 2\ \ 3} \\ 8.9 \\ 2\ 3.8 \\ 4.7 \\ 2\ 2.1 \\ +\ 1\ 8.6 \\ \hline 7\ 8.1 \end{array}$$

Thus, $d = 78.1$.

Check. To check we can repeat the addition. We can also check by rounding:

$$8.9 + 23.8 + 4.7 + 22.1 + 18.6 \approx 9 + 24 + 5 + 22 + 19 = 79 \approx 78.1$$

State. The distance around the figure is 78.1 cm.

31. *Familiarize*. This is a multistep problem. First we find the sum s of the two 0.8 cm segments. Then we use this length to find d.

Translate and Solve.

Length of one small segment	plus	Length of other small segment	is	Total length.
↓	↓	↓	↓	↓
0.8	$+$	0.8	$=$	s

To solve we carry out the addition.

$$\begin{array}{r} {\scriptstyle 1} \\ 0.\ 8 \\ +\ 0.\ 8 \\ \hline 1.\ 6 \end{array}$$

Thus, $s = 1.6$.

Now we find d.

Total length of smaller segments	plus	length of d	is	3.91 cm
↓	↓	↓	↓	↓
1.6	$+$	d	$=$	3.91

To solve we subtract 1.6 on both sides of the equation.

$$d = 3.91 - 1.6$$
$$d = 2.31$$

$$\begin{array}{r} 3.9\ 1 \\ -\ 1.6\ 0 \\ \hline 2.3\ 1 \end{array}$$

Check. We repeat the calculations.

State. The length d is 2.31 cm.

33. *Familiarize*. We visualize the situation. We let t = the number of degrees by which the temperature of the bath water exceeds normal body temperature.

Translate. We have a "how-much-more" situation.

Normal body temperature	plus	Additional degrees	is	Bath water temperature
↓	↓	↓	↓	↓
98.6	$+$	t	$=$	100

Solve. To solve we subtract 98.6 on both sides of the equation.

$$t = 100 - 98.6$$
$$t = 1.4$$

$$\begin{array}{r} {\scriptstyle 9\ \ 9\ 10} \\ \cancel{1\ 0\ 0}.\ \cancel{0} \\ -\ \ \ 9\ 8.\ 6 \\ \hline 1.\ 4 \end{array}$$

Check. To check we add 1.4 to 98.6 to get 100. This checks.

State. The temperature of the bath water is 1.4°F above normal body temperature.

35. *Familiarize*. This is a multistep problem. We will find the correct change and the amount of the actual change and compare them to determine if the change was correct. We let c = the correct change and a = the actual change.

Translate. We write two equations.

$20	less	Cost of book	is	Correct change
↓	↓	↓	↓	↓
20	$-$	10.75	$=$	c

Amount in $5 bills	plus	Amount in $1 bills	plus	Amount in dimes	plus
↓	↓	↓	↓	↓	↓
5	$+$	$(1 + 1 + 1)$	$+$	0.10	$+$

Amount in nickels	is	Actual change
↓	↓	↓
$(0.05 + 0.05)$	$=$	a

Solve. To solve the first equation we carry out the subtraction.

$$\begin{array}{r} {\scriptstyle 1\ \ 9\ \ 9\ 10} \\ \cancel{2\ 0.\ 0\ 0} \\ -\ 1\ 0.\ 7\ 5 \\ \hline 9.\ 2\ 5 \end{array}$$

Thus, $c = 9.25$.

To solve the second equation we carry out the addition.

$$5 + 1 + 1 + 1 + 0.10 + 0.05 + 0.05 = 8.20$$

Thus, $a = 8.20$.

The correct change is different from the actual change.

Check. We repeat the calculations.

State. The change was not correct.

37.
$$\begin{array}{r} \overset{1\ 1\ 1}{4\ 5\ 6\ 9} \\ +\ 1\ 7\ 6\ 6 \\ \hline 6\ 3\ 3\ 5 \end{array}$$

39.
$$\begin{aligned} \frac{5}{6} + \frac{7}{10} &= \frac{5}{6} \cdot \frac{5}{5} + \frac{7}{10} \cdot \frac{3}{3} \\ &= \frac{25}{30} + \frac{21}{30} \\ &= \frac{25 + 21}{30} = \frac{46}{30} \\ &= \frac{2 \cdot 23}{2 \cdot 15} = \frac{2}{2} \cdot \frac{23}{15} \\ &= \frac{23}{15} \end{aligned}$$

41.
$$\begin{array}{r} \overset{3\ 15}{\cancel{4}\ \cancel{5}\ 6\ 9} \\ -\ 1\ 7\ 6\ 6 \\ \hline 2\ 8\ 0\ 3 \end{array}$$

43.
$$\begin{aligned} \frac{5}{6} - \frac{7}{10} &= \frac{5}{6} \cdot \frac{5}{5} - \frac{7}{10} \cdot \frac{3}{3} \\ &= \frac{25}{30} - \frac{21}{30} \\ &= \frac{25 - 21}{30} = \frac{4}{30} \\ &= \frac{2 \cdot 2}{2 \cdot 15} = \frac{2}{2} \cdot \frac{2}{15} \\ &= \frac{2}{15} \end{aligned}$$

Chapter 5

Multiplication and Division: Decimal Notation

Exercise Set 5.1

1.
```
    8. 6    (1 decimal place)
  ×   7    (0 decimal places)
  6 0. 2    (1 decimal place)
```

3.
```
    0. 8 4    (2 decimal places)
  ×     8    (0 decimal places)
    6. 7 2    (2 decimal places)
```

5.
```
      6. 3    (1 decimal place)
  × 0. 0 4    (2 decimal places)
    0. 2 5 2  (3 decimal places)
```

7.
```
        8 7    (0 decimal places)
  × 0. 0 0 6   (3 decimal places)
    0. 5 2 2   (3 decimal places)
```

9. 10×23.76

```
        23.7.6
           └┘
```

1 zero   Move 1 place to the right.

$10 \times 23.76 = 237.6$

11. 1000×583.686852

```
        583.686.852
            └──┘
```

3 zeros   Move 3 places to the right.

$1000 \times 583.686852 = 583,686.852$

13. 7.8×100

```
        7.80.
          └┘
```

2 zeros   Move 2 places to the right.

$7.8 \times 100 = 780$

15. 0.1×89.23

```
        8.9.23
          └┘
```

1 decimal place   Move 1 place to the left.

$0.1 \times 89.23 = 8.923$

17. 0.001×97.68

```
        0.097.68
          └─┘
```

3 decimal places   Move 3 places to the left.

$0.001 \times 97.68 = 0.09768$

19. 78.2×0.01

```
        0.78.2
          └┘
```

2 decimal places   Move 2 places to the left.

$78.2 \times 0.01 = 0.782$

21.
```
      3 2. 6    (1 decimal place)
  ×     1 6    (0 decimal places)
      1 9 5 6
      3 2 6 0
      5 2 1. 6  (1 decimal place)
```

23.
```
      0. 9 8 4    (3 decimal places)
  ×       3. 3    (1 decimal place)
        2 9 5 2
      2 9 5 2 0
      3. 2 4 7 2  (4 decimal places)
```

25.
```
      3 7 4    (0 decimal places)
  ×     2. 4   (1 decimal place)
      1 4 9 6
      7 4 8 0
      8 9 7. 6  (1 decimal place)
```

27.
```
      7 4 9    (0 decimal places)
  ×   0. 4 3   (2 decimal places)
      2 2 4 7
      2 9 9 6 0
      3 2 2. 0 7  (2 decimal places)
```

29.
```
      0. 8 7    (2 decimal places)
  ×     6 4    (0 decimal places)
        3 4 8
      5 2 2 0
      5 5. 6 8  (2 decimal places)
```

31.
```
      4 6. 5 0    (2 decimal places)
  ×       7 5    (0 decimal places)
      2 3 2 5 0
      3 2 5 5 0 0
      3 4 8 7. 5 0  (2 decimal places)
```

Since the last decimal place is 0, we could also write this answer as 3487.5.

33.
```
        8 1. 7    (1 decimal place)
  ×   0. 6 1 2    (3 decimal places)
        1 6 3 4
        8 1 7 0
      4 9 0 2 0 0
      5 0. 0 0 0 4  (4 decimal places)
```

35.
```
        1 0. 1 0 5    (3 decimal places)
  ×   1 1. 3 2 4      (3 decimal places)
          4 0 4 2 0
        2 0 2 1 0 0
      3 0 3 1 5 0 0
    1 0 1 0 5 0 0 0
  1 0 1 0 5 0 0 0 0
  1 1 4. 4 2 9 0 2 0  (6 decimal places)
```

or 114.42902

37.   1 2. 3   (1 decimal place)
          × 1. 0 8   (2 decimal places)
          ─────────
              9 8 4
          1 2 3 0 0
          ─────────
          1 3. 2 8 4   (3 decimal places)

39.   3 2. 4   (1 decimal place)
          ×   2. 8   (1 decimal place)
          ─────────
            2 5 9 2
            6 4 8 0
          ─────────
            9 0. 7 2   (2 decimal places)

41.   0. 0 0 3 4 2   (5 decimal places)
          ×   0. 8 4   (2 decimal places)
          ───────────────
                1 3 6 8
              2 7 3 6 0
          ───────────────
          0. 0 0 2 8 7 2 8   (7 decimal places)

43.   0. 3 4 7   (3 decimal places)
          ×   2. 0 9   (2 decimal places)
          ─────────────
            3 1 2 3
          6 9 4 0 0
          ─────────────
          0. 7 2 5 2 3   (5 decimal places)

45.   3. 0 0 5   (3 decimal places)
          × 0. 6 2 3   (3 decimal places)
          ─────────────
            9 0 1 5
            6 0 1 0 0
          1 8 0 3 0 0 0
          ─────────────
          1. 8 7 2 1 1 5   (6 decimal places)

47. 1000 × 45.678   45.678.
                                 └──┘↑

3 zeros   Move 3 places to the right.
1000 × 45.678 = 45,678

49. Move 2 places to the right.

$28.88.¢

Change from $ sign in front to ¢ sign at end.
       $28.88 = 2888¢

51. Move 2 places to the right.

$0.66.¢

Change from $ sign in front to ¢ sign at end.
       $0.66 = 66¢

53. Move 2 places to the left.

$0.34.¢

Change from ¢ sign at end to $ sign in front.
       34¢ = $0.34

55. Move 2 places to the left.

$34.45.¢

Change from ¢ sign at end to $ sign in front.
       3345¢ = $34.45

57. $4.03 trillion = $4.03 × 1,$\underbrace{000,000,000,000}_{12 \text{ zeros}}$

$4.030000000000.
 └──────────────┘↑

Move 12 places to the right.
$4.03 trillion = $4,030,000,000,000

59. $4.7 million = $4.7 × 1,$\underbrace{000,000}_{6 \text{ zeros}}$

$4.700000.
 └──────┘↑

Move 6 places to the right.
$4.7 million = $4,700,000

61. $2\frac{1}{3} \cdot 4\frac{4}{5} = \frac{7}{3} \cdot \frac{24}{5} = \frac{7 \cdot 3 \cdot 8}{3 \cdot 5}$

$\qquad = \frac{3}{3} \cdot \frac{7 \cdot 8}{5} = \frac{56}{5}$

$\qquad = 11\frac{1}{5}$

63.
```
           3 4 2
     24 ) 8 2 0 8
          7 2 0 0
          ───────
          1 0 0 8
            9 6 0
          ───────
              4 8
              4 8
          ───────
               0
```
The answer is 342.

65.
```
            4 5 6 6
      7 ) 3 1, 9 6 2
          2 8 0 0 0
          ─────────
            3 9 6 2
            3 5 0 0
          ─────────
              4 6 2
              4 2 0
          ─────────
                4 2
                4 2
          ─────────
                 0
```
The answer is 4566.

67. (1 trillion) · (1 billion)

$= 1,\underbrace{000,000,000,000}_{12 \text{ zeros}} \times 1,\underbrace{000,000,000}_{9 \text{ zeros}}$

$= 1,\underbrace{000,000,000,000,000,000,000}_{21 \text{ zeros}}$

$= 10^{21}$

Exercise Set 5.2

```
      2.99
1.  2 ) 5.98
      4 00
      ─────
      1 98
      1 80
      ─────
        18
        18
        ───
         0
```

Divide as though dividing whole numbers. Place the decimal point directly above the decimal point in the dividend.

```
      23.78
3.  4 ) 95.12
      8 000
      ─────
      1 512
      1 200
      ─────
        3 12
        2 80
        ─────
          32
          32
          ───
           0
```

Divide as though dividing whole numbers. Place the decimal point directly above the decimal point in the dividend.

```
       7.48
5.  12 ) 89.76
       8 4 00
       ─────
         5 76
         4 80
         ─────
           96
           96
           ───
            0
```

```
       7.2
7.  33 ) 237.6
       2 31 0
       ─────
          66
          66
          ───
           0
```

```
      1.143
9.  8 ) 9.144
      8 000
      ─────
      1 144
        800
      ─────
        344
        320
        ─────
         24
         24
         ───
          0
```

```
       4.041
11. 3 ) 12.123
       12 000
       ──────
          123
          120
       ──────
            3
            3
            ─
            0
```

```
      0.07
13. 5 ) 0.35
        35
        ──
         0
```

```
        70.
15. 0.12∧) 8.40∧
        8 40
        ────
           0
```

Multiply the divisor by 100 (move the decimal point 2 places). Multiply the same way in the dividend (move 2 places). Then divide.

```
        20.
17. 3.4∧) 68.0∧
        6 80
        ────
           0
```

Put a decimal point at the end of the whole number. Multiply the divisor by 10 (move the decimal point 1 place). Multiply the same way in the dividend (move 1 place), adding an extra 0. Then divide.

```
       0.4
19. 15 ) 6.0
        60
        ──
         0
```

Put a decimal point at the end of the whole number. Write an extra 0 to the right of the decimal point. Then divide.

```
       0.41
21. 36 ) 14.76
       1 440
       ─────
          36
          36
          ──
           0
```

```
        8.5
23. 3.2∧) 27.2∧0
        2 56
        ─────
          1 60      Write an extra 0.
          1 60
          ─────
             0
```

```
        9.3
25. 4.2∧) 39.0∧6
        37 8 0
        ──────
           1 26
           1 26
           ─────
              0
```

```
      0.625
27. 8 ) 5.000
      4 8
      ───
        20       Write an extra 0.
        16
        ──
         40       Write an extra 0.
         40
         ──
          0
```

```
        0.26
29. 0.47∧) 0.12∧22
         9 40
         ─────
          2 82
          2 82
          ─────
             0
```

```
        15.625
31. 4.8∧) 75.0∧000
         48 0
         ─────
         27 0
         24 0
         ─────
          3 00
          2 88
          ─────
          1 20
            96
          ─────
          2 40
          2 40
          ─────
             0
```

33.

$$0.032_\wedge \overline{)0.074_\wedge 88}$$

$$\begin{array}{r} 2.34 \\ \hline 6400 \\ \hline 1088 \\ 960 \\ \hline 128 \\ 128 \\ \hline 0 \end{array}$$

35.

$$82 \overline{)38.54}$$

$$\begin{array}{r} 0.47 \\ \hline 3280 \\ \hline 574 \\ 574 \\ \hline 0 \end{array}$$

37. $\dfrac{213.4567}{1000}$

0.213.4567 ↰

3 zeros   Move 3 places to the left.

$\dfrac{213.4567}{1000} = 0.2134567$

39. $\dfrac{213.4567}{10}$

21.3.4567 ↰

1 zero   Move 1 place to the left.

$\dfrac{213.4567}{10} = 21.34567$

41. $\dfrac{1.0237}{0.001}$

1.023.7 ↱

3 decimal places   Move 3 places to the right.

$\dfrac{1.0237}{0.001} = 1023.7$

43. $4.2 \cdot x = 39.06$

$\dfrac{4.2 \cdot x}{4.2} = \dfrac{39.06}{4.2}$ Dividing on both sides by 4.2

$x = 9.3$

$$4.2_\wedge \overline{)39.0_\wedge 6}$$

$$\begin{array}{r} 09.3 \\ \hline 3780 \\ \hline 126 \\ 126 \\ \hline 0 \end{array}$$

The solution is 9.3.

45. $1000 \cdot y = 9.0678$

$\dfrac{1000 \cdot y}{1000} = \dfrac{9.0678}{1000}$ Dividing on both sides by 1000

$y = 0.0090678$ Moving the decimal point 3 places to the left

The solution is 0.0090678.

47. $1048.8 = 23 \cdot t$

$\dfrac{1048.8}{23} = \dfrac{23 \cdot t}{23}$ Dividing on both sides by 23

$45.6 = t$

$$23 \overline{)1048.8}$$

$$\begin{array}{r} 45.6 \\ \hline 9200 \\ \hline 1288 \\ 1150 \\ \hline 138 \\ 138 \\ \hline 0 \end{array}$$

The solution is 45.6.

49. $14 \times (82.6 + 67.9) = 14 \times (150.5)$ Doing the calculation inside the parentheses

$\qquad = 2107$ Multiplying

51. $0.003 + 3.03 \div 0.01 = 0.003 + 303$ Dividing first

$\qquad = 303.003$ Adding

53. $42 \times (10.6 + 0.024)$

$= 42 \times 10.624$ Doing the calculation inside the parentheses

$= 446.208$ Multiplying

55. $4.2 \times 5.7 + 0.7 \div 3.5$

$= 23.94 + 0.2$ Doing the multiplications and divisions in order from left to right

$= 24.14$ Adding

57. $9.0072 + 0.04 \div 0.1^2$

$= 9.0072 + 0.04 \div 0.01$ Evaluating the exponential expression

$= 9.0072 + 4$ Dividing

$= 13.0072$ Adding

59. $(8 - 0.04)^2 \div 4 + 8.7 \times 0.4$

$= (7.96)^2 \div 4 + 8.7 \times 0.4$ Doing the calculation inside the parentheses

$= 63.3616 \div 4 + 8.7 \times 0.4$ Evaluating the exponential expression

$= 15.8404 + 3.48$ Doing the multiplications and divisions in order from left to right

$= 19.3204$ Adding

61. $86.13 + 95.7 \div (9.6 - 0.03)$

$= 86.13 + 95.7 \div 9.57$ Doing the calculation inside the parentheses

$= 86.13 + 10$ Dividing

$= 96.13$ Adding

63. $4 \div 0.4 + 0.1 \times 5 - 0.1^2$

$= 4 \div 0.4 + 0.1 \times 5 - 0.01$ Evaluating the exponential expression

$= 10 + 0.5 - 0.01$ Doing the multiplications and divisions in order from left to right

$= 10.49$ Adding and subtracting in order from left to right

65. $5.5^2 \times [(6 - 4.2) \div 0.06 + 0.12]$

$= 5.5^2 \times [1.8 \div 0.06 + 0.12]$ Doing the calculation in the innermost parentheses first

$= 5.5^2 \times [30 + 0.12]$ Doing the calculation inside the parentheses

$= 5.5^2 \times 30.12$

$= 30.25 \times 30.12$ Evaluating the exponential expression

$= 911.13$ Multiplying

67. $200 \times \{[(4 - 0.25) \div 2.5] - (4.5 - 4.025)\}$

$= 200 \times \{[3.75 \div 2.5] - 0.475\}$ Doing the calculations in the innermost parentheses first

$= 200 \times \{1.5 - 0.475\}$ Again, doing the calculations in the innermost parentheses

$= 200 \times 1.025$ Subtracting inside the parentheses

$= 205$ Multiplying

69. We add the numbers and then divide by the number of addends.

$(\$1276.59 + \$1350.49 + \$1123.78 + \$1402.56) \div 4$

$= \$5153.42 \div 4$

$= \$1288.355$

71. $10\frac{1}{2} + 4\frac{5}{8} = 10\frac{4}{8} + 4\frac{5}{8}$

$\qquad = 14\frac{9}{8} = 15\frac{1}{8}$

73. $\dfrac{36}{42} = \dfrac{6 \cdot 6}{6 \cdot 7} = \dfrac{6}{6} \cdot \dfrac{6}{7} = \dfrac{6}{7}$

75.

$$
\begin{array}{r}
1\,9 \\
3\,\overline{\smash{)}\,5\,7} \\
3\,\overline{\smash{)}\,1\,7\,1} \\
2\,\overline{\smash{)}\,3\,4\,2} \\
2\,\overline{\smash{)}\,6\,8\,4}
\end{array}
$$

$684 = 2 \cdot 2 \cdot 3 \cdot 3 \cdot 19$

Exercise Set 5.3

1. $\dfrac{3}{5} = \dfrac{3}{5} \cdot \dfrac{2}{2}$ We use $\dfrac{2}{2}$ for 1 to get a denominator of 10.

$\quad = \dfrac{6}{10} = 0.6$

3. $\dfrac{13}{40} = \dfrac{13}{40} \cdot \dfrac{25}{25}$ We use $\dfrac{25}{25}$ for 1 to get a denominator of 1000.

$\quad = \dfrac{325}{1000} = 0.325$

5. $\dfrac{1}{5} = \dfrac{1}{5} \cdot \dfrac{2}{2} = \dfrac{2}{10} = 0.2$

7. $\dfrac{17}{20} = \dfrac{17}{20} \cdot \dfrac{5}{5} = \dfrac{85}{100} = 0.85$

9. $\dfrac{19}{40} = \dfrac{19}{40} \cdot \dfrac{25}{25} = \dfrac{475}{1000} = 0.475$

11. $\dfrac{39}{40} = \dfrac{39}{40} \cdot \dfrac{25}{25} = \dfrac{975}{1000} = 0.975$

13. $\dfrac{13}{25} = \dfrac{13}{25} \cdot \dfrac{4}{4} = \dfrac{52}{100} = 0.52$

15. $\dfrac{2502}{125} = \dfrac{2502}{125} \cdot \dfrac{8}{8} = \dfrac{20,016}{1000} = 20.016$

17. $\dfrac{1}{4} = \dfrac{1}{4} \cdot \dfrac{25}{25} = \dfrac{25}{100} = 0.25$

19. $\dfrac{23}{40} = \dfrac{23}{40} \cdot \dfrac{25}{25} = \dfrac{575}{1000} = 0.575$

21. $\dfrac{18}{25} = \dfrac{18}{25} \cdot \dfrac{4}{4} = \dfrac{72}{100} = 0.72$

23. $\dfrac{19}{16} = \dfrac{19}{16} \cdot \dfrac{625}{625} = \dfrac{11,875}{10,000} = 1.1875$

25. $\dfrac{4}{15} = 4 \div 15$

$$
\begin{array}{r}
0.\,2\,6\,6 \\
15\,\overline{\smash{)}\,4.\,0\,0\,0} \\
3\,0 \\
\hline
1\,0\,0 \\
9\,0 \\
\hline
1\,0\,0 \\
9\,0 \\
\hline
1\,0
\end{array}
$$

Since 10 keeps reappearing as a remainder, the digits repeat and

$\dfrac{4}{15} = 0.2666\ldots$ or $0.2\overline{6}$.

27. $\dfrac{1}{3} = 1 \div 3$

$$
\begin{array}{r}
0.\,3\,3\,3 \\
3\,\overline{\smash{)}\,1.\,0\,0\,0} \\
9 \\
\hline
1\,0 \\
9 \\
\hline
1\,0 \\
9 \\
\hline
1
\end{array}
$$

Since 1 keeps reappearing as a remainder, the digits repeat and

$\dfrac{1}{3} = 0.333\ldots$ or $0.\overline{3}$.

29. $\dfrac{4}{3} = 4 \div 3$

```
        1. 3 3
    3 | 4. 0 0
        3
        ――
        1 0
          9
        ――
          1 0
            9
        ――
            1
```

Since 1 keeps reappearing as a remainder, the digits repeat and

$\dfrac{4}{3} = 1.333\ldots$ or $1.\overline{3}$.

31. $\dfrac{7}{6} = 7 \div 6$

```
        1. 1 6 6
    6 | 7. 0 0 0
        6
        ――
        1 0
          6
        ――
          4 0
          3 6
        ――
            4 0
            3 6
        ――
              4
```

Since 4 keeps reappearing as a remainder, the digits repeat and

$\dfrac{7}{6} = 1.166\ldots$ or $1.1\overline{6}$.

33. $\dfrac{4}{7} = 4 \div 7$

```
          0. 5 7 1 4 2 8 5
    7 | 4. 0 0 0 0 0 0 0
        3 5
        ――
          5 0
          4 9
        ――
            1 0
              7
        ――
              3 0
              2 8
        ――
                2 0
                1 4
        ――
                  6 0
                  5 6
        ――
                    4 0
                    3 5
        ――
                      5
```

Since 5 reappears as a remainder, the sequence repeats and

$\dfrac{4}{7} = 0.571428571428\ldots$ or $0.\overline{571428}$.

35. $\dfrac{11}{12} = 11 \div 12$

```
          0. 9 1 6 6
    1 2 | 1 1. 0 0 0 0
          1 0 8
          ――――
              2 0
              1 2
          ――――
                8 0
                7 2
          ――――
                  8 0
                  7 2
          ――――
                    8
```

Since 8 keeps reappearing as a remainder, the digits repeat and $\dfrac{11}{12} = 0.91666\ldots$ or $0.91\overline{6}$.

37. Round 0. 2 [6] 6 6 ... to the nearest tenth.

⬇ ↑＿＿＿ Hundredths digit is 5 or higher.

0. 3   Round up.

Round 0. 2 6 [6] 6 ... to the nearest hundredth.

⬇ ↑＿＿＿ Thousandths digit is 5 or higher.

0. 2 7   Round up.

Round 0. 2 6 6 [6] ... to the nearest thousandth.

⬇ ↑＿＿＿ Ten-thousandths digit is 5 or higher.

0. 2 6 7   Round up.

39. Round 0. 3 [3] 3 3 ... to the nearest tenth.

⬇ ↑＿＿＿ Hundredths digit is 4 or lower.

0. 3   Round down.

Round 0. 3 3 [3] 3 ... to the nearest hundredth.

⬇ ↑＿＿＿ Thousandths digit is 4 or lower.

0. 3 3   Round down.

Round 0. 3 3 3 [3] ... to the nearest thousandth.

⬇ ↑＿＿＿ Ten-thousandths digit is 4 or lower.

0. 3 3 3   Round down.

41. Round 1. 3 [3] 3 3 ... to the nearest tenth.

⬇ ↑＿＿＿ Hundredths digit is 4 or lower.

1. 3   Round down.

Round 1. 3 3 [3] 3 ... to the nearest hundredth.

⬇ ↑＿＿＿ Thousandths digit is 4 or lower.

1. 3 3   Round down.

Round 1. 3 3 3 [3] ... to the nearest thousandth.

⬇ ↑＿＿＿ Ten-thousandths digit is 4 or lower.

1. 3 3 3   Round down.

43. Round 1. 1 [6] 6 6 ... to the nearest tenth.

⬇ ↑＿＿＿ Hundredths digit is 5 or higher.

1. 2   Round up.

Round 1. 1 $\underline{6}$ $\boxed{6}$ 6 ... to the nearest hundredth.
        ↓ ↑   Thousandths digit is 5 or higher.

1. 1 7   Round up.

Round 1. 1 6 6 $\boxed{6}$ ... to the nearest thousandth.
          ↑   Ten-thousandths digit is 5 or higher.

1. 1 6 7   Round up.

45. $0.\overline{571428}$

Round to the nearest tenth.

0.5 $\boxed{7}$ 1428571428...
    ↓ ↑   Hundredths digit is 5 or higher.

0.6   Round up.

Round to the nearest hundredth.

0.5$\underline{7}$ $\boxed{1}$ 428571428...
     ↓ ↑   Thousandths digit is 4 or lower.

0.57   Round down.

Round to the nearest thousandth.

0.571 $\boxed{4}$ 28571428...
       ↑   Ten-thousandths digit is 4 or lower.

0.571   Round down.

47. Round 0. $\underline{9}$ $\boxed{1}$ 6 6 ... to the nearest tenth.
        ↓ ↑   Hundredths digit is 4 or lower.

0. 9   Round down.

Round 0. 9 $\underline{1}$ $\boxed{6}$ 6 ... to the nearest hundredth.
         ↓ ↑   Thousandths digit is 5 or higher.

0. 9 2   Round up.

Round 0. 9 1 $\underline{6}$ $\boxed{6}$ ... to the nearest thousandth.
          ↑   Ten-thousandths digit is 5 or higher.

0. 9 1 7   Round up

49. Round 0. $\underline{1}$ $\boxed{8}$ 1 8 ... to the nearest tenth.
        ↓ ↑   Hundredths digit is 5 or higher.

0. 2   Round up.

Round 0. 1 $\underline{8}$ $\boxed{1}$ 8 ... to the nearest hundredth.
         ↓ ↑   Thousandths digit is 4 or lower.

0. 1 8   Round down.

Round 0. 1 8 $\underline{1}$ $\boxed{8}$ ... to the nearest thousandth.
          ↑   Ten-thousandths digit is 5 or higher.

0. 1 8 2   Round up.

51. Round 0. 2 $\underline{7}$ $\boxed{7}$ 7 ... to the nearest tenth.
        ↓ ↑   Hundredths digit is 5 or higher.

0. 3   Round up.

Round 0. 2 $\underline{7}$ $\boxed{7}$ 7 ... to the nearest hundredth.
        ↓ ↑   Thousandths digit is 5 or higher.

0. 2 8   Round up.

Round 0. 2 7 $\underline{7}$ $\boxed{7}$ ... to the nearest thousandth.
          ↑   Ten-thousandths digit is 5 or higher.

0. 2 7 8   Round up.

53. We will use the first method discussed in the text.

$$\frac{7}{8} \times 12.64 = \frac{7}{8} \times \frac{1264}{100} = \frac{7 \cdot 1264}{8 \cdot 100}$$

$$= \frac{7 \cdot 2 \cdot 2 \cdot 2 \cdot 2 \cdot 79}{2 \cdot 2 \cdot 2 \cdot 2 \cdot 2 \cdot 5 \cdot 5}$$

$$= \frac{2 \cdot 2 \cdot 2 \cdot 2}{2 \cdot 2 \cdot 2 \cdot 2} \cdot \frac{7 \cdot 79}{2 \cdot 5 \cdot 5}$$

$$= 1 \cdot \frac{7 \cdot 79}{2 \cdot 5 \cdot 5}$$

$$= \frac{7 \cdot 79}{2 \cdot 5 \cdot 5} = \frac{553}{50}, \text{ or } 11.06$$

55. $2\frac{3}{4} + 5.65 = 2.75 + 5.65$ Writing $2\frac{3}{4}$ using decimal notation

           $= 8.4$ Adding

57. We will use the second method discussed in the text.

$$\frac{47}{9} \times 79.95 = 5.\overline{2} \times 79.95$$

$$\approx 5.222 \times 79.95 = 417.4989$$

Note that this answer is not as accurate as those found using either of the other methods, due to rounding.

59. $\frac{1}{2} - 0.5 = 0.5 - 0.5$ Writing $\frac{1}{2}$ using decimal notation

        $= 0$

61. $4.875 - 2\frac{1}{16} = 4.875 - 2.0625$ Writing $2\frac{1}{16}$ using decimal notation

              $= 2.8125$

63. We will use the third method discussed in the text.

$$\frac{5}{6} \times 0.0765 + \frac{5}{4} \times 0.1124 = \frac{5}{6} \times \frac{0.0765}{1} + \frac{5}{4} \times \frac{0.1124}{1}$$

$$= \frac{5 \times 0.0765}{6 \times 1} + \frac{5 \times 0.1124}{4 \times 1}$$

$$= \frac{0.3825}{6} + \frac{0.562}{4}$$

$$= 0.06375 + 0.1405$$

$$= 0.20425$$

65. We use the rules for order of operations, doing the multiplication first and then the division. Then we add.

$$\frac{4}{5} \times 384.8 + 24.8 \div \frac{8}{3} = 307.84 + 24.8 \cdot \frac{3}{8}$$

$$= 307.84 + 9.3$$

$$= 317.14$$

67. We do the multiplications in order from left to right. Then we subtract.

$$\frac{7}{8} \times 0.86 - 0.76 \times \frac{3}{4} = 0.7525 - 0.76 \times \frac{3}{4}$$
$$= 0.7525 - 0.57$$
$$= 0.1825$$

69. $3.375 \times 5\frac{1}{3} = 3.375 \times \frac{16}{3}$ Writing $5\frac{1}{3}$ using fractional notation

$$= 18 \qquad \text{Multiplying}$$

71. $6.84 \div 2\frac{1}{2} = 6.84 \div 2.5$ Writing $2\frac{1}{2}$ using decimal notation

$$= 2.736 \qquad \text{Dividing}$$

73. $9 \cdot 2\frac{1}{3} = \frac{9}{1} \cdot \frac{7}{3} = \frac{9 \cdot 7}{1 \cdot 3} = \frac{3 \cdot 3 \cdot 7}{1 \cdot 3} = \frac{3}{3} \cdot \frac{3 \cdot 7}{1} = 21$

75.
$$\begin{array}{r} 20 = 19\frac{5}{5} \\ -16\frac{3}{5} = -16\frac{3}{5} \\ \hline 3\frac{2}{5} \end{array}$$

77. a) Find the prime factorization of each number.

$$25 = 5 \cdot 5, \quad 65 = 5 \cdot 13$$

b) Create a product by writing factors, using each the greatest number of times it occurs in any one factorization. The LCM is $5 \cdot 5 \cdot 13$, or 325.

Exercise Set 5.4

1. We are estimating the sum

$$\$109.95 + \$249.95.$$

We round both numbers to the nearest ten. The estimate is

$$\$110 + \$250 = \$360.$$

Answer (d) is correct.

3. We are estimating the difference

$$\$299 - \$249.95.$$

We round both numbers to the nearest ten. The estimate is

$$\$300 - \$250 = \$50.$$

Answer (c) is correct.

5. We are estimating the product

$$9 \times \$299.$$

We round $299 to the nearest ten. The estimate is

$$9 \times \$300 = \$2700.$$

Answer (a) is correct.

7. We are estimating the quotient

$$\$1700 \div \$299.$$

Rounding $299, we get $300. Since $1700 is close to $1800, which is a multiple of $300, we estimate

$$\$1800 \div \$300,$$

so the answer is about 6.

Answer (c) is correct.

9. This is about $0.0 + 1.3 + 0.3$, so the answer is about 1.6.

11. This is about $6 + 0 + 0$, so the answer is about 6.

13. This is about $52 + 1 + 7$, so the answer is about 60.

15. This is about $2.7 - 0.4$, so the answer is about 2.3.

17. This is about $200 - 20$, so the answer is about 180.

19. This is about 50×8, rounding 49 to the nearest ten and 7.89 to the nearest one, so the answer is about 400. Answer (a) is correct.

21. This is about 100×0.08, rounding 98.4 to the nearest ten and 0.083 to the nearest hundredth, so the answer is about 8. Answer (c) is correct.

23. This is about $4 \div 4$, so the answer is about 1. Answer (b) is correct.

25. This is about $75 \div 25$, so the answer is about 3. Answer (b) is correct.

27.
$$\begin{array}{r} 3 \\ 3\overline{\smash{\big)}\,9} \\ 3\overline{\smash{\big)}\,27} \\ 2\overline{\smash{\big)}\,54} \\ 2\overline{\smash{\big)}\,108} \end{array}$$
$$108 = 2 \cdot 2 \cdot 3 \cdot 3 \cdot 3$$

29.

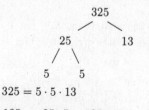

$$325 = 5 \cdot 5 \cdot 13$$

31. $\frac{125}{400} = \frac{25 \cdot 5}{25 \cdot 16} = \frac{25}{25} \cdot \frac{5}{16} = \frac{5}{16}$

33. $\frac{72}{81} = \frac{9 \cdot 8}{9 \cdot 9} = \frac{9}{9} \cdot \frac{8}{9} = \frac{8}{9}$

35. We round each factor to the nearest ten. The estimate is $180 \times 60 = 10,800$. The estimate is close to the result given, so the decimal point was placed correctly.

Exercise Set 5.5

1. *Familiarize.* Repeated addition fits this situation. We let C = the cost of 8 pairs of socks.

$$\underbrace{\boxed{\$4.95} + \boxed{\$4.95} + \cdots + \boxed{\$4.95}}_{\text{8 addends}}$$

Translate.

$$\underbrace{\text{Price per pair}}_{4.95} \quad \underset{\times}{\text{times}} \quad \underbrace{\text{Number of pairs}}_{8} \quad \underset{=}{\text{is}} \quad \underbrace{\text{Total cost}}_{C}$$

Solve. We carry out the multiplication.
$$\begin{array}{r} 4.\,9\,5 \\ \times \quad 8 \\ \hline 3\,9.\,6\,0 \end{array}$$

Thus, $C = 39.60$.

Check. We obtain a partial check by rounding and estimating:
$$4.95 \times 8 \approx 5 \times 8 = 40 \approx 39.00.$$

State. Eight pairs of socks cost $39.60.

3. *Familiarize.* Repeated addition fits this situation. We let c = the cost of 17.7 gal of gasoline.

Translate.

$$\underbrace{\text{Cost per gallon}}_{1.099} \quad \underset{\cdot}{\text{times}} \quad \underbrace{\text{Number of gallons}}_{17.7} \quad \underset{=}{\text{is}} \quad \underbrace{\text{Total cost}}_{c}$$

Solve. We carry out the multiplication.
$$\begin{array}{r} 1.\,0\,9\,9 \\ \times \quad 1\,7.\,7 \\ \hline 7\,6\,9\,3 \\ 7\,6\,9\,3\,0 \\ 1\,0\,9\,9\,0\,0 \\ \hline 1\,9.\,4\,5\,2\,3 \end{array}$$

Thus, $c = 19.4523$.

Check. We obtain a partial check by rounding and estimating:
$$1.099 \times 17.7 \approx 1 \times 18 = 18 \approx 19.4523.$$

State. We round $19.4523 to the nearest cent and find that the cost of the gasoline is $19.45.

5. *Familiarize.* Think of this as a rectangular array with 250 objects (miles) arranged in 4 rows. We want to find how many objects (miles) are in each row. We let m represent this number.

Translate. We think (Total miles) ÷ (Number of hours) = (Number of miles in 1 hour).

Solve. To solve the equation we carry out the division.

$$\begin{array}{r} 6\,2.5 \\ 4\,\overline{)2\,5\,0.0} \\ 2\,4\,0 \\ \hline 1\,0 \\ 8 \\ \hline 2\,0 \\ 2\,0 \\ \hline 0 \end{array}$$

Thus, $m = 62.5$.

Check. We obtain a partial check by rounding and estimating:
$$250 \div 4 \approx 300 \div 5 = 60 \approx 62.5.$$

State. The car went 62.5 mi in 1 hour.

7. *Familiarize.* Think of this as a rectangular array containing $47.60 arranged in 8 rows. We want to find how many dollars are in each row. We let s represent this number.

Translate. We think (Total cost) ÷ (Number of students) = (Each student's share).

Solve. To solve the equation we carry out the division.

$$\begin{array}{r} 5.9\,5 \\ 8\,\overline{)4\,7.6\,0} \\ 4\,0\,0\,0 \\ \hline 7\,6\,0 \\ 7\,2\,0 \\ \hline 4\,0 \\ 4\,0 \\ \hline 0 \end{array}$$

Thus, $s = 5.95$.

Check. We obtain a partial check by rounding and estimating:
$$47.60 \div 8 \approx 50 \div 10 = 5 \approx 5.95$$

State. Each person's share is $5.95.

9. *Familiarize.* We draw a picture, letting A = the area.

Translate. We use the formula $A = l \cdot w$.
$$A = 800.4 \times 312.6$$

Solve. We carry out the multiplication.
$$\begin{array}{r} 3\,1\,2.6 \\ \times\,8\,0\,0.4 \\ \hline 1\,2\,5\,0\,4 \\ 2\,5\,0\,0\,8\,0\,0\,0 \\ \hline 2\,5\,0,\,2\,0\,5.\,0\,4 \end{array}$$

Thus, $A = 250,205.04$.

Check. We obtain a partial check by rounding and estimating:
$$800.4 \times 312.6 \approx 800 \times 300 = 240,000 \approx 250,205.04$$

State. The area is 250,205.04 sq ft.

11. Familiarize. Repeated addition fits this situation. We let S = the income from the sale of 100,000 lottery tickets.

Translate.

Price per ticket	times	Number of tickets	is	Total income
↓	↓	↓	↓	↓
1.50	×	100,000	=	S

Solve. We carry out the multiplication. Since 100,000 has 5 zeros, we move the decimal point in 1.50 five places to the right.

1.50000.

$1.50 \times 100,000 = 150,000$

Check. We repeat the calculation. The answer checks.

State. The income is $150,000.

13. Familiarize. Repeated addition fits this situation. We let t = the total savings in 1 year.

Translate.

Number of weeks	times	Savings per week	is	Total savings
↓	↓	↓	↓	↓
52	·	6.72	=	t

Solve. To solve the equation we carry out the multiplication.

```
      6. 7 2
  ×      5 2
   1 3 4 4
 3 3 6 0 0
 3 4 9. 4 4
```

Check. We can obtain a partial check by rounding and estimating:

$$52 \cdot 6.72 \approx 50 \cdot 7 = 350 \approx 349.44.$$

State. The family would save $349.44 in 1 year.

15. Familiarize. Think of a rectangular array containing $11,178.72 arranged in 24 rows. We want to find how many dollars are in each row. We let p represent this number.

Translate. We think (Amount of loan) ÷ (Number of payments) = (Amount of each payment).

Solve. To solve the equation we carry out the division.

```
              4 6 5. 7 8
    2 4 | 1 1, 1 7 8. 7 2
          9 6 0 0 0 0
          1 5 7 8 7 2
          1 4 4 0 0 0
            1 3 8 7 2
            1 2 0 0 0
              1 8 7 2
              1 6 8 0
                1 9 2
                1 9 2
                    0
```

Thus $p = 465.78$.

Check. We obtain a partial check by rounding and estimating:

$$11,178.72 \div 24 \approx 10,000 \div 20 = 500 \approx 465.78.$$

State. Each payment is $465.78.

17. Familiarize. Repeated addition fits this situation. We let d = the distance the plane travels.

Translate.

Number of km flown in 1 hr	times	Number of hours flown	is	Total distance
↓	↓	↓	↓	↓
147.9	×	6	=	d

Solve. To solve the equation we carry out the multiplication.

```
    1 4 7. 9
  ×       6
    8 8 7. 4
```

Thus $d = 887.4$.

Check. We obtain a partial check by rounding and estimating:

$$147.9 \times 6 \approx 150 \times 6 = 900 \approx 887.4.$$

State. The plane travels 887.4 km.

19. Familiarize. This is a two-step problem. First we find the mileage cost of driving 120 miles at 27¢ per mile. We let c represent this number.

Translate and Solve.

Cost per mile	times	Number of miles	is	Mileage cost
↓	↓	↓	↓	↓
0.27	·	120	=	c

To solve the equation we carry out the multiplication.

```
    1 2 0
  × 0. 2 7
    8 4 0
  2 4 0 0
  3 2. 4 0
```

Thus $c = 32.40$.

Next we add the rental cost for one day to the mileage cost to find the total cost. We let y represent this number.

Cost for one day	plus	Mileage cost	is	Total cost for one day
↓	↓	↓	↓	↓
24.95	+	32.40	=	y

To solve the equation we carry out the addition.

```
    2 4. 9 5
  + 3 2. 4 0
    5 7. 3 5
```

Thus $y = 57.35$.

Check. To check, we first subtract the daily charge from the total cost to get the total mileage cost:

$57.35 - 24.95 = 32.40$

Then we divide the total mileage cost by the cost per mile

$32.4 \div 0.27 = 120$

The number 57.35 checks.

State. The total cost of driving 120 miles in 1 day is $57.35.

21. *Familiarize.* This is a two-step problem. First, we find the number of miles that have been driven between fillups. This is a "how-much-more" situation. We let $n =$ the number of miles driven.

Translate and Solve.

First odometer reading	plus	Number of miles driven	is	Second odometer reading
↓	↓	↓	↓	↓
26,342.8	+	n	=	26,736.7

To solve the equation we subtract 26,342.8 on both sides.

$n = 26,736.7 - 26,342.8$
$n = 393.9$

$$
\begin{array}{r}
2\,6,7\,3\,6.7 \\
-\ 2\,6,3\,4\,2.8 \\
\hline
3\,9\,3.9
\end{array}
$$

Second, we divide the total number of miles driven by the number of gallons. This gives us $m =$ the number of miles per gallon.

$393.9 \div 19.5 = m$

To find the number m, we divide.

$$
\begin{array}{r}
2\,0.2 \\
19.5_\wedge \overline{)\,3\,9\,3.\,9_\wedge 0} \\
3\,9\,0\,0 \\
\hline
3\,9\ 0 \\
3\,9\ 0 \\
\hline
0
\end{array}
$$

Thus, $m = 20.2$.

Check. To check, we first multiply the number of miles per gallon times the number of gallons:

$19.5 \times 20.2 = 393.9$

Then we add 393.9 to 26,342.8:

$26,342.8 + 393.9 = 26,736.7$

The number 20.2 checks.

State. The driver gets 20.2 miles per gallon.

23. *Familiarize.* We visualize a rectangular array consisting of 748.45 objects with 62.5 objects in each row. We want to find n, the number of rows.

Translate. We think (Total number of pounds) ÷ (Pounds per cubic foot) = (Number of cubic feet).

$748.45 \div 62.5 = n$

Solve. We carry out the division.

$$
\begin{array}{r}
1\,1.9\,7\,5\,2 \\
6\,2.5_\wedge \overline{)\,7\,4\,8.\,4_\wedge 5\,0\,0\,0} \\
6\,2\,5\,0\,0 \\
\hline
1\,2\,3\,4\,5 \\
6\,2\,5\,0 \\
\hline
6\,0\,9\,5 \\
5\,6\,2\,5 \\
\hline
4\,7\,0\,0 \\
4\,3\,7\,5 \\
\hline
3\,2\,5\,0 \\
3\,1\,2\,5 \\
\hline
1\,2\,5\,0 \\
1\,2\,5\,0 \\
\hline
0
\end{array}
$$

Thus, $n = 11.9752$.

Check. We obtain a partial check by rounding and estimating:

$748.45 \div 62.5 \approx 700 \div 70 = 10 \approx 11.9752$

State. The tank holds 11.9752 cubic feet of water.

25. *Familiarize.* This is a two-step problem. First, we find the number of games that can be played in one hour. Think of an array containing 60 minutes (1 hour = 60 minutes) with 1.5 minutes in each row. We want to find how many rows there are. We let g represent this number.

Translate and Solve. We think (Number of minutes) ÷ (Number of minutes per game) = (Number of games).

$60 \div 1.5 = g$

To solve the equation we carry out the division.

$$
\begin{array}{r}
4\,0. \\
1.5_\wedge \overline{)\,6\,0.\,0_\wedge} \\
6\,0\,0 \\
\hline
0 \\
0 \\
\hline
0
\end{array}
$$

Thus, $g = 40$.

Second, we find the cost t of playing 40 video games. Repeated addition fits this situation. (We express 25¢ as $0.25.)

Cost of one game	times	Number of games played	is	Total cost
↓	↓	↓	↓	↓
0.25	×	40	=	t

To solve the equation we carry out the multiplication.

$$
\begin{array}{r}
0.\,2\,5 \\
\times\quad 4\,0 \\
\hline
1\,0.\,0\,0
\end{array}
$$

Thus, $t = 10$.

Check. To check, we first divide the total cost by the cost per game to find the number of games played:

$10 \div 0.25 = 40$

Then we multiply 40 by 1.5 to find the total time:

$1.5 \times 40 = 60$

The number 10 checks.

State. It costs $10 to play video games for one hour.

27. Familiarize. This is a two-step problem. First, we find how many minutes there are in 2 hr. We let m represent this number. Repeated addition fits this situation (Remember that 1 hr = 60 min.)

Translate and Solve.

Number of minutes in 1 hour	times	Number of hours	is	Total number of minutes
↓	↓	↓	↓	↓
60	·	2	=	m

To solve the equation we carry out the multiplication.

$$\begin{array}{r} 6\,0 \\ \times \quad 2 \\ \hline 1\,2\,0 \end{array}$$

Thus, $m = 120$.

Next, we find how many calories are burned in 120 minutes. We let t represent this number. Repeated addition fits this situation also.

Number of calories burned in 1 minute	times	Number of minutes	is	Total number of calories burned
↓	↓	↓	↓	↓
8.6	×	120	=	t

To solve the equation we carry out the multiplication.

$$\begin{array}{r} 1\,2\,0 \\ \times \quad 8.\,6 \\ \hline 7\,2\,0 \\ 9\,6\,0\,0 \\ \hline 1\,0\,3\,2.\,0 \end{array}$$

Thus, $t = 1032$.

Check. To check, we first divide the total number of calories by the number of calories burned in one minute to find the total number of minutes the person mowed:

$$1032 \div 8.6 = 120$$

Then we divide 120 by 60 to find the number of hours:

$$120 \div 60 = 2$$

The number 1032 checks.

State. In 2 hr of mowing, 1032 calories would be burned.

29. Familiarize. The batting average is a fraction whose numerator is the number of hits and whose denominator is the number of at bats. We let a = the batting average.

Translate. We think (Number of hits) ÷ (Number of at bats) = (Batting average).

$$58 \div 138 = a$$

Solve. We carry out the division.

$$\begin{array}{r} 0.\,4\,2\,0\,2 \\ 138\overline{)\,5\,8.\,0\,0\,0\,0} \\ 5\,5\,2 \\ \hline 2\,8\,0 \\ 2\,7\,6 \\ \hline 4\,0\,0 \\ 2\,7\,6 \\ \hline 1\,2\,4 \end{array}$$

We stop dividing at this point, because we will round to the nearest thousandth. Thus, $a \approx 0.420$.

Check. We can obtain a partial check by rounding and estimating:

$$58 \div 138 \approx 60 \div 150 = 0.4 \approx 0.420.$$

State. 0.420 of the at bats were hits.

31. Familiarize. We let h = the hourly wage.

Translate. We think (Total earnings) ÷ (Number of hours) = (Hourly wage).

$$\$310.37 \div 40 = h$$

Solve. We carry out the division.

$$\begin{array}{r} 7.\,7\,5\,9 \\ 40\overline{)\,3\,1\,0.\,3\,7\,0} \\ 2\,8\,0\,0\,0 \\ \hline 3\,0\,3\,7 \\ 2\,8\,0\,0 \\ \hline 2\,3\,7 \\ 2\,0\,0 \\ \hline 3\,7\,0 \\ 3\,6\,0 \\ \hline 1\,0 \end{array}$$

We stop dividing at this point, because we will round to the nearest cent. Thus, $h \approx \$7.76$.

Check. We can obtain a partial check by rounding and estimating:

$$310.37 \div 40 \approx 320 \div 40 = 8 \approx 7.76.$$

State. The hourly wage was about $7.76.

33. Familiarize. We let d = the amount earned each day. (We will assume the student earned the same amount each day.)

Translate. We think (Total earnings) ÷ (Number of days) = (Amount earned each day).

$$78.27 \div 9 = d$$

Solve. We carry out the division.

$$\begin{array}{r} 8.\,6\,9\,6 \\ 9\overline{)\,7\,8.\,2\,7\,0} \\ 7\,2\,0\,0 \\ \hline 6\,2\,7 \\ 5\,4\,0 \\ \hline 8\,7 \\ 8\,1 \\ \hline 6\,0 \\ 5\,4 \\ \hline 6 \end{array}$$

Thus, $d = 8.69\overline{6}$.

Check. We can obtain a partial check by rounding and estimating:

$$78.27 \div 9 \approx 80 \div 10 = 8 \approx 8.69\overline{6}.$$

State. We round $8.69\overline{6}$ and find the amount earned each day to be about \$8.70.

35. **Familiarize.** We let $c =$ the cost per pound.

Translate. We think (Total cost) $\div$ (Number of pounds) = (Cost per pound).

$$\$31.92 \div 8 = c$$

Solve. We carry out the division.

```
      3. 9 9
  8 ⟌3 1. 9 2
    2 4 0 0
    ───────
      7 9 2
      7 2 0
    ───────
        7 2
        7 2
    ───────
         0
```

Thus, $c = \$3.99$.

Check. We can obtain a partial check by rounding and estimating:

$$31.92 \div 8 \approx 32 \div 8 = 4 \approx 3.99.$$

State. The ham costs \$3.99 per pound.

37. **Familiarize.** We let $c =$ the cost per acre.

Translate. We think (Total cost) $\div$ (Number of acres) = (Cost per acre).

$$\$906,816 \div 47.5 = c$$

Solve. We carry out the division.

```
            1 9 0 9 0. 8 6 3
  4 7.5∧⟌9 0 6, 8 1 6. 0∧0 0 0
        4 7 5 0 0 0 0
        ─────────────
        4 3 1 8 1 6 0
        4 2 7 5 0 0 0
        ─────────────
          4 3 1 6 0
          4 2 7 5 0
        ─────────────
            4 1 0 0
            3 8 0 0
        ─────────────
            3 0 0 0
            2 8 5 0
        ─────────────
            1 5 0 0
            1 4 2 5
        ─────────────
              7 5
```

We stop dividing at this point because we will round to the nearest cent.

Thus, $c \approx \$19,090.86$.

Check. We can obtain a partial check by rounding and estimating:

$$906,816 \div 47.5 \approx 900,000 \div 45 = 20,000 \approx 19,090.86.$$

State. The land cost about \$19,090.86 per acre.

39. **Familiarize.** This is a three-step problem. We will find the area S of a standard soccer field and the area F of a standard football field using the formula Area $= l \cdot w$.

Then we will find E, the amount by which the area of a soccer field exceeds the area of a football field.

Translate and Solve.

$$S = l \cdot w = 114.9 \times 74.4 = 8548.56$$
$$F = l \cdot w = 120 \times 53.3 = 6396$$

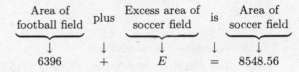

Area of football field	plus	Excess area of soccer field	is	Area of soccer field
↓	↓	↓	↓	↓
6396	+	E	=	8548.56

To solve the equation we subtract 6396 on both sides.

$$E = 8548.56 - 6396$$
$$E = 2152.56$$

```
        4 14
    8 5̸ 4̸ 8. 5 6
  − 6 3 9 6. 0 0
  ─────────────
    2 1 5 2. 5 6
```

Check. We can obtain a partial check by rounding and estimating:

$$114.9 \times 74.4 \approx 110 \times 75 = 8250 \approx 8548.56$$
$$120 \times 53.3 \approx 120 \times 50 = 6000 \approx 6396$$
$$8250 - 6000 = 2250 \approx 2152.56$$

State. The area of a soccer field is 2152.56 sq yd greater than the area of a football field.

41. **Familiarize.** This is a three-step problem. First we find $m =$ the number of months in 30 years.

Translate and Solve.

Number of years	times	Number of months in a year	is	Total number of months
↓	↓	↓	↓	↓
30	×	12	=	m

We carry out the multiplication.

```
    1 2
  × 3 0
  ─────
  3 6 0
```

Thus, $m = 360$.

Next we find $a =$ the amount paid back.

Monthly payment	times	Number of months	is	Amount paid back
↓	↓	↓	↓	↓
880.52	×	360	=	a

We carry out the multiplication.

```
        8 8 0. 5 2
    ×       3 6 0
    ───────────────
    5 2 8 3 1 2 0
  2 6 4 1 5 6 0 0
  ───────────────
  3 1 6, 9 8 7. 2 0
```

Thus, $a = 316,987.20$.

Finally, we find $p =$ the amount by which the amount paid back exceeds the amount of the loan. This is a "how-much-more" situation.

$$\underbrace{\text{Amount} \atop \text{of loan}} \text{ plus } \underbrace{\text{Excess} \atop \text{amount}} \text{ is } \underbrace{\text{Amount} \atop \text{paid back}}$$

$$120,000 \quad + \quad p \quad = \quad 316,987.20$$

We subtract 120,000 on both sides.

$$p = 316,987.20 - 120,000$$
$$p = 196,987.20$$

$$\begin{array}{r} 3\,1\,6,\,9\,8\,7.\,2\,0 \\ -\,1\,2\,0,\,0\,0\,0.\,0\,0 \\ \hline 1\,9\,6,\,9\,8\,7.\,2\,0 \end{array}$$

Check. We repeat the calculations.

State. You pay back \$316,987.20. This is \$196,987.20 more than the amount of the loan.

43.
$$\begin{array}{r} 24\frac{1}{4} = \ 24\frac{3}{12} = \ 23\frac{15}{12} \\ -10\frac{2}{3} = -10\frac{8}{12} = -10\frac{8}{12} \\ \hline 13\frac{7}{12} \end{array}$$

45.
$$\begin{array}{r} 24\frac{1}{4} = \ 24\frac{3}{12} \\ +10\frac{2}{3} = +10\frac{8}{12} \\ \hline 34\frac{11}{12} \end{array}$$

47.
$$24\frac{1}{4} \cdot 10\frac{2}{3} = \frac{97}{4} \cdot \frac{32}{3} = \frac{97 \cdot 32}{4 \cdot 3}$$
$$= \frac{97 \cdot 4 \cdot 8}{4 \cdot 3} = \frac{4}{4} \cdot \frac{97 \cdot 8}{3}$$
$$= 1 \cdot \frac{97 \cdot 8}{3} = \frac{776}{3}$$
$$= 258\frac{2}{3}$$

Chapter 6

Ratio and Proportion

Exercise Set 6.1

---

1. The ratio of 4 to 5 is $\dfrac{4}{5}$.

3. The ratio of 178 to 572 is $\dfrac{178}{572}$.

5. The ratio of 0.4 to 12 is $\dfrac{0.4}{12}$.

7. The ratio of 3.8 to 7.4 is $\dfrac{3.8}{7.4}$.

9. The ratio of 56.78 to 98.35 is $\dfrac{56.78}{98.35}$.

11. The ratio of $8\frac{3}{4}$ to $9\frac{5}{6}$ is $\dfrac{8\frac{3}{4}}{9\frac{5}{6}}$.

13. The ratio of lawyers to people is $\dfrac{36.1}{1000}$.

The ratio of people to lawyers is $\dfrac{1000}{36.1}$.

15. The ratio of \$2.7 billion to \$13.1 billion is $\dfrac{2.7}{13.1}$.

The ratio of \$13.1 billion to \$2.7 billion is $\dfrac{13.1}{2.7}$.

17. The ratio of length to width is $\dfrac{478}{213}$.

The ratio of width to length is $\dfrac{213}{478}$.

19. The ratio of 4 to 6 is $\dfrac{4}{6}=\dfrac{2\cdot2}{2\cdot3}=\dfrac{2}{2}\cdot\dfrac{2}{3}=\dfrac{2}{3}$.

21. The ratio of 18 to 24 is $\dfrac{18}{24}=\dfrac{3\cdot6}{4\cdot6}=\dfrac{3}{4}\cdot\dfrac{6}{6}=\dfrac{3}{4}$.

23. The ratio of 4.8 to 10 is $\dfrac{4.8}{10}=\dfrac{4.8}{10}\cdot\dfrac{10}{10}=\dfrac{48}{100}=\dfrac{4\cdot12}{4\cdot25}=$
$\dfrac{4}{4}\cdot\dfrac{12}{25}=\dfrac{12}{25}$.

25. The ratio of 2.8 to 3.6 is $\dfrac{2.8}{3.6}=\dfrac{2.8}{3.6}\cdot\dfrac{10}{10}=\dfrac{28}{36}=\dfrac{4\cdot7}{4\cdot9}=$
$\dfrac{4}{4}\cdot\dfrac{7}{9}=\dfrac{7}{9}$.

27. The ratio is $\dfrac{20}{30}=\dfrac{2\cdot10}{3\cdot10}=\dfrac{2}{3}\cdot\dfrac{10}{10}=\dfrac{2}{3}$.

29. The ratio is $\dfrac{56}{100}=\dfrac{4\cdot14}{4\cdot25}=\dfrac{4}{4}\cdot\dfrac{14}{25}=\dfrac{14}{25}$.

31. The ratio is $\dfrac{128}{256}=\dfrac{1\cdot128}{2\cdot128}=\dfrac{1}{2}\cdot\dfrac{128}{128}=\dfrac{1}{2}$.

33. The ratio is $\dfrac{0.48}{0.64}=\dfrac{0.48}{0.64}\cdot\dfrac{100}{100}=\dfrac{48}{64}=\dfrac{3\cdot16}{4\cdot16}=\dfrac{3}{4}\cdot\dfrac{16}{16}=\dfrac{3}{4}$.

35. The ratio is $\dfrac{54}{100}=\dfrac{2\cdot27}{2\cdot50}=\dfrac{2}{2}\cdot\dfrac{27}{50}=\dfrac{27}{50}$.

37. The ratio is $\dfrac{6.4}{20.2}=\dfrac{6.4}{20.2}\cdot\dfrac{10}{10}=\dfrac{64}{202}=\dfrac{2\cdot32}{2\cdot101}=\dfrac{2}{2}\cdot\dfrac{32}{101}=$
$\dfrac{32}{101}$.

39. We find the cross products:

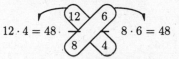

$$12\cdot4=48 \qquad 8\cdot6=48$$

Since the cross products are equal, $\dfrac{12}{8}=\dfrac{6}{4}$.

41.
$$\begin{array}{r} 50 \\ 4\overline{)200} \\ \underline{200} \\ 0 \\ \underline{0} \\ 0 \end{array}$$

43.
$$\begin{array}{r} 14.5 \\ 16\overline{)232.0} \\ \underline{160} \\ 72 \\ \underline{64} \\ 80 \\ \underline{80} \\ 0 \end{array}$$

45. Use a calculator.

The ratio of people to sheep is $\dfrac{13,339,000}{145,304,000}$, or about 0.09.

The ratio of sheep to people is $\dfrac{145,304,000}{13,339,000}$, or about 10.89.

Exercise Set 6.2

---

1. $\dfrac{120 \text{ km}}{3 \text{ hr}}$, or $40\ \dfrac{\text{km}}{\text{hr}}$

3. $\dfrac{440 \text{ m}}{40 \text{ sec}}$, or $11\ \dfrac{\text{m}}{\text{sec}}$

5. $\dfrac{342 \text{ yd}}{2.25 \text{ days}}$, or $152\ \dfrac{\text{yd}}{\text{day}}$

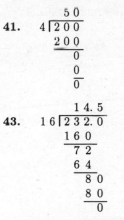

7. $\dfrac{500 \text{ mi}}{20 \text{ hr}} = 25 \dfrac{\text{mi}}{\text{hr}}$

$\dfrac{20 \text{ hr}}{500 \text{ mi}} = 0.04 \dfrac{\text{hr}}{\text{mi}}$

9. $\dfrac{623 \text{ gal}}{1000 \text{ sq ft}} = 0.623 \dfrac{\text{gal}}{\text{sq ft}}$

11. $\dfrac{\$5.75}{10 \text{ min}} = \dfrac{575\cent}{10 \text{ min}} = 57.5 \dfrac{\cent}{\text{min}}$

13. $\dfrac{186,000 \text{ mi}}{1 \text{ sec}} = 186,000 \dfrac{\text{mi}}{\text{sec}}$

15. $\dfrac{310 \text{ km}}{2.5 \text{ hr}} = 124 \dfrac{\text{km}}{\text{hr}}$

17. $\dfrac{2660 \text{ mi}}{4.75 \text{ hr}} = 560 \dfrac{\text{mi}}{\text{hr}}$

19. Unit price $= \dfrac{\text{Price}}{\text{Number of units}} = \dfrac{\$80.75}{8.5 \text{ yd}} = 9.5 \dfrac{\text{dollars}}{\text{yd}}$, or $\$9.50/\text{yd}$

21. We need to find the number of ounces in 2 pounds:

$2 \text{ lb} = 2 \times 1 \text{ lb} = 2 \times 16 \text{ oz} = 32 \text{ oz}$

Unit price $= \dfrac{\text{Price}}{\text{Number of units}} = \dfrac{\$8.59}{32 \text{ oz}} = \dfrac{859\cent}{32 \text{ oz}} \approx 26.84 \dfrac{\cent}{\text{oz}}$

23. $\dfrac{\$2.89}{\frac{2}{3} \text{ lb}} = 2.89 \times \dfrac{3}{2} \times \dfrac{\text{dollars}}{\text{lb}} \approx 4.34 \dfrac{\text{dollars}}{\text{lb}}$, or $\$4.34/\text{lb}$

25. Compare the unit prices.

For Brand A: $\dfrac{\$1.79}{18 \text{ oz}} \approx \$0.099/\text{oz}$

For Brand B: $\dfrac{\$1.65}{16 \text{ oz}} \approx \$0.103/\text{oz}$

Thus, Brand A has the lower unit price.

27. Compare the unit prices. Recall that 1 qt = 32 oz, so 2 qt = 2×1 qt = 2×32 oz = 64 oz.

For Brand A: $\dfrac{\$2.69}{64 \text{ oz}} \approx \$0.042/\text{oz}$

For Brand B: $\dfrac{\$1.97}{48 \text{ oz}} \approx \$0.041/\text{oz}$

Thus, Brand B has the lower unit price.

29. Compare the unit prices.

For Brand A: $\dfrac{\$3.96}{3 \text{ bars}} = \$1.32/\text{bar}$

For Brand B: $\dfrac{\$2.67}{2 \text{ bars}} = \$1.335/\text{bar}$

Thus, Brand A has the lower unit price.

31. Compare the unit prices.

For Brand A: $\dfrac{\$1.29}{7 \text{ oz}} \approx \$0.184/\text{oz}$

For Brand B: $\dfrac{\$1.19}{6\frac{1}{8} \text{ oz}} \approx \$0.194/\text{oz}$

Thus, Brand A has the lower unit price.

33. Six 10-oz bottles contain 6×10 oz = 60 oz of sparkling water; four 12-oz bottles contain 4×12 oz = 48 oz of sparkling water.

Compare the unit prices.

Six 10-oz bottles: $\dfrac{\$3.09}{60 \text{ oz}} = \$0.0515/\text{oz}$

Four 12-oz bottles: $\dfrac{\$2.39}{48 \text{ oz}} \approx \$0.0498/\text{oz}$

Thus, four 12-oz bottles have the lower unit price.

35. Compare the unit prices.

For Package A: $\dfrac{63\cent}{3.9 \text{ oz}} \approx 16.2\cent/\text{oz}$

For Package B: $\dfrac{99\cent}{5.9 \text{ oz}} \approx 16.8\cent/\text{oz}$

Thus, Package A has the lower unit price.

37. **Familiarize.** We visualize the situation. We let $p =$ the number by which the number of piano players exceeds the number of guitar players, in millions.

18.9 million	p
20.6 million	

Translate. This is a "how-much-more" situation.

$$\underbrace{\begin{matrix} \text{Number} \\ \text{of guitar} \\ \text{players} \end{matrix}}_{\downarrow} + \underbrace{\begin{matrix} \text{Additional} \\ \text{number of} \\ \text{piano players} \end{matrix}}_{\downarrow} = \underbrace{\begin{matrix} \text{Number} \\ \text{of piano} \\ \text{players} \end{matrix}}_{\downarrow}$$

$$18.9 \quad + \quad p \quad = \quad 20.6$$

Solve. To solve the equation we subtract 18.9 on both sides.

$p = 20.6 - 18.9$
$p = 1.7$

$$\begin{array}{r} \overset{1\ \ 9\ 16}{2\ \cancel{0}.\ \cancel{6}} \\ -\ 1\ 8.\ 9 \\ \hline 1.\ 7 \end{array}$$

Check. We repeat the calculation.

State. There are 1.7 million more piano players than guitar players.

39. $1\underline{00} \times 678.19 \qquad 678.19.$

2 zeros   Move 2 places to the right.

$100 \times 678.19 = 67,819$

41.
$$\begin{array}{r} 6\ 9.\ 2 \\ \times\ 8\ 4.\ 3 \\ \hline 2\ 0\ 7\ 6 \\ 2\ 7\ 6\ 8\ 0 \\ 5\ 5\ 3\ 6\ 0\ 0 \\ \hline 5\ 8\ 3\ 3.\ 5\ 6 \end{array}$$

43. a) For the 6-oz container: $\dfrac{65\cent}{6 \text{ oz}} \approx 10.83\cent/\text{oz}$

For the 5.5-oz container: $\dfrac{60\cent}{5.5 \text{ oz}} \approx 10.91\cent/\text{oz}$

b) For the \$0.89 roll: $\dfrac{\$0.89}{78 \text{ ft}^2} = \dfrac{89\cancel{c}}{78 \text{ ft}^2} \approx 1.14\cancel{c}/\text{ft}^2$

For the \$0.79 roll: $\dfrac{\$0.79}{65 \text{ ft}^2} = \dfrac{79\cancel{c}}{65 \text{ ft}^2} \approx 1.22\cancel{c}/\text{ft}^2$

Exercise Set 6.3

1. We can use cross-products:

$$5 \cdot 9 = 45 \qquad \begin{matrix} 5 & 7 \\ 6 & 9 \end{matrix} \qquad 6 \cdot 7 = 42$$

Since the cross-products are not the same, $45 \neq 42$, we know that the numbers are not proportional.

3. We can use cross-products:

$$1 \cdot 20 = 20 \qquad \begin{matrix} 1 & 10 \\ 2 & 20 \end{matrix} \qquad 2 \cdot 10 = 20$$

Since the cross-products are the same, $20 = 20$, we know that $\dfrac{1}{2} = \dfrac{10}{20}$, so the numbers are proportional.

5. We can use cross-products:

$$2.4 \cdot 2.7 = 6.48 \qquad \begin{matrix} 2.4 & 1.8 \\ 3.6 & 2.7 \end{matrix} \qquad 3.6 \cdot 1.8 = 6.48$$

Since the cross-products are the same, $6.48 = 6.48$, we know that $\dfrac{2.4}{3.6} = \dfrac{1.8}{2.7}$, so the numbers are proportional.

7. We can use cross-products:

$$5\tfrac{1}{3} \cdot 9\tfrac{1}{2} = 50\tfrac{2}{3} \qquad \begin{matrix} 5\tfrac{1}{3} & 2\tfrac{1}{5} \\ 8\tfrac{1}{4} & 9\tfrac{1}{2} \end{matrix} \qquad 8\tfrac{1}{4} \cdot 2\tfrac{1}{5} = 18\tfrac{3}{20}$$

Since the cross-products are not the same, $50\tfrac{2}{3} \neq 18\tfrac{3}{20}$, we know that the numbers are not proportional.

9.
$$\dfrac{18}{4} = \dfrac{x}{10}$$

$18 \cdot 10 = 4 \cdot x$ Finding cross-products

$\dfrac{18 \cdot 10}{4} = x$ Dividing by 4

$\dfrac{180}{4} = x$ Multiplying

$45 = x$ Dividing

11.
$$\dfrac{x}{8} = \dfrac{9}{6}$$

$6 \cdot x = 8 \cdot 9$ Finding cross-products

$x = \dfrac{8 \cdot 9}{6}$ Dividing by 6

$x = \dfrac{72}{6}$ Multiplying

$x = 12$ Dividing

13.
$$\dfrac{t}{12} = \dfrac{5}{6}$$
$$6 \cdot t = 12 \cdot 5$$
$$t = \dfrac{12 \cdot 5}{6}$$
$$t = \dfrac{60}{6}$$
$$t = 10$$

15.
$$\dfrac{2}{5} = \dfrac{8}{n}$$
$$2 \cdot n = 5 \cdot 8$$
$$n = \dfrac{5 \cdot 8}{2}$$
$$n = \dfrac{40}{2}$$
$$n = 20$$

17.
$$\dfrac{n}{15} = \dfrac{10}{30}$$
$$30 \cdot n = 15 \cdot 10$$
$$n = \dfrac{15 \cdot 10}{30}$$
$$n = \dfrac{150}{30}$$
$$n = 5$$

19.
$$\dfrac{16}{12} = \dfrac{24}{x}$$
$$16 \cdot x = 12 \cdot 24$$
$$x = \dfrac{12 \cdot 24}{16}$$
$$x = \dfrac{288}{16}$$
$$x = 18$$

21.
$$\dfrac{6}{11} = \dfrac{12}{x}$$
$$6 \cdot x = 11 \cdot 12$$
$$x = \dfrac{11 \cdot 12}{6}$$
$$x = \dfrac{132}{6}$$
$$x = 22$$

23.
$$\dfrac{20}{7} = \dfrac{80}{x}$$
$$20 \cdot x = 7 \cdot 80$$
$$x = \dfrac{7 \cdot 80}{20}$$
$$x = \dfrac{560}{20}$$
$$x = 28$$

25. $\dfrac{12}{9} = \dfrac{x}{7}$

$12 \cdot 7 = 9 \cdot x$

$\dfrac{12 \cdot 7}{9} = x$

$\dfrac{84}{9} = x$

$\dfrac{28}{3} = x$ Simplifying

$9\dfrac{1}{3} = x$ Writing a mixed numeral

27. $\dfrac{x}{13} = \dfrac{2}{9}$

$9 \cdot x = 13 \cdot 2$

$x = \dfrac{13 \cdot 2}{9}$

$x = \dfrac{26}{9}$, or $2\dfrac{8}{9}$

29. $\dfrac{t}{0.16} = \dfrac{0.15}{0.40}$

$0.40 \times t = 0.16 \times 0.15$

$t = \dfrac{0.16 \times 0.15}{0.40}$

$t = \dfrac{0.024}{0.40}$

$t = 0.06$

31. $\dfrac{100}{25} = \dfrac{20}{n}$

$100 \cdot n = 25 \cdot 20$

$n = \dfrac{25 \cdot 20}{100}$

$n = \dfrac{500}{100}$

$n = 5$

33. $\dfrac{7}{\frac{1}{4}} = \dfrac{28}{x}$

$7 \cdot x = \dfrac{1}{4} \cdot 28$

$x = \dfrac{\frac{1}{4} \cdot 28}{7}$

$x = \dfrac{7}{7}$

$x = 1$

35. $\dfrac{\frac{1}{4}}{\frac{1}{2}} = \dfrac{\frac{1}{2}}{x}$

$\dfrac{1}{4} \cdot x = \dfrac{1}{2} \cdot \dfrac{1}{2}$

$x = \dfrac{\frac{1}{2} \cdot \frac{1}{2}}{\frac{1}{4}}$

$x = \dfrac{\frac{1}{4}}{\frac{1}{4}}$

$x = 1$

37. $\dfrac{1}{2} = \dfrac{7}{x}$

$1 \cdot x = 2 \cdot 7$

$x = \dfrac{2 \cdot 7}{1}$

$x = 14$

39. $\dfrac{\frac{2}{7}}{\frac{3}{4}} = \dfrac{\frac{5}{6}}{y}$

$\dfrac{2}{7} \cdot y = \dfrac{3}{4} \cdot \dfrac{5}{6}$

$y = \dfrac{3}{4} \cdot \dfrac{5}{6} \cdot \dfrac{7}{2}$ Dividing by $\dfrac{2}{7}$

$y = \dfrac{3}{4} \cdot \dfrac{5}{2 \cdot \cancel{3}} \cdot \dfrac{7}{2}$

$y = \dfrac{5 \cdot 7}{4 \cdot 2 \cdot 2}$

$y = \dfrac{35}{16}$, or $2\dfrac{3}{16}$

41. $\dfrac{2\frac{1}{2}}{3\frac{1}{3}} = \dfrac{x}{4\frac{1}{4}}$

$2\dfrac{1}{2} \cdot 4\dfrac{1}{4} = 3\dfrac{1}{3} \cdot x$

$\dfrac{5}{2} \cdot \dfrac{17}{4} = \dfrac{10}{3} \cdot x$

$\dfrac{3}{10} \cdot \dfrac{5}{2} \cdot \dfrac{17}{4} = x$ Dividing by $\dfrac{10}{3}$

$\dfrac{3}{\cancel{5} \cdot 2} \cdot \dfrac{\cancel{5}}{2} \cdot \dfrac{17}{4} = x$

$\dfrac{3 \cdot 17}{2 \cdot 2 \cdot 4} = x$

$\dfrac{51}{16} = x$, or

$3\dfrac{3}{16} = x$

43. $\dfrac{1.28}{3.76} = \dfrac{4.28}{y}$

$1.28 \times y = 3.76 \times 4.28$

$y = \dfrac{3.76 \times 4.28}{1.28}$

$y = \dfrac{16.0928}{1.28}$

$y = 12.5725$

45. $\dfrac{10\frac{3}{8}}{12\frac{2}{3}} = \dfrac{5\frac{3}{4}}{y}$

$10\dfrac{3}{8} \cdot y = 12\dfrac{2}{3} \cdot 5\dfrac{3}{4}$

$\dfrac{83}{8} \cdot y = \dfrac{38}{3} \cdot \dfrac{23}{4}$

$y = \dfrac{38}{3} \cdot \dfrac{23}{4} \cdot \dfrac{8}{83}$ Dividing by $\dfrac{83}{3}$

$y = \dfrac{38}{3} \cdot \dfrac{23}{4} \cdot \dfrac{2 \cdot 4}{83}$

$y = \dfrac{38 \cdot 23 \cdot 2}{3 \cdot 83}$

$y = \dfrac{1748}{249}$, or $7\dfrac{5}{249}$

47. We find the cross-products:

$3 \cdot 6 = 18$ $\dfrac{3}{4} \times \dfrac{5}{6}$ $4 \cdot 5 = 20$

Since $18 \neq 20$, we have $\dfrac{3}{4} \neq \dfrac{5}{6}$.

49. We find the cross-products:

$7 \cdot 9 = 63$ $\dfrac{7}{8} \times \dfrac{7}{9}$ $8 \cdot 7 = 56$

Since $63 \neq 56$, we have $\dfrac{7}{8} \neq \dfrac{7}{9}$.

51. $\dfrac{1728}{5643} = \dfrac{836.4}{x}$

$1728 \cdot x = 5643 \cdot 836.4$

$x = \dfrac{5643 \cdot 836.4}{1728}$

$x \approx 2731.4$ Using a calculator to multiply and divide

Exercise Set 6.4

1. Let d = the distance traveled in 42 days.

Distance $\rightarrow \dfrac{234}{14} = \dfrac{d}{42} \leftarrow$ Distance
Time $\rightarrow$ $\qquad\qquad\leftarrow$ Time

Solve: $234 \cdot 42 = 14 \cdot d$ Finding cross-products

$\dfrac{234 \cdot 42}{14} = d$ Dividing by 14

$\dfrac{234 \cdot 3 \cdot 14}{14} = d$ Factoring

$234 \cdot 3 = d$ Simplifying

$702 = d$ Dividing

The car would travel 702 km in 42 days.

3. Let x = the cost of 9 sweatshirts.

Sweatshirts $\rightarrow \dfrac{2}{18.80} = \dfrac{9}{x} \leftarrow$ Sweatshirts
Dollars $\rightarrow$ $\qquad\qquad\leftarrow$ Dollars

Solve: $2 \cdot x = 18.80 \cdot 9$ Finding cross-products

$x = \dfrac{18.80 \cdot 9}{2}$ Dividing by 2

$x = \dfrac{2 \cdot 9.40 \cdot 9}{2}$ Factoring

$x = 9.40 \cdot 9$ Simplifying

$x = 84.60$ Dividing

Thus, 9 sweatshirts cost $84.60.

5. Let E = the earned run average.

Earned runs $\rightarrow \dfrac{E}{9} = \dfrac{73}{211} \leftarrow$ Earned runs
Innings $\rightarrow$ $\qquad\qquad\leftarrow$ Innings

Solve: $211 \cdot E = 9 \cdot 73$

$E = \dfrac{9 \cdot 73}{211}$

$E = \dfrac{657}{211}$ Multiplying

$E \approx 3.11$ Dividing and rounding to the nearest hundredth

The earned run average was about 3.11.

7. Let D = the number of deer in the game preserve.

Deer tagged originally $\rightarrow \dfrac{318}{D} = \dfrac{56}{168} \leftarrow$ Tagged deer caught later
Deer in game preserve $\rightarrow$ $\qquad\qquad\leftarrow$ Deer caught later

Solve: $318 \cdot 168 = 56 \cdot D$

$\dfrac{318 \cdot 168}{56} = D$

$\dfrac{2 \cdot 3 \cdot 53 \cdot 2 \cdot 2 \cdot 2 \cdot 3 \cdot 7}{2 \cdot 2 \cdot 2 \cdot 7} = D$

$2 \cdot 3 \cdot 53 \cdot 3 = D$

$954 = D$

We estimate that there are 954 deer in the game preserve.

9. Let w = the weight of 40 books.

Books $\rightarrow \dfrac{24}{37} = \dfrac{40}{w} \leftarrow$ Books
Weight $\rightarrow$ $\qquad\qquad\leftarrow$ Weight

Solve: $24 \cdot w = 37 \cdot 40$

$w = \dfrac{37 \cdot 40}{24}$

$w = \dfrac{37 \cdot 5 \cdot 8}{3 \cdot 8}$

$w = \dfrac{185}{3}$, or $61\dfrac{2}{3}$

The weight of 40 books is $61\dfrac{2}{3}$ lb.

11. Let t = the number of trees required to produce 391 pounds of coffee.

$$\text{Trees} \rightarrow \frac{14}{17} = \frac{t}{391} \leftarrow \text{Trees}$$
$$\text{Pounds} \rightarrow \quad\quad\quad \leftarrow \text{Pounds}$$

Solve: $14 \cdot 391 = 17 \cdot t$

$$\frac{14 \cdot 391}{17} = t$$
$$\frac{14 \cdot 17 \cdot 23}{17} = t$$
$$14 \cdot 23 = t$$
$$322 = t$$

Thus, 322 trees are required to produce 391 pounds of coffee.

13. Let z = the number of pounds of zinc in the alloy.

$$\text{Zinc} \rightarrow \frac{3}{13} = \frac{z}{520} \leftarrow \text{Zinc}$$
$$\text{Copper} \rightarrow \quad\quad\quad \leftarrow \text{Copper}$$

Solve: $3 \cdot 520 = 13 \cdot z$

$$\frac{3 \cdot 520}{13} = z$$
$$\frac{3 \cdot 13 \cdot 2 \cdot 2 \cdot 2 \cdot 5}{13} = z$$
$$3 \cdot 2 \cdot 2 \cdot 2 \cdot 5 = z$$
$$120 = z$$

There are 120 lb of zinc in the alloy.

15. Let d = the number of defective bulbs in a lot of 22,000.

$$\text{Defective bulbs} \rightarrow \frac{18}{200} = \frac{d}{22,000} \leftarrow \text{Defective bulbs}$$
$$\text{Bulbs in lot} \rightarrow \quad\quad\quad\quad \leftarrow \text{Bulbs in lot}$$

Solve: $18 \cdot 22,000 = 200 \cdot d$

$$\frac{18 \cdot 22,000}{200} = d$$
$$\frac{18 \cdot 110 \cdot 200}{200} = d$$
$$18 \cdot 110 = d$$
$$1980 = d$$

There would be 1980 defective bulbs in a lot of 22,000.

17. Let d = the actual distance between the cities.

$$\text{Map distance} \rightarrow \frac{1}{16.6} = \frac{3.5}{d} \leftarrow \text{Map distance}$$
$$\text{Actual distance} \rightarrow \quad\quad\quad \leftarrow \text{Actual distance}$$

Solve: $1 \cdot d = 16.6 \cdot 3.5$
$$d = 58.1$$

The cities are 58.1 mi apart.

19. Let w = the number of inches of water to which $5\frac{1}{2}$ ft of snow will melt.

$$\text{Snow} \rightarrow \frac{1\frac{1}{2}}{2} = \frac{5\frac{1}{2}}{w} \leftarrow \text{Snow}$$
$$\text{Water} \rightarrow \quad\quad\quad \leftarrow \text{Water}$$

Solve: $1\frac{1}{2} \cdot w = 2 \cdot 5\frac{1}{2}$

$$\frac{3}{2} \cdot w = \frac{2}{1} \cdot \frac{11}{2} \quad \text{Writing fractional}$$
$$\text{notation}$$
$$w = \frac{2}{1} \cdot \frac{11}{2} \cdot \frac{2}{3} \quad \text{Dividing by } \frac{3}{2}$$
$$w = \frac{2 \cdot 11 \cdot 2}{1 \cdot 2 \cdot 3}$$
$$w = \frac{11 \cdot 2}{1 \cdot 3} \quad \text{Simplifying}$$
$$w = \frac{22}{3}, \text{ or } 7\frac{1}{3}$$

Thus, $5\frac{1}{2}$ ft of snow will melt to $7\frac{1}{3}$ in. of water.

21. Let s = the amount the student will spend during the academic year if spending continues at the given rate. Note that the length of the academic year is $2 \cdot 16$ weeks, or 32 weeks.

$$\text{Weeks} \rightarrow \frac{3}{80} = \frac{32}{s} \leftarrow \text{Weeks}$$
$$\text{Spending} \rightarrow \quad\quad\quad \leftarrow \text{Spending}$$

Solve: $3 \cdot s = 80 \cdot 32$
$$s = \frac{80 \cdot 32}{3}$$
$$s = \frac{2560}{3}$$
$$s \approx 853.33$$

At the given rate the student would spend about $853.33 during the academic year. Thus, the budget will not be adequate.

To find when the money will be exhausted, we let w = the number of weeks it will take to spend $800 if spending continues at the given rate.

$$\text{Weeks} \rightarrow \frac{3}{80} = \frac{w}{800} \leftarrow \text{Weeks}$$
$$\text{Spending} \rightarrow \quad\quad\quad \leftarrow \text{Spending}$$

Solve: $3 \cdot 800 = 80 \cdot w$
$$\frac{3 \cdot 800}{80} = w$$
$$\frac{3 \cdot 80 \cdot 10}{80 \cdot 1} = w$$
$$30 = w$$

The money will be gone in 30 weeks.

To find how much more will be needed to complete the year we subtract the amount available from the amount needed, if spending continues at the given rate.

$$\$853.33 - \$800 = \$53.33$$

The student will need $53.33 more to complete the year.

23. Let R = the earned runs.

$$\text{Earned runs} \rightarrow \frac{2.63}{9} = \frac{R}{7356} \leftarrow \text{Earned runs}$$
$$\text{Innings} \rightarrow \quad\quad\quad \leftarrow \text{Innings}$$

Solve: $2.63 \cdot 7356 = 9 \cdot R$

$$\frac{2.63 \cdot 7356}{9} = R$$

$$\frac{19,346.28}{9} = R$$

$$2150 \approx R$$

Cy Young gave up about 2150 earned runs.

Exercise Set 6.5

1. The ratio of h to 5 is the same as the ratio of 45 to 9. We have the proportion

$$\frac{h}{5} = \frac{45}{9}.$$

Solve: $9 \cdot h = 5 \cdot 45$ Finding cross-products

$\qquad h = \dfrac{5 \cdot 45}{9}$ Dividing by 9 on both sides

$\qquad h = 25$ Simplifying

The missing length h is 25.

3. The ratio of x to 2 is the same as the ratio of 2 to 3. We have the proportion

$$\frac{x}{2} = \frac{2}{3}.$$

Solve: $3 \cdot x = 2 \cdot 2$ Finding cross-products

$\qquad x = \dfrac{2 \cdot 2}{3}$ Dividing by 3 on both sides

$\qquad x = \dfrac{4}{3}$, or $1\dfrac{1}{3}$

The missing length x is $\dfrac{4}{3}$, or $1\dfrac{1}{3}$. We could also have used $\dfrac{x}{2} = \dfrac{1}{1\frac{1}{2}}$ to find x.

5. First we find x. The ratio of x to 9 is the same as the ratio of 6 to 8. We have the proportion

$$\frac{x}{9} = \frac{6}{8}.$$

Solve: $8 \cdot x = 9 \cdot 6$

$\qquad x = \dfrac{9 \cdot 6}{8}$

$\qquad x = \dfrac{27}{4}$, or $6\dfrac{3}{4}$

The missing length x is $\dfrac{27}{4}$, or $6\dfrac{3}{4}$.

Next we find y. The ratio of y to 12 is the same as the ratio of 6 to 8. We have the proportion

$$\frac{y}{12} = \frac{6}{8}.$$

Solve: $8 \cdot y = 12 \cdot 6$

$\qquad y = \dfrac{12 \cdot 6}{8}$

$\qquad y = 9$

The missing length y is 9.

7. First we find x. The ratio of x to 2.5 is the same as the ratio of 2.1 to 0.7. We have the proportion

$$\frac{x}{2.5} = \frac{2.1}{0.7}.$$

Solve: $0.7 \cdot x = 2.5 \cdot 2.1$

$\qquad x = \dfrac{2.5 \cdot 2.1}{0.7}$

$\qquad x = 7.5$

The missing length x is 7.5.

Next we find y. The ratio of y to 2.4 is the same as the ratio of 2.1 to 0.7. We have the proportion

$$\frac{y}{2.4} = \frac{2.1}{0.7}.$$

Solve: $0.7 \cdot y = 2.4 \cdot 2.1$

$\qquad y = \dfrac{2.4 \cdot 2.1}{0.7}$

$\qquad y = 7.2$

The missing length y is 7.2.

9. If we use the sun's rays to represent the third side of a triangle in a drawing of the situation, we see that we have similar triangles. We let $s =$ the length of a shadow cast by a person 2 m tall.

Sun's rays   8 m   Sun's rays   2 m
5 m s

The ratio of s to 5 is the same as the ratio of 2 to 8. We have the proportion

$$\frac{s}{5} = \frac{2}{8}.$$

Solve: $8 \cdot s = 5 \cdot 2$

$\qquad s = \dfrac{5 \cdot 2}{8}$

$\qquad s = \dfrac{5}{4}$, or 1.25

The length of a shadow cast by a person 2 m tall is 1.25 m.

11. If we use the sun's rays to represent the third side of a triangle in a drawing of the situation, we see that we have similar triangles. We let $h =$ the height of the tree.

Sun's rays h
Sun's rays   4 ft
3 ft   27 ft

The ratio of h to 4 is the same as the ratio of 27 to 3. We have the proportion

$$\frac{h}{4} = \frac{27}{3}.$$

Solve: $3 \cdot h = 4 \cdot 27$

$\qquad h = \dfrac{4 \cdot 27}{3}$

$\qquad h = 36$

The tree is 36 ft tall.

13. *Familiarize*. This is a multistep problem.

First we find the total cost of the purchases. We let $c =$ this amount.

Translate and Solve.

Price of book	plus	Price of CD	plus	Price of sweatshirt	is	Total cost
↓	↓	↓	↓	↓	↓	↓
$49.95	+	$14.88	+	$29.95	=	c

To solve the equation we carry out the addition.

$$
\begin{array}{r}
\overset{2\ \ 2\ \ 1}{4\,9.\,9\,5} \\
1\,4.\,8\,8 \\
+\,2\,9.\,9\,5 \\
\hline
9\,4.\,7\,8
\end{array}
$$

Thus, $c = \$94.78$.

Now we find how much more money the student needs to make these purchases. We let $m =$ this amount.

Money student has	plus	How much more money	is	Total cost of purchases
↓	↓	↓	↓	↓
$34.97	+	m	=	$94.78

To solve the equation we subtract 34.97 on both sides.

$m = 94.78 - 34.97$
$m = 59.81$

$$
\begin{array}{r}
\overset{\ \ \ \ \ \ 13}{\overset{8\ \ \cancel{3}\,17}{\cancel{9}\,\cancel{4}.\,7\,8}} \\
-\,3\,4.\,9\,7 \\
\hline
5\,9.\,8\,1
\end{array}
$$

Check. We repeat the calculations.

State. The student needs $59.81 more to make the purchases.

15.

$$
\begin{array}{r}
\overset{7\ \ 7\ \ 1}{\overset{3\ \ 3}{8\,0.\,8\,9\,2}} \\
\times\ \ \ \ \ \ 8.\,4 \\
\hline
3\,2\,3\,5\,6\,8 \\
6\,4\,7\,1\,3\,6\,0 \\
\hline
6\,7\,9.\,4\,9\,2\,8
\end{array}
$$

17. $1\underline{00} \times 274.568 \qquad 274.56.8$

  2 zeros   Move 2 places to the right.

  $100 \times 274.568 = 27,456.8$

19. Since the ratio of d to 25 ft is the same as the ratio of 40 ft to 10 ft, we have the proportion

$$\frac{d}{25} = \frac{40}{10}.$$

Solve: $10 \cdot d = 25 \cdot 40$

$$d = \frac{25 \cdot 40}{10}$$

$$d = 100$$

The distance across the river is 100 ft.

Chapter 7

Percent Notation

Exercise Set 7.1

1.
$$90\% = \frac{90}{100} \qquad \text{A ratio of 90 to 100}$$
$$90\% = 90 \times \frac{1}{100} \qquad \text{Replacing \% with } \times \frac{1}{100}$$
$$90\% = 90 \times 0.01 \qquad \text{Replacing \% with } \times 0.01$$

3.
$$12.5\% = \frac{12.5}{100} \qquad \text{A ratio of 12.5 to 100}$$
$$12.5\% = 12.5 \times \frac{1}{100} \qquad \text{Replacing \% with } \times \frac{1}{100}$$
$$12.5\% = 12.5 \times 0.01 \qquad \text{Replacing \% with } \times 0.01$$

5. 67%

a) Replace the percent symbol with $\times 0.01$.

67×0.01

b) Move the decimal point two places to the left.

0.67.

Thus, 67% = 0.67.

7. 45.6%

a) Replace the percent symbol with $\times 0.01$.

45.6×0.01

b) Move the decimal point two places to the left.

0.45.6

Thus, 45.6% = 0.456.

9. 59.01%

a) Replace the percent symbol with $\times 0.01$.

59.01×0.01

b) Move the decimal point two places to the left.

0.59.01

Thus, 59.01% = 0.5901.

11. 10%

a) Replace the percent symbol with $\times 0.01$.

10×0.01

b) Move the decimal point two places to the left.

0.10.

Thus, 10% = 0.1.

13. 1%

a) Replace the percent symbol with $\times 0.01$.

1×0.01

b) Move the decimal point two places to the left.

0.01.

Thus, 1% = 0.01.

15. 200%

a) Replace the percent symbol with $\times 0.01$.

200×0.01

b) Move the decimal point two places to the left.

2.00.

Thus, 200% = 2.

17. 0.1%

a) Replace the percent symbol with $\times 0.01$.

0.1×0.01

b) Move the decimal point two places to the left.

0.00.1

Thus, 0.1% = 0.001.

19. 0.09%

a) Replace the percent symbol with $\times 0.01$.

0.09×0.01

b) Move the decimal point two places to the left.

0.00.09

Thus, 0.09% = 0.0009.

21. 0.18%

a) Replace the percent symbol with $\times 0.01$.

0.18×0.01

b) Move the decimal point two places to the left.

0.00.18

Thus, 0.18% = 0.0018.

23. 23.19%

a) Replace the percent symbol with $\times 0.01$.

23.19×0.01

b) Move the decimal point two places to the left.

0.23.19

Thus, 23.19% = 0.2319.

25. 90%

    a) Replace the percent symbol with ×0.01.

       90 × 0.01

    b) Move the decimal point two places to the left.

       0.90.

        ↑⎵

    Thus, 90% = 0.9.

27. 10.8%

    a) Replace the percent symbol with ×0.01.

       10.8 × 0.01

    b) Move the decimal point two places to the left.

       0.10.8

        ↑⎵

    Thus, 10.8% = 0.108.

29. 45.8%

    a) Replace the percent symbol with ×0.01.

       45.8 × 0.01

    b) Move the decimal point two places to the left.

       0.45.8

        ↑⎵

    Thus, 45.8% = 0.458.

31. 0.47

    a) Move the decimal point two places to the right.

       0.47.

        ⎵↑

    b) Write a percent symbol: 47%

    Thus, 0.47 = 47%.

33. 0.03

    a) Move the decimal point two places to the right.

       0.03.

        ⎵↑

    b) Write a percent symbol: 3%

    Thus, 0.03 = 3%.

35. 1.00

    a) Move the decimal point two places to the right.

       1.00.

        ⎵↑

    b) Write a percent symbol: 100%

    Thus, 1.00 = 100%.

37. 0.334

    a) Move the decimal point two places to the right.

       0.33.4

        ⎵↑

    b) Write a percent symbol: 33.4%

    Thus, 0.334 = 33.4%.

39. 0.75

    a) Move the decimal point two places to the right.

       0.75.

        ⎵↑

    b) Write a percent symbol: 75%

    Thus, 0.75 = 75%.

41. 0.4

    a) Move the decimal point two places to the right.

       0.40.

        ⎵↑

    b) Write a percent symbol: 40%

    Thus, 0.4 = 40%.

43. 0.006

    a) Move the decimal point two places to the right.

       0.00.6

        ⎵↑

    b) Write a percent symbol: 0.6%

    Thus, 0.006 = 0.6%.

45. 0.017

    a) Move the decimal point two places to the right.

       0.01.7

        ⎵↑

    b) Write a percent symbol: 1.7%

    Thus, 0.017 = 1.7%.

47. 0.2718

    a) Move the decimal point two places to the right.

       0.27.18

        ⎵↑

    b) Write a percent symbol: 27.18%

    Thus, 0.2718 = 27.18%.

49. 0.0239

    a) Move the decimal point two places to the right.

       0.02.39

        ⎵↑

    b) Write a percent symbol: 2.39%

    Thus, 0.0239 = 2.39%.

51. 0.025

    a) Move the decimal point two places to the right.

       0.02.5

        ⎵↑

    b) Write a percent symbol: 2.5%

    Thus, 0.025 = 2.5%.

53. 0.684

a) Move the decimal point two places to the right.

0.68.4

b) Write a percent symbol: 68.4%

Thus, 0.684 = 68.4%.

55. To convert $\dfrac{100}{3}$ to a mixed numeral, we divide.

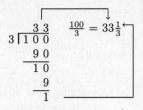

$\dfrac{100}{3} = 33\dfrac{1}{3}$

57. To convert $\dfrac{75}{8}$ to a mixed numeral, we divide.

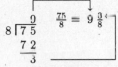

$\dfrac{75}{8} = 9\dfrac{3}{8}$

59. To convert $\dfrac{2}{3}$ to decimal notation, we divide.

```
  0.6 6
3 ) 2.0 0
    1 8
    ‾‾‾
      2 0
      1 8
      ‾‾‾
        2
```

Since 2 keeps reappearing as a remainder, the digits repeat and

$$\dfrac{2}{3} = 0.66\ldots \quad \text{or} \quad 0.\overline{6}.$$

61. To convert $\dfrac{5}{6}$ to decimal notation, we divide.

```
  0.8 3
6 ) 5.0 0
    4 8
    ‾‾‾
      2 0
      1 8
      ‾‾‾
        2
```

Since 2 keeps reappearing as a remainder, the digits repeat and

$$\dfrac{5}{6} = 0.833\ldots \quad \text{or} \quad 0.8\overline{3}.$$

63. Multiply by 100. (This is equivalent to moving the decimal point two places to the right.)

Exercise Set 7.2

1. We use the definition of percent as a ratio.

$$\dfrac{41}{100} = 41\%$$

3. We use the definition of percent as a ratio.

$$\dfrac{5}{100} = 5\%$$

5. We multiply by 1 to get 100 in the denominator.

$$\dfrac{2}{10} = \dfrac{2}{10} \cdot \dfrac{10}{10} = \dfrac{20}{100} = 20\%$$

7. We multiply by 1 to get 100 in the denominator.

$$\dfrac{3}{10} = \dfrac{3}{10} \cdot \dfrac{10}{10} = \dfrac{30}{100} = 30\%$$

9. $\dfrac{1}{2} = \dfrac{1}{2} \cdot \dfrac{50}{50} = \dfrac{50}{100} = 50\%$

11. Find decimal notation by division.

```
  0.6 2 5
8 ) 5.0 0 0
    4 8
    ‾‾‾
      2 0
      1 6
      ‾‾‾
        4 0
        4 0
        ‾‾‾
          0
```

$$\dfrac{5}{8} = 0.625$$

Convert to percent notation.

0.62.5

$$\dfrac{5}{8} = 62.5\%, \text{ or } 62\dfrac{1}{2}\%$$

13. $\dfrac{4}{5} = \dfrac{4}{5} \cdot \dfrac{20}{20} = \dfrac{80}{100} = 80\%$

15. Find decimal notation by division.

```
  0.6 6 6
3 ) 2.0 0 0
    1 8
    ‾‾‾
      2 0
      1 8
      ‾‾‾
        2 0
        1 8
        ‾‾‾
          2
```

We get a repeating decimal: $\dfrac{2}{3} = 0.66\overline{6}$

Convert to percent notation.

0.66.$\overline{6}$

$$\dfrac{2}{3} = 66.\overline{6}\%, \text{ or } 66\dfrac{2}{3}\%$$

17.
$$6 \overline{)\begin{array}{c} 0.1\,6\,6 \\ 1.0\,0\,0 \end{array}}$$
$$\underline{6}$$
$$4\,0$$
$$\underline{3\,6}$$
$$4\,0$$
$$\underline{3\,6}$$
$$4$$

We get a repeating decimal: $\frac{1}{6} = 0.16\overline{6}$

Convert to percent notation.

$$0.16.\overline{6}$$
$$\llcorner \uparrow$$

$\frac{1}{6} = 16.\overline{6}\%$, or $16\frac{2}{3}\%$

19. $\frac{4}{25} = \frac{4}{25} \cdot \frac{4}{4} = \frac{16}{100} = 16\%$

21. $\frac{1}{20} = \frac{1}{20} \cdot \frac{5}{5} = \frac{5}{100} = 5\%$

23. $\frac{17}{50} = \frac{17}{50} \cdot \frac{2}{2} = \frac{34}{100} = 34\%$

25. $\frac{9}{25} = \frac{9}{25} \cdot \frac{4}{4} = \frac{36}{100} = 36\%$

27. $\frac{3}{100} = 3\%$ Using the definition of percent as a ratio

29. $\frac{16}{25} = \frac{16}{25} \cdot \frac{4}{4} = \frac{64}{100} = 64\%$

31. $85\% = \frac{85}{100}$ Definition of percent

$$= \frac{5 \cdot 17}{5 \cdot 20}$$
$$= \frac{5}{5} \cdot \frac{17}{20} \left.\begin{array}{c} \\ \\ \\ \end{array}\right\} \text{Simplifying}$$
$$= \frac{17}{20}$$

33. $62.5\% = \frac{62.5}{100}$ Definition of percent

$$= \frac{62.5}{100} \cdot \frac{10}{10}$$ Multiplying by 1 to eliminate the decimal point in the numerator

$$= \frac{625}{1000}$$

$$= \frac{5 \cdot 125}{8 \cdot 125}$$
$$= \frac{5}{8} \cdot \frac{125}{125} \left.\begin{array}{c} \\ \\ \\ \end{array}\right\} \text{Simplifying}$$
$$= \frac{5}{8}$$

35. $33\frac{1}{3}\% = \frac{100}{3}\%$ Converting from mixed numeral to fractional notation

$$= \frac{100}{3} \times \frac{1}{100}$$ Definition of percent

$$= \frac{100 \cdot 1}{3 \cdot 100}$$ Multiplying

$$= \frac{1}{3} \cdot \frac{100}{100} \left.\begin{array}{c} \\ \\ \\ \end{array}\right\}$$
$$= \frac{1}{3} \qquad\qquad\quad \text{Simplifying}$$

37. $16.\overline{6}\% = 16\frac{2}{3}\%$ $\left(16.\overline{6} = 16\frac{2}{3}\right)$

$$= \frac{50}{3}\%$$ Converting from mixed numeral to fractional notation

$$= \frac{50}{3} \times \frac{1}{100}$$ Definition of percent

$$= \frac{50 \cdot 1}{3 \cdot 50 \cdot 2}$$ Multiplying

$$= \frac{1}{2 \cdot 3} \cdot \frac{50}{50} \left.\begin{array}{c} \\ \\ \\ \end{array}\right\}$$
$$= \frac{1}{6} \qquad\qquad\quad \text{Simplifying}$$

39. $7.25\% = \frac{7.25}{100} = \frac{7.25}{100} \cdot \frac{100}{100}$

$$= \frac{725}{10,000} = \frac{29 \cdot 25}{400 \cdot 25} = \frac{29}{400} \cdot \frac{25}{25}$$

$$= \frac{29}{400}$$

41. $0.8\% = \frac{0.8}{100} = \frac{0.8}{100} \cdot \frac{10}{10}$

$$= \frac{8}{1000} = \frac{1 \cdot 8}{125 \cdot 8} = \frac{1}{125} \cdot \frac{8}{8}$$

$$= \frac{1}{125}$$

43. $25\frac{3}{8}\% = \frac{203}{8}\%$

$$= \frac{203}{8} \times \frac{1}{100}$$ Definition of percent

$$= \frac{203}{800}$$

45. $78\frac{2}{9}\% = \frac{704}{9}\%$

$$= \frac{704}{9} \times \frac{1}{100}$$ Definition of percent

$$= \frac{4 \cdot 176 \cdot 1}{9 \cdot 4 \cdot 25}$$

$$= \frac{4}{4} \cdot \frac{176 \cdot 1}{9 \cdot 25}$$

$$= \frac{176}{225}$$

47. $64\frac{7}{11}\% = \frac{711}{11}\%$

$= \frac{711}{11} \times \frac{1}{100}$

$= \frac{711}{1100}$

49. $150\% = \frac{150}{100} = \frac{3 \cdot 50}{2 \cdot 50} = \frac{3}{2} \cdot \frac{50}{50} = \frac{3}{2}$

51. $0.0325\% = \frac{0.0325}{100} = \frac{0.0325}{100} \cdot \frac{10,000}{10,000} = \frac{325}{1,000,000} =$

$\frac{25 \cdot 13}{25 \cdot 40,000} = \frac{25}{25} \cdot \frac{13}{40,000} = \frac{13}{40,000}$

53. Note that $33.\overline{3}\% = 33\frac{1}{3}\%$ and proceed as in Exercise 35;

$33.\overline{3}\% = \frac{1}{3}$.

55. $55\% = \frac{55}{100} = \frac{5 \cdot 11}{5 \cdot 20} = \frac{5}{5} \cdot \frac{11}{20} = \frac{11}{20}$

57. $38\% = \frac{38}{100} = \frac{2 \cdot 19}{2 \cdot 50} = \frac{2}{2} \cdot \frac{19}{50} = \frac{19}{50}$

59. $11\% = \frac{11}{100}$

61. $24\% = \frac{24}{100} = \frac{4 \cdot 6}{4 \cdot 25} = \frac{4}{4} \cdot \frac{6}{25} = \frac{6}{25}$

63. $27.5\% = \frac{27.5}{100} = \frac{27.5}{100} \cdot \frac{10}{10} = \frac{275}{1000} = \frac{25 \cdot 11}{25 \cdot 40} = \frac{25}{25} \cdot \frac{11}{40} = \frac{11}{40}$

65. $\frac{1}{8} = 1 \div 8$

$$
\begin{array}{r}
0.1\,2\,5 \\
8\,\overline{)1.0\,0\,0} \\
\underline{8} \\
2\,0 \\
\underline{1\,6} \\
4\,0 \\
\underline{4\,0} \\
0
\end{array}
$$

$\frac{1}{8} = 0.125 = 12\frac{1}{2}\%$, or 12.5%

$\frac{1}{6} = 1 \div 6$

$$
\begin{array}{r}
0.1\,6\,6 \\
6\,\overline{)1.0\,0\,0} \\
\underline{6} \\
4\,0 \\
\underline{3\,6} \\
4\,0 \\
\underline{3\,6} \\
4
\end{array}
$$

We get a repeating decimal: $0.1\overline{6}$

0.16.$\overline{6}$ $0.1\overline{6} = 16.\overline{6}\%$

$\frac{1}{6} = 0.1\overline{6} = 16.\overline{6}\%$, or $16\frac{2}{3}\%$

$20\% = \frac{20}{100} = \frac{1}{5} \cdot \frac{20}{20} = \frac{1}{5}$

0.20. $20\% = 0.2$

$\frac{1}{5} = 0.2 = 20\%$

0.25. $0.25 = 25\%$

$25\% = \frac{25}{100} = \frac{1}{4} \cdot \frac{25}{25} = \frac{1}{4}$

$\frac{1}{4} = 0.25 = 25\%$

$33\frac{1}{3}\% = \frac{100}{3}\% = \frac{100}{3} \times \frac{1}{100} = \frac{100}{300} = \frac{1}{3} \cdot \frac{100}{100} = \frac{1}{3}$

0.33.$\overline{3}$ $33.\overline{3}\% = 0.33\overline{3}$, or $0.\overline{3}$

$\frac{1}{3} = 0.\overline{3} = 33\frac{1}{3}\%$, or $33.\overline{3}\%$

$37.5\% = \frac{37.5}{100} = \frac{37.5}{100} \cdot \frac{10}{10} = \frac{375}{1000} = \frac{3}{8} \cdot \frac{125}{125} = \frac{3}{8}$

0.37.5 $37.5\% = 0.375$

$\frac{3}{8} = 0.375 = 37\frac{1}{2}\%$, or 37.5%

$40\% = \frac{40}{100} = \frac{2}{5} \cdot \frac{20}{20} = \frac{2}{5}$

0.40. $40\% = 0.4$

$\frac{2}{5} = 0.4 = 40\%$

67. 0.50. $0.5 = 50\%$

$50\% = \frac{50}{100} = \frac{1}{2} \cdot \frac{50}{50} = \frac{1}{2}$

$\frac{1}{2} = 0.5 = 50\%$

$\frac{1}{3} = 1 \div 3$

$$
\begin{array}{r}
0.3 \\
3\,\overline{)1.0} \\
\underline{9} \\
1
\end{array}
$$

We get a repeating decimal: $0.\overline{3}$

0.33.$\overline{3}$ $0.\overline{3} = 33.\overline{3}\%$

$\frac{1}{3} = 0.\overline{3} = 33.\overline{3}\%,$ or $33\frac{1}{3}\%$

$25\% = \frac{25}{100} = \frac{25}{25} \cdot \frac{1}{4} = \frac{1}{4}$

0.25. $25\% = 0.25$

$\frac{1}{4} = 0.25 = 25\%$

$16\frac{2}{3}\% = \frac{50}{3}\% = \frac{50}{3} \times \frac{1}{100} = \frac{50 \cdot 1}{3 \cdot 2 \cdot 50} = \frac{50}{50} \cdot \frac{1}{6} = \frac{1}{6}$

$\frac{1}{6} = 1 \div 6$

$$\begin{array}{r} 0.1\,6 \\ 6\,\overline{\big)\,1.0\,0} \\ \underline{6} \\ 4\,0 \\ \underline{3\,6} \\ 4 \end{array}$$

We get a repeating decimal: $0.1\overline{6}$

$\frac{1}{6} = 0.1\overline{6} = 16\frac{2}{3}\%,$ or $16.\overline{6}\%$

0.12.5 $0.125 = 12.5\%$

$12.5\% = \frac{12.5}{100} = \frac{12.5}{100} \cdot \frac{10}{10} = \frac{125}{1000} = \frac{125}{125} \cdot \frac{1}{8} = \frac{1}{8}$

$\frac{1}{8} = 0.125 = 12.5\%,$ or $12\frac{1}{2}\%$

$\frac{3}{4} = \frac{3}{4} \cdot \frac{25}{25} = \frac{75}{100} = 75\%$

0.75. $75\% = 0.75$

$\frac{3}{4} = 0.75 = 75\%$

$0.8\overline{3} = 0.83.\overline{3}$ $0.8\overline{3} = 83.\overline{3}\%$

$83.\overline{3}\% = 83\frac{1}{3}\% = \frac{250}{3}\% = \frac{250}{3} \times \frac{1}{100} = \frac{5 \cdot 50}{3 \cdot 2 \cdot 50} =$
$\frac{5}{6} \cdot \frac{50}{50} = \frac{5}{6}$

$\frac{5}{6} = 0.8\overline{3} = 83.\overline{3}\%,$ or $83\frac{1}{3}\%$

$\frac{3}{8} = 3 \div 8$

$$\begin{array}{r} 0.3\,7\,5 \\ 8\,\overline{\big)\,3.0\,0\,0} \\ \underline{2\,4} \\ 6\,0 \\ \underline{5\,6} \\ 4\,0 \\ \underline{4\,0} \\ 0 \end{array}$$

$\frac{3}{8} = 0.375$

0.37.5 $0.375 = 37.5\%$

$\frac{3}{8} = 0.375 = 37.5\%,$ or $37\frac{1}{2}\%$

69. $10 \cdot x = 725$

$\frac{10 \cdot x}{10} = \frac{725}{10}$

$x = 72.5$

71. $0.05 \times b = 20$

$\frac{0.05 \times b}{0.05} = \frac{20}{0.05}$

$b = 400$

73. $\frac{24}{37} = \frac{15}{x}$

$24 \cdot x = 37 \cdot 15$ Finding cross-products

$x = \frac{37 \cdot 15}{24}$

$x = 23.125$

75. Use a calculator.

$\frac{41}{369} = 0.11.\overline{1} = 11.\overline{1}\%$

77. $\frac{14}{9}\% = \frac{14}{9} \times \frac{1}{100} = \frac{2 \cdot 7 \cdot 1}{9 \cdot 2 \cdot 50} = \frac{2}{2} \cdot \frac{7}{450} = \frac{7}{450}$

To find decimal notation for $\frac{7}{450}$ we divide.

$$\begin{array}{r} 0.0\,1\,5\,5 \\ 4\,5\,0\,\overline{\big)\,7.0\,0\,0\,0} \\ \underline{4\,5\,0} \\ 2\,5\,0\,0 \\ \underline{2\,2\,5\,0} \\ 2\,5\,0\,0 \\ \underline{2\,2\,5\,0} \\ 2\,5\,0 \end{array}$$

We get a repeating decimal: $\frac{14}{9}\% = 0.01\overline{5}$

Exercise Set 7.3

1. What is 32% of 78?

$$\begin{array}{ccccc} \downarrow & \downarrow & \downarrow & \downarrow & \downarrow \\ a & = 32\% & \times & 78 \end{array}$$

3. 89 is what percent of 99?

$$\begin{array}{cccc} \downarrow\downarrow & & \downarrow & \downarrow\downarrow \\ 89 = & n & \times & 99 \end{array}$$

5. 13 is 25% of what?

$$\begin{array}{ccccc} \downarrow & \downarrow & \downarrow & \downarrow & \downarrow \\ 13 = & 25\% & \times & & b \end{array}$$

7. What is 85% of 276?

Translate: $a = 85\% \cdot 276$

Solve: The letter is by itself. To solve the equation we convert 85% to decimal notation and multiply.

$$\begin{array}{r} 2\,7\,6 \\ \times\ 0.8\,5 \\ \hline 1\,3\,8\,0 \\ 2\,2\,0\,8\,0 \\ \hline a = 2\,3\,4.6\,0 \end{array} \quad (85\% = 0.85)$$

234.6 is 85% of 276. The answer is 234.6.

9. 150% of 30 is what?

Translate: $150\% \times 30 = a$

Solve: Convert 150% to decimal notation and multiply.

$$\begin{array}{r} 3\,0 \\ \times\ 1.5 \\ \hline 1\,5\,0 \\ 3\,0\,0 \\ \hline a = 4\,5.0 \end{array} \quad (150\% = 1.5)$$

150% of 30 is 45. The answer is 45.

11. What is 6% of $300?

Translate: $a = 6\% \cdot \$300$

Solve: Convert 6% to decimal notation and multiply.

$$\begin{array}{r} \$\,3\,0\,0 \\ \times\ 0.0\,6 \\ \hline a = \$\,1\,8.0\,0 \end{array} \quad (6\% = 0.06)$$

$18 is 6% of $300. The answer is $18.

13. 3.8% of 50 is what?

Translate: $3.8\% \cdot 50 = a$

Solve: Convert 3.8% to decimal notation and multiply.

$$\begin{array}{r} 5\,0 \\ \times\ 0.0\,3\,8 \\ \hline 4\,0\,0 \\ 1\,5\,0\,0 \\ \hline a = 1.9\,0\,0 \end{array} \quad (3.8\% = 0.038)$$

3.8% of 50 is 1.9. The answer is 1.9.

15. $39 is what percent of $50?

Translate: $39 = n \times 50$

Solve: To solve the equation we divide on both sides by 50 and convert the answer to percent notation.

$$n \cdot 50 = 39$$
$$\frac{n \cdot 50}{50} = \frac{39}{50}$$
$$n = 0.78 = 78\%$$

$39 is 78% of $50. The answer is 78%.

17. 20 is what percent of 10?

Translate: $20 = n \times 10$

Solve: To solve the equation we divide on both sides by 10 and convert the answer to percent notation.

$$n \cdot 10 = 20$$
$$\frac{n \cdot 10}{10} = \frac{20}{10}$$
$$n = 2 = 200\%$$

20 is 200% of 10. The answer is 200%.

19. What percent of $300 is $150?

Translate: $n \times 300 = 150$

Solve: $n \cdot 300 = 150$

$$\frac{n \cdot 300}{300} = \frac{150}{300}$$
$$n = 0.5 = 50\%$$

50% of $300 is $150. The answer is 50%.

21. What percent of 80 is 100?

Translate: $n \times 80 = 100$

Solve: $n \cdot 80 = 100$

$$\frac{n \cdot 80}{80} = \frac{100}{80}$$
$$n = 1.25 = 125\%$$

125% of 80 is 100. The answer is 125%.

23. 20 is 50% of what?

Translate: $20 = 50\% \times b$

Solve: To solve the equation we divide on both sides by 50%:

$$20 \div 50\% = b$$
$$20 \div 0.5 = b \quad (50\% = 0.5)$$
$$40 = b$$

$$\begin{array}{r} 4\,0. \\ 0.5_\wedge\overline{)2\,0.0_\wedge} \\ 2\,0\,0 \\ \hline 0 \\ 0 \\ \hline 0 \end{array}$$

20 is 50% of 40. The answer is 40.

25. 40% of what is $16?

Translate: $40\% \times b = 16$

Solve: To solve the equation we divide on both sides by 40%:

$b = 16 \div 40\%$

$b = 16 \div 0.4 \quad (40\% = 0.4)$

$b = 40$

$$0.4\,_\wedge\!\overline{)\,1\,6.\,0\,_\wedge}\frac{40.}{}$$

$$\begin{array}{r} 1\ 6\ 0 \\ \hline 0 \\ 0 \\ \hline 0 \end{array}$$

40% of $40 is $16. The answer is $40.

27. 56.32 is 64% of what?

Translate: $56.32 = 64\% \times b$

Solve: $56.32 \div 64\% = b$

$56.32 \div 0.64 = b$

$88 = b$

$$0.64\,_\wedge\!\overline{)\,5\,6.\,3\,2\,_\wedge}\frac{88.}{}$$

$$\begin{array}{r} 5\ 1\ 2\ 0 \\ \hline 5\ 1\ 2 \\ 5\ 1\ 2 \\ \hline 0 \end{array}$$

56.32 is 64% of 88. The answer is 88.

29. 70% of what is 14?

Translate: $70\% \times b = 14$

Solve: $b = 14 \div 70\%$

$b = 14 \div 0.7$

$b = 20$

$$0.7\,_\wedge\!\overline{)\,1\,4.\,0\,_\wedge}\frac{20.}{}$$

$$\begin{array}{r} 1\ 4\ 0 \\ \hline 0 \\ 0 \\ \hline 0 \end{array}$$

70% of 20 is 14. The answer is 20.

31. What is $62\frac{1}{2}\%$ of 10?

Translate: $a = 62\frac{1}{2}\% \times 10$

Solve: $a = 0.625 \times 10 \quad (62\frac{1}{2}\% = 0.625)$

$a = 6.25 \qquad$ Multiplying

6.25 is $62\frac{1}{2}\%$ of 10. The answer is 6.25.

33. What is 8.3% of $10,200?

Translate: $a = 8.3\% \times 10,200$

Solve: $a = 8.3\% \times 10,200$

$a = 0.083 \times 10,200 \quad (8.3\% = 0.083)$

$a = 846.6 \qquad$ Multiplying

$846.60 is 8.3% of $10,200. The answer is $846.60.

35. $\underline{0.09} = \dfrac{9}{100}$

2 decimal places   2 zeros

37. $0.\underline{875} = \dfrac{875}{1000}$

3 decimal places   3 zeros

$\dfrac{875}{1000} = \dfrac{7 \cdot 125}{8 \cdot 125} = \dfrac{7}{8} \cdot \dfrac{125}{125} = \dfrac{7}{8}$

Thus, $0.875 = \dfrac{875}{1000}$, or $\dfrac{7}{8}$.

39. $\dfrac{89}{100} \qquad 0.89.$

2 zeros   Move 2 places

$\dfrac{89}{100} = 0.89$

41. $\dfrac{3}{10} \qquad 0.3.$

1 zero   Move 1 place

$\dfrac{3}{10} = 0.3$

43. Estimate: Round 7.75% to 8% and $10,880 to $11,000. Then translate:

What is 8% of $11,000?

$\downarrow \qquad \downarrow \ \downarrow \ \downarrow \qquad \downarrow$

$a \quad = 8\% \times \quad 11,000$

We convert 8% to decimal notation and multiply.

$$\begin{array}{r} 1\ 1,0\ 0\ 0 \\ \times \quad 0.0\ 8 \\ \hline 8\ 8\ 0.0\ 0 \end{array} \quad (8\% = 0.08)$$

$880 is about 7.75% of $10,880. (Answers may vary.)

Calculate: First we translate.

What is 7.75% of $10,880?

$\downarrow \qquad \downarrow \ \downarrow \ \downarrow \qquad \downarrow$

$a \quad = 7.75\% \times \quad 10,880$

Use a calculator to multiply:

$0.0775 \times 10,880 = 843.2$

$843.20 is 7.75% of $10,880.

Exercise Set 7.4

1. What is 37% of 74?

$\quad\ \downarrow \qquad\quad \downarrow \qquad \downarrow$

amount   number   base

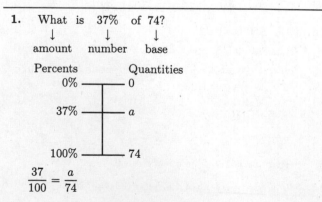

$\dfrac{37}{100} = \dfrac{a}{74}$

3.   4.3   is   what   percent   of 5.9?
         ↓   ↓   ↓
      amount   number   base

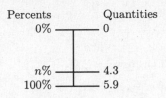

Percents   Quantities
   0% ──────── 0

   n% ──────── 4.3
  100% ──────── 5.9

$$\frac{n}{100} = \frac{4.3}{5.9}$$

5.   14   is   25%   of what?
         ↓   ↓   ↓
      amount   number   base

Percents   Quantities
   0% ──────── 0
  25% ──────── 14

  100% ──────── b

$$\frac{25}{100} = \frac{14}{b}$$

7.   What   is   76%   of 90?
          ↓   ↓   ↓
       amount   number   base

Percents   Quantities
   0% ──────── 0

  76% ──────── a
 100% ──────── 90

Translate: $\dfrac{76}{100} = \dfrac{a}{90}$

Solve: $76 \cdot 90 = 100 \cdot a$ Finding cross-products

$$\frac{76 \cdot 90}{100} = a$$ Dividing by 100

$$\frac{6840}{100} = a$$

$$68.4 = a$$ Simplifying

68.4 is 76% of 90. The answer is 68.4.

9.   70%   of 660   is what?
          ↓   ↓   ↓
       number   base   amount

Percents   Quantities
   0% ──────── 0

  70% ──────── a
 100% ──────── 660

Translate: $\dfrac{70}{100} = \dfrac{a}{660}$

Solve: $70 \cdot 660 = 100 \cdot a$ Finding cross-products

$$\frac{70 \cdot 660}{100} = a$$ Dividing by 100

$$\frac{46,200}{100} = a$$

$$462 = a$$ Simplifying

70% of 660 is 462. The answer is 462.

11.   What   is   4%   of 1000?
           ↓   ↓   ↓
        amount   number   base

Percents   Quantities
   0% ──────── 0
   4% ──────── a

 100% ──────── 1000

Translate: $\dfrac{4}{100} = \dfrac{a}{1000}$

Solve: $4 \cdot 1000 = 100 \cdot a$

$$\frac{4 \cdot 1000}{100} = a$$

$$\frac{4000}{100} = a$$

$$40 = a$$

40 is 4% of 1000. The answer is 40.

13.   4.8%   of 60   is what?
           ↓   ↓   ↓
        number   base   amount

Percents   Quantities
   0% ──────── 0
  4.8% ──────── a

 100% ──────── 60

Translate: $\dfrac{4.8}{100} = \dfrac{a}{60}$

Solve: $4.8 \cdot 60 = 100 \cdot a$

$$\frac{4.8 \cdot 60}{100} = a$$

$$\frac{288}{100} = a$$

$$2.88 = a$$

4.8% of 60 is 2.88. The answer is 2.88.

15. $24   is   what   percent of $96?

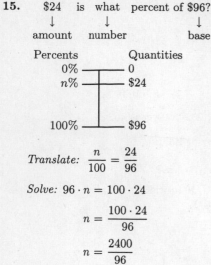

Translate: $\dfrac{n}{100} = \dfrac{24}{96}$

Solve: $96 \cdot n = 100 \cdot 24$

$n = \dfrac{100 \cdot 24}{96}$

$n = \dfrac{2400}{96}$

$n = 25$

$24 is 25% of $96. The answer is 25%.

17.   102   is   what   percent of 100?

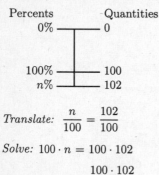

Translate: $\dfrac{n}{100} = \dfrac{102}{100}$

Solve: $100 \cdot n = 100 \cdot 102$

$n = \dfrac{100 \cdot 102}{100}$

$n = \dfrac{10,200}{100}$

$n = 102$

102 is 102% of 100. The answer is 102%.

19.   What   percent of $480 is $120?

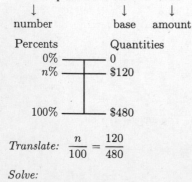

Translate: $\dfrac{n}{100} = \dfrac{120}{480}$

Solve:

$480 \cdot n = 100 \cdot 120$

$n = \dfrac{100 \cdot 120}{480}$

$n = \dfrac{12,000}{480}$

$n = 25$

25% of $480 is $120. The answer is 25%.

21.   What   percent of 160 is   150?

↓ number   ↓ base   ↓ amount

Percents   Quantities

0% ——— 0

n% ——— 150
100% ——— 160

Translate: $\dfrac{n}{100} = \dfrac{150}{160}$

Solve: $160 \cdot n = 100 \cdot 150$

$n = \dfrac{100 \cdot 150}{160}$

$n = \dfrac{15,000}{160}$

$n = 93.75$

93.75% of 160 is 150. The answer is 93.75%.

23. $18   is   25%   of what?

↓ amount   ↓ number   ↓ base

Percents   Quantities

0% ——— 0
25% ——— $18

100% ——— b

Translate: $\dfrac{25}{100} = \dfrac{18}{b}$

Solve: $25 \cdot b = 100 \cdot 18$

$b = \dfrac{100 \cdot 18}{25}$

$b = \dfrac{1800}{25}$

$b = 72$

$18 is 25% of $72. The answer is $72.

25.   60%   of what is $54?

↓ number   ↓ base   ↓ amount

Percents   Quantities

0% ——— 0

60% ——— 54

100% ——— b

Translate: $\dfrac{60}{100} = \dfrac{54}{b}$

Solve: $60 \cdot b = 100 \cdot 54$

$$b = \frac{100 \cdot 54}{60}$$

$$b = \frac{5400}{60}$$

$$b = 90$$

60% of 90 is 54. The answer is 90.

27. $\underset{\underset{\text{amount}}{\downarrow}}{65.12}$ is $\underset{\underset{\text{number}}{\downarrow}}{74\%}$ of $\underset{\underset{\text{base}}{\downarrow}}{\text{what?}}$

Percents   Quantities
0% ——— 0

74% ——— 65.12
100% ——— b

Translate: $\dfrac{74}{100} = \dfrac{65.12}{b}$

Solve: $74 \cdot b = 100 \cdot 65.12$

$$b = \frac{100 \cdot 65.12}{74}$$

$$b = \frac{6512}{74}$$

$$b = 88$$

65.12 is 74% of 88. The answer is 88.

29. $\underset{\underset{\text{number}}{\downarrow}}{80\%}$ of $\underset{\underset{\text{base}}{\downarrow}}{\text{what}}$ is $\underset{\underset{\text{amount}}{\downarrow}}{16?}$

Percents   Quantities
0% ——— 0

80% ——— 16
100% ——— b

Translate: $\dfrac{80}{100} = \dfrac{16}{b}$

Solve: $80 \cdot b = 100 \cdot 16$

$$b = \frac{100 \cdot 16}{80}$$

$$b = \frac{1600}{80}$$

$$b = 20$$

80% of 20 is 16. The answer is 20.

31. $\underset{\underset{\text{amount}}{\downarrow}}{\text{What}}$ is $\underset{\underset{\text{number}}{\downarrow}}{62\frac{1}{2}\%}$ of $\underset{\underset{\text{base}}{\downarrow}}{40?}$

Percents   Quantities
0% ——— 0

$62\frac{1}{2}\%$ ——— a

100% ——— 40

Translate: $\dfrac{62\frac{1}{2}}{100} = \dfrac{a}{40}$

Solve: $62\dfrac{1}{2} \cdot 40 = 100 \cdot a$

$$\frac{125}{2} \cdot \frac{40}{1} = 100 \cdot a$$

$$2500 = 100 \cdot a$$

$$\frac{2500}{100} = a$$

$$25 = a$$

25 is $62\frac{1}{2}\%$ of 40. The answer is 25.

33. $\underset{\underset{\text{amount}}{\downarrow}}{\text{What}}$ is $\underset{\underset{\text{number}}{\downarrow}}{9.4\%}$ of $\underset{\underset{\text{base}}{\downarrow}}{\$8300?}$

Percents   Quantities
0% ——— 0
9.4% ——— a

100% ——— 8300

Translate: $\dfrac{9.4}{100} = \dfrac{a}{8300}$

Solve: $9.4 \cdot 8300 = 100 \cdot a$

$$\frac{9.4 \cdot 8300}{100} = a$$

$$\frac{78,020}{100} = a$$

$$780.2 = a$$

$780.20 is 9.4% of $8300. The answer is $780.20.

35. $\dfrac{x}{188} = \dfrac{2}{47}$

$$47 \cdot x = 188 \cdot 2$$

$$x = \frac{188 \cdot 2}{47}$$

$$x = \frac{4 \cdot 47 \cdot 2}{47}$$

$$x = 8$$

37. $\dfrac{4}{7} = \dfrac{x}{14}$

$$4 \cdot 14 = 7 \cdot x$$

$$\frac{4 \cdot 14}{7} = x$$

$$\frac{4 \cdot 2 \cdot 7}{7} = x$$

$$8 = x$$

39.

$$\frac{5000}{t} = \frac{3000}{60}$$

$$5000 \cdot 60 = 3000 \cdot t$$

$$\frac{5000 \cdot 60}{3000} = t$$

$$\frac{5 \cdot 1000 \cdot 3 \cdot 20}{3 \cdot 1000} = t$$

$$100 = t$$

41. Estimate: Round 8.85% to 9%, and $12,640 to $12,600.

What is   9%   of $12,600?

↓   ↓   ↓

amount   number   base

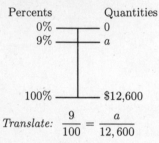

Percents	Quantities
0%	0
9%	a
100%	$12,600

Translate: $\dfrac{9}{100} = \dfrac{a}{12,600}$

Solve: $9 \cdot 12,600 = 100 \cdot a$

$$\frac{9 \cdot 12,600}{100} = a$$

$$\frac{113,400}{100} = a$$

$$1134 = a$$

$1134 is about 8.85% of $12,640. (Answers may vary.)

Calculate:

What is 8.85% of $12,640?

↓   ↓   ↓

amount   number   base

Percents	Quantities
0%	0
8.85%	a
100%	$12,640

Translate: $\dfrac{8.85}{100} = \dfrac{a}{12,640}$

Solve: $8.85 \cdot 12,640 = 100 \cdot a$

$$\frac{8.85 \cdot 12,640}{100} = a$$

$$\frac{111,864}{100} = a \qquad \text{Use a calculator to multiply and divide.}$$

$$1118.64 = a$$

$1118.64 is 8.85% of $12,640.

Exercise Set 7.5

1. We first find how many bowlers would be expected to be left-handed.

Method 1: Solve using an equation.

Restate: What is 17% of 160?

↓   ↓   ↓   ↓   ↓

Translate: a = 17% × 160

Solve: We convert 17% to decimal notation and multiply.

$$\begin{array}{r} 1\,6\,0 \\ \times\ 0.1\,7 \\ \hline 1\ 1\ 2\ 0 \\ 1\ 6\ 0\ 0 \\ \hline 2\,7.2\,0 \end{array} \qquad (17\% = 0.17)$$

You would expect 27 bowlers to be left-handed.

Method 2: Solve using a proportion.

Restate: What is   17%   of 160?

↓   ↓   ↓

amount   number   base

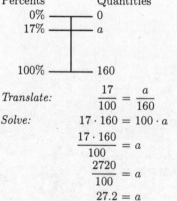

Percents	Quantities
0%	0
17%	a
100%	160

Translate: $\dfrac{17}{100} = \dfrac{a}{160}$

Solve: $17 \cdot 160 = 100 \cdot a$

$$\frac{17 \cdot 160}{100} = a$$

$$\frac{2720}{100} = a$$

$$27.2 = a$$

You would expect 27 bowlers to be left-handed.

We can subtract to find the number of bowlers who would not be left-handed.

$$160 - 27 = 133$$

You would expect that 133 bowlers would not be left-handed.

3. We first find the number of moviegoers in the 12-29 age group.

Method 1: Solve using an equation.

Restate: What is 67% of 800?

↓   ↓   ↓   ↓   ↓

Translate: a = 67% × 800

Solve: We convert 67% to decimal notation and multiply.

$$\begin{array}{r} 8\,0\,0 \\ \times\ 0.6\,7 \\ \hline 5\ 6\ 0\ 0 \\ 4\ 8\ 0\ 0\ 0 \\ \hline 5\,3\,6.0\,0 \end{array} \qquad (67\% = 0.67)$$

Thus, 536 were in the 12-29 age group.

Method 2: Solve using a proportion.

Restate: What is   67%   of 800?
              ↓   ↓   ↓
           amount number   base

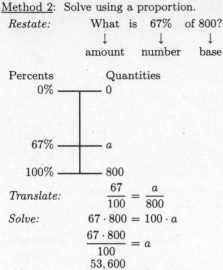

Percents   Quantities
   0% ——————— 0

  67% ——————— a

 100% ——————— 800

Translate: $\dfrac{67}{100} = \dfrac{a}{800}$

Solve: $67 \cdot 800 = 100 \cdot a$

            $\dfrac{67 \cdot 800}{100} = a$

            $\dfrac{53,600}{100} = a$

               $536 = a$

Thus, 536 were in the 12-29 age group.

We subtract to find the number of moviegoers who were not in this age group.

$$800 - 536 = 264$$

264 moviegoers were not in the 12-29 age group.

5. We first find the percent that are hits.

Method 1: Solve using an equation.

Restate: 13 is what percent of 40?

Translate: 13 = n × 40

Solve: To solve the equation we divide on both sides by 40.

$$13 \div 40 = n$$
$$0.325 = n$$
$$32.5\% = n$$

Thus, 32.5% are hits.

Method 2: Solve using a proportion.

Restate: 13   is what percent of 40?
              ↓   ↓   ↓
           amount   number   base

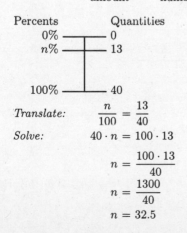

Percents   Quantities
   0% ——————— 0
  n% ——————— 13

 100% ——————— 40

Translate: $\dfrac{n}{100} = \dfrac{13}{40}$

Solve: $40 \cdot n = 100 \cdot 13$

            $n = \dfrac{100 \cdot 13}{40}$

            $n = \dfrac{1300}{40}$

            $n = 32.5$

Thus, 32.5% are hits.

We can subtract to find the percent that are not hits.

$$100\% - 32.5\% = 67.5\%$$

Thus, 67.5% are not hits.

7. We first find how many milliliters are acid.

Method 1: Solve using an equation.

Restate: What is 3% of 680?
              ↓   ↓ ↓ ↓   ↓
Translate: a = 3% × 680

Solve: We convert 3% to decimal notation and multiply.

$$\begin{array}{r} 6\,8\,0 \\ \times\ 0.0\,3 \\ \hline 2\,0.4\,0 \end{array} \quad (3\% = 0.03)$$

Thus, 20.4 milliliters are acid.

Method 2: Solve using a proportion.

Restate: What is   3%   of 680?
              ↓   ↓   ↓
           amount   number   base

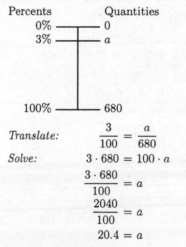

Percents   Quantities
   0% ——————— 0
   3% ——————— a

 100% ——————— 680

Translate: $\dfrac{3}{100} = \dfrac{a}{680}$

Solve: $3 \cdot 680 = 100 \cdot a$

            $\dfrac{3 \cdot 680}{100} = a$

            $\dfrac{2040}{100} = a$

            $20.4 = a$

Thus, 20.4 milliliters are acid.

We subtract to find how many milliliters are water.

$$680 - 20.4 = 659.6$$

Thus, 659.6 milliliters are water.

9. Method 1: Solve using an equation.

Restate: 2190 is what percent of 8760?

Translate: 2190 = n × 8760

Solve: We divide on both sides by 8760.

$$2190 \div 8760 = n$$
$$0.25 = n$$
$$25\% = n$$

Thus, 2190 is 25% of 8760.

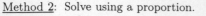

Method 2: Solve using a proportion.

Restate: 2190   is what percent of 8760?
                      ↓   ↓   ↓
                  amount   number   base

Percents   Quantities
  0% ——————— 0
  n% ——————— 2190

 100% ——————— 8760

Translate: $\dfrac{n}{100} = \dfrac{2190}{8760}$

Solve: $8760 \cdot n = 100 \cdot 2190$

$$n = \frac{100 \cdot 2190}{8760}$$

$$n = \frac{219,000}{8760}$$

$$n = 25$$

Thus, 2190 is 25% of 8760.

11. First we find the total weight of the nuts.

$$1800 \text{ lb} + 1500 \text{ lb} + 700 \text{ lb} = 4000 \text{ lb}$$

Then we find what percent are peanuts.

Method 1: Solve using an equation.

Restate: What percent of 4000 is 1800?
                      ↓   ↓   ↓   ↓   ↓
Translate: n   × 4000 = 1800

Solve: We divide on both sides by 4000.

$$n = 1800 \div 4000$$
$$n = 0.45$$
$$n = 45\%$$

Thus, 45% are peanuts.

Method 2: Solve using a proportion.

Restate: What percent of 4000 is   1800?
                      ↓   ↓   ↓
                  number   base   amount

Percents   Quantities
  0% ——————— 0

  n% ——————— 1800

 100% ——————— 4000

Translate: $\dfrac{n}{100} = \dfrac{1800}{4000}$

Solve: $4000 \cdot n = 100 \cdot 1800$

$$n = \frac{100 \cdot 1800}{4000}$$

$$n = \frac{180,000}{4000}$$

$$n = 45$$

Thus, 45% are peanuts.

Next we find the percent that are cashews.

Method 1: Solve using an equation.

Restate: What percent of 4000 is 1500?
                      ↓   ↓   ↓   ↓   ↓
Translate: n   × 4000 = 1500

Solve: We divide on both sides by 4000.

$$n = 1500 \div 4000$$
$$n = 0.375$$
$$n = 37.5\%$$

Thus, 37.5% are cashews.

Method 2: Solve using a proportion.

Restate: What percent of 4000 is   1500?
                      ↓   ↓   ↓
                  number   base   amount

Percents   Quantities
  0% ——————— 0

  n% ——————— 1500

 100% ——————— 4000

Translate: $\dfrac{n}{100} = \dfrac{1500}{4000}$

Solve: $4000 \cdot n = 100 \cdot 1500$

$$n = \frac{100 \cdot 1500}{4000}$$

$$n = \frac{150,000}{4000}$$

$$n = 37.5$$

Thus, 37.5% are cashews.

Finally, we find the percent that are almonds. First we observe that peanuts and cashews make up 45% + 37.5%, or 82.5%, of the total weight. Then we subtract.

$$100\% - 82.5\% = 17.5\%$$

Thus, 17.5% are almonds.

13.

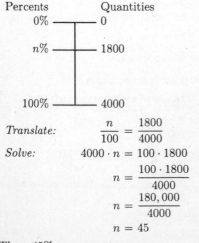

a) First find the increase by subtracting.

$$\begin{array}{rl} 2\,1\,6 & \text{New amount} \\ -\,2\,0\,0 & \text{Original amount} \\ \hline 1\,6 & \text{Increase} \end{array}$$

The increase is $16.

b) Now we ask: $16 is what percent of $200 (the original amount)?

Method 1: Solve using an equation.

$16 is what percent of $200?

$$16 = n \times 200 \quad \text{Translating}$$

$16 \div 200 = n \quad$ Dividing on both sides by 200

$0.08 = n$

$8\% = n$

The percent of increase was 8%.

Method 2: Solve using a proportion.

$16 is what percent of $200?

↓   ↓   ↓

amount   number   base

$$\frac{n}{100} = \frac{16}{200} \quad \text{Translating}$$

$$200 \cdot n = 16 \cdot 100 \quad \text{Finding cross-products}$$

$$n = \frac{16 \cdot 100}{200} \quad \text{Dividing by 200}$$

$$n = \frac{1600}{200}$$

$$n = 8$$

The percent of increase was 8%.

15.

$70	
$56	$14

100%	
	?%

a) First find the decrease by subtracting.

$$\begin{array}{rl} 7\,0 & \text{Original price} \\ -\,5\,6 & \text{New price} \\ \hline 1\,4 & \text{Decrease} \end{array}$$

The decrease was $14.

b) Now we ask: $14 is what percent of $70 (the original price)?

Method 1: Solve using an equation.

$14 is what percent of $70?

$$14 = n \times 70 \quad \text{Translating}$$

$14 \div 70 = n \quad$ Dividing on both sides by 70

$0.2 = n$

$20\% = n$

The percent of decrease was 20%.

Method 2: Solve using a proportion.

$14 is what percent of $70?

↓   ↓   ↓

amount   number   base

$$\frac{n}{100} = \frac{14}{70} \quad \text{Translating}$$

$$70 \cdot n = 100 \cdot 14 \quad \text{Finding cross-products}$$

$$n = \frac{100 \cdot 14}{70} \quad \text{Dividing by 70}$$

$$n = \frac{1400}{70}$$

$$n = 20$$

The percent of decrease was 20%.

17.

$18,600	$?

100%	5%

a) First we find the increase. We ask:

What is 5% of $18,600?

Method 1: Solve using an equation.

What is 5% of $18,600?

$$a = 5\% \times 18,600$$

We convert to decimal notation and multiply:

$$a = 0.05 \times 18,600 = 930$$

The increase is $930.

Method 2: Solve using a proportion.

What is 5% of $18,600?

↓   ↓   ↓

amount   number   base

$$\frac{5}{100} = \frac{a}{18,600}$$

$$5 \cdot 18,600 = 100 \cdot a \quad \text{Finding cross-products}$$

$$\frac{5 \cdot 18,600}{100} = a$$

$$\frac{93,000}{100} = a$$

$$930 = a$$

The increase is $930.

b) The new salary is $18,600 + $930 = $19,530.

19.

$18,000	

	$?

100%	

70%	30%

a) First we find the decrease. We ask:

What is 30% of $18,000?

Method 1: Solve using an equation.

What is 30% of $18,000?

$$a = 30\% \times 18,000$$

We convert to decimal notation and multiply:

$$a = 0.3 \times 18,000 = 5400$$

The decrease is $5400.

Method 2: Solve using a proportion.

   What   is   30%   of $18,000?
    ↓   ↓   ↓
 amount   number   base

$$\frac{30}{100} = \frac{a}{18,000}$$

$30 \cdot 18,000 = 100 \cdot a$ Finding cross-products

$$\frac{30 \cdot 18,000}{100} = a$$

$$\frac{540,000}{100} = a$$

$$5400 = a$$

The decrease is $5400.

b) The value one year later is $18,000 - $5400 = $12,600.

21. a) To find the population in 1996 we first find the increase. We ask:

   What is 1.6% of 5.6 billion?

Method 1: Solve using an equation.

 What is 1.6% of 5.6 billion?
  ↓   ↓   ↓   ↓   ↓
  a = 1.6% ×   5.6

We convert to decimal notation and multiply:

$$a = 0.016 \times 5.6 = 0.0896$$

The increase is 0.0896 billion.

Method 2: Solve using a proportion.

   What   is   1.6%   of 5.6 billion?
    ↓   ↓   ↓
 amount   number   base

$$\frac{1.6}{100} = \frac{a}{5.6}$$

$1.6 \cdot 5.6 = 100 \cdot a$ Finding cross-products

$$\frac{1.6 \cdot 5.6}{100} = a$$

$$\frac{8.96}{100} = a$$

$$0.0896 = a$$

The increase is 0.0896 billion.

b) The population in 1996 will be 5.6 billion + 0.0896 billion = 5.6896 billion ≈ 5.7 billion.

a) To find the population in 1997 we first find the increase. We ask:

   What is 1.6% of 5.7 billion?

Method 1: Solve using an equation.

 What is 1.6% of 5.7 billion?
  ↓   ↓   ↓   ↓   ↓
  a = 1.6% ×   5.7

We convert to decimal notation and multiply:

$$a = 0.016 \times 5.7 = 0.0912$$

The increase is 0.0912 billion.

Method 2: Solve using a proportion.

   What   is   1.6%   of 5.7 billion?
    ↓   ↓   ↓
 amount   number   base

$$\frac{1.6}{100} = \frac{a}{5.7}$$

$1.6 \cdot 5.7 = 100 \cdot a$ Finding cross-products

$$\frac{1.6 \cdot 5.7}{100} = a$$

$$\frac{9.12}{100} = a$$

$$0.0912 = a$$

The increase is 0.0912 billion.

b) The population in 1997 will be 5.7 billion + 0.0912 billion = 5.7912 billion ≈ 5.8 billion.

a) To find the population in 1998 we first find the increase. We ask:

   What is 1.6% of 5.8 billion?

Method 1: Solve using an equation.

 What is 1.6% of 5.8 billion?
  ↓   ↓   ↓   ↓   ↓
  a = 1.6% ×   5.8

We convert to decimal notation and multiply:

$$a = 0.016 \times 5.8 = 0.0928$$

The increase is 0.0928 billion.

Method 2: Solve using a proportion.

   What   is   1.6%   of 5.8 billion?
    ↓   ↓   ↓
 amount   number   base

$$\frac{1.6}{100} = \frac{a}{5.8}$$

$1.6 \cdot 5.8 = 100 \cdot a$ Finding cross-products

$$\frac{1.6 \cdot 5.8}{100} = a$$

$$\frac{9.28}{100} = a$$

$$0.0928 = a$$

The increase is 0.0928 billion.

b) The population in 1998 will be 5.8 billion + 0.0928 billion = 5.8928 billion ≈ 5.9 billion.

23. Since the car depreciates 30% in the first year, its value after the first year is 100% − 30%, or 70%, of the original value. To find the decrease in value, we ask:

   $11,480 is 70% of what?

Method 1: Solve using an equation.

$11,480 is 70% of what?
↓   ↓   ↓   ↓   ↓
$11,480 = 70% × b

$$11,480 = 0.7 \times b$$
$$11,480 \div 0.7 = b \qquad \text{Dividing on both sides by 0.7}$$
$$16,400 = b$$

The original cost was $16,400.

Method 2: Solve using a proportion.

$11,480 is   70%   of what?
↓   ↓   ↓
amount   number   base

$$\frac{70}{100} = \frac{11,480}{b}$$
$$70 \cdot b = 100 \cdot 11,480$$
$$b = \frac{100 \cdot 11,480}{70}$$
$$b = \frac{1,148,000}{70}$$
$$b = 16,400$$

The original cost was $16,400.

25. To find the maximum heartrate for a 25 year old person, we first subtract 25 from 220:

$$220 - 25 = 195$$

Then we ask:

What is 85% of 195?

Method 1: Solve using an equation.

What is 85% of 195?
↓   ↓   ↓   ↓   ↓
a = 85% × 195

We convert to decimal notation and multiply:

$$a = 0.85 \times 195 = 165.75 \approx 166$$

The maximum heartrate for a 25 year old person is 166 beats per minute.

Method 2: Solve using a proportion.

What   is   85%   of 195?
↓   ↓   ↓
amount   number   base

$$\frac{85}{100} = \frac{a}{195}$$
$$85 \cdot 195 = 100 \cdot a$$
$$\frac{85 \cdot 195}{100} = a$$
$$\frac{16,575}{100} = a$$
$$165.75 = a$$
$$166 \approx a$$

The maximum heartrate for a 25 year old person is 166 beats per minute.

To find the maximum heartrate for a 36 year old person, we first subtract 36 from 220:

$$220 - 36 = 184$$

Then we ask:

What is 85% of 184?

Method 1: Solve using an equation.

What is 85% of 184?
↓   ↓   ↓   ↓   ↓
a = 85% ×   184

We convert to decimal notation and multiply:

$$a = 0.85 \times 184 = 156.4 \approx 156$$

The maximum heartrate for a 36 year old person is 156 beats per minute.

Method 2: Solve using a proportion.

What   is   85%   of 184?
↓   ↓   ↓
amount   number   base

$$\frac{85}{100} = \frac{a}{184}$$
$$85 \cdot 184 = 100 \cdot a$$
$$\frac{85 \cdot 184}{100} = a$$
$$\frac{15,640}{100} = a$$
$$156.4 = a$$
$$156 \approx a$$

The maximum heartrate for a 36 year old person is 156 beats per minute.

To find the maximum heartrate for a 48 year old person, we first subtract 48 from 220:

$$220 - 48 = 172$$

Then we ask:

What is 85% of 172?

Method 1: Solve using an equation.

What is 85% of 172?
↓   ↓   ↓   ↓   ↓
a = 85% ×   172

We convert to decimal notation and multiply:

$$a = 0.85 \times 172 = 146.2 \approx 146$$

The maximum heartrate for a 48 year old person is 146 beats per minute.

Method 2: Solve using a proportion.

What   is   85%   of 172?
↓   ↓   ↓
amount   number   base

$$\frac{85}{100} = \frac{a}{172}$$
$$85 \cdot 172 = 100 \cdot a$$
$$\frac{85 \cdot 172}{100} = a$$
$$\frac{14,620}{100} = a$$
$$146.2 = a$$
$$146 \approx a$$

The maximum heartrate for a 48 year old person is 146 beats per minute.

To find the maximum heartrate for a 60 year old person, we first subtract 60 from 220:

$$220 - 60 = 160$$

Then we ask:

What is 85% of 160?

Method 1: Solve using an equation.

What is 85% of 160?
$$\downarrow \quad \downarrow \quad \downarrow \quad \downarrow \quad \downarrow$$
$$a \quad = 85\% \times \quad 160$$

We convert to decimal notation and multiply:

$$a = 0.85 \times 160 = 136$$

The maximum heartrate for a 60 year old person is 136 beats per minute.

Method 2: Solve using a proportion.

What   is   85%   of 160?
$$\downarrow \qquad\quad \downarrow \qquad\quad \downarrow$$
amount   number   base

$$\frac{85}{100} = \frac{a}{160}$$
$$85 \cdot 160 = 100 \cdot a$$
$$\frac{85 \cdot 160}{100} = a$$
$$\frac{13,600}{100} = a$$
$$136 = a$$

The maximum heartrate for a 60 year old person is 136 beats per minute.

To find the maximum heartrate for a 76 year old person, we first subtract 76 from 220:

$$220 - 76 = 144$$

Then we ask:

What is 85% of 144?

Method 1: Solve using an equation.

What is 85% of 144?
$$\downarrow \quad \downarrow \quad \downarrow \quad \downarrow \quad \downarrow$$
$$a \quad = 85\% \times \quad 144$$

We convert to decimal notation and multiply:

$$a = 0.85 \times 144 = 122.4 \approx 122$$

The maximum heartrate for a 76 year old person is 122 beats per minute.

Method 2: Solve using a proportion.

What   is   85%   of 144?
$$\downarrow \qquad\quad \downarrow \qquad\quad \downarrow$$
amount   number   base

$$\frac{85}{100} = \frac{a}{144}$$
$$85 \cdot 144 = 100 \cdot a$$
$$\frac{85 \cdot 144}{100} = a$$
$$\frac{12,240}{100} = a$$
$$122.4 = a$$
$$122 \approx a$$

The maximum heartrate for a 76 year old person is 122 beats per minute.

27. First we find the area of the strike zone:

$$A = l \cdot w = 40 \text{ in.} \times 15 \text{ in.} = 600 \text{ sq in.}$$

When a 2-in. border is added to the outside of the strike zone, the dimensions of the larger zone are 19 in. by 44 in. The area of this zone is:

$$A = l \cdot w = 44 \text{ in.} \times 19 \text{ in.} = 836 \text{ sq in.}$$

To find the percent of increase, we first find the amount of the increase by subtracting:

$$836 \text{ sq in.} - 600 \text{ sq in.} = 236 \text{ sq in.}$$

Then we ask:

236 is what percent of 600?

Method 1: Solve using an equation.

236 is what percent of 600?
$$\downarrow \downarrow \qquad \downarrow \qquad \downarrow \downarrow$$
$$236 = \qquad n \qquad \times 600$$

To solve we divide on both sides by 600.

$$n = 236 \div 600 = 0.39\overline{3} = 39.\overline{3}\%$$

The area of the strike zone is increased by $39.\overline{3}\%$.

Method 2: Solve using a proportion.

236   is what percent of 600?
$$\downarrow \qquad\qquad \downarrow \qquad\qquad \downarrow$$
amount   number   base

$$\frac{n}{100} = \frac{236}{600}$$
$$600 \cdot n = 100 \cdot 236$$
$$n = \frac{100 \cdot 236}{600}$$
$$n = 39.\overline{3} = 39.\overline{3}\%$$

The area of the strike zone is increased by $39.\overline{3}\%$.

29. a) We find $1000 increased by 15%. First we find the amount of the increase.

What is 15% of $1000?
$$\downarrow \quad \downarrow \quad \downarrow \quad \downarrow \quad \downarrow$$
$$a \quad = 15\% \times \quad 1000$$

We multiply.

$$0.15 \times 1000 = 150$$

The increase is $150. Then $1000 increased by 15% is $1000 + $150, or $1150.

Next we find $1150 decreased by 15%. First we find the amount of the decrease.

What is 15% of $1150?

$$a \quad = 15\% \times 1150$$

We multiply.

$$0.15 \times 1150 = 172.5.$$

The decrease is $172.50. Then $1150 decreased by 15% is $1150 - $172.50, or $977.50.

b) We find $1000 decreased by 15%. First we find the amount of the decrease, or 15% of $1000. In part (a) we found this to be $150. Then $1000 decreased by 15% is $1000 - $150, or $850.

Next we find $850 increased by 15%. First we find the amount of the increase.

What is 15% of $850?

$$a \quad = 15\% \times 850$$

We multiply.

$$0.15 \times 850 = 127.5.$$

The increase is $127.50. Then $850 increased by 15% is $850 + $127.50, or $977.50.

Neither (a) nor (b) is higher. They are the same.

31. a) Note that 4 ft, 8 in. = 48 in. + 8 in. = 56 in.

56 inches is 84.4% of what?

$$56 \quad = 84.4\% \times b$$

To solve the equation we divide on both sides by 84.4%:

$$56 \div 84.4\% = b$$
$$56 \div 0.844 = b$$
$$66 \approx b$$

```
                  6 6 .3
0. 8 4 4⌐| 5 6 .0 0 0 ∧0
           5 0 6 4
           5 3 6 0
           5 0 6 4
             2 9 6 0
             2 5 3 2
               4 2 8
```

Her final adult height is about 66 in., or 5 ft, 6 in.

Exercise Set 7.6

1. a) We find the sales tax using an equation.

Sales tax = Sales tax rate × Purchase price

$$t \quad = \quad 8.25\% \quad \times \quad 248$$

This tells us what to do. We multiply.

```
      2 4 8
  × 0. 0 8 2 5
      1 2 4 0
      4 9 6 0
    1 9 8 4 0 0
    2 0. 4 6 0 0
```

The sales tax is $20.46.

b) The total price is purchase price plus sales tax, or

```
    2 4 8.0 0
  +   2 0.4 6
    2 6 8.4 6
```

The total price is $268.46.

3. a) We find the sales tax using an equation.

Sales tax = Sales tax rate × Purchase price

$$t \quad = \quad 6\% \quad \times \quad 189.95$$

This tells us what to do. We multiply.

```
    1 8 9.9 5
  ×     0. 0 6
  1 1. 3 9 7 0
```

The sales tax is $11.397, or about $11.40.

b) The total price is purchase price plus sales tax, or

```
    1 8 9.9 5
  +   1 1.4 0
    2 0 1.3 5
```

The total price is $201.35.

5. *Restate:*   Sales tax   is   what percent   of   purchase price?

Translate: 48   =   r   ×   960

Solve: We divide on both sides by 960.

$$48 \div 960 = r$$
$$0.05 = r$$
$$5\% = r$$

The sales tax rate is 5%.

7. *Restate:*   Sales tax   is   what percent   of   purchase price?

Translate: 35.80   =   r   ×   895

Solve: We divide on both sides by 895.

$$35.80 \div 895 = r$$
$$0.04 = r$$
$$4\% = r$$

The sales tax rate is 4%.

9. *Restate:*   Sales tax is 5% of what?

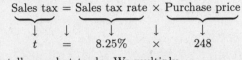

Translate: 100   = 5% ×   p

Solve:　We divide on both sides by 0.05.

$$100 \div 0.05 = p$$
$$2000 = p$$

$$0.05_\wedge \overline{)100.00_\wedge}$$
$$\underline{10000}$$
$$0$$

$$2000.$$

The purchase price is \$2000.

11. *Restate:*　Sales tax is 3.5% of what?

Translate:　$28 = 3.5\% \times p$,　or
　　　　　　　$28 = 0.035 \times p$

Solve:　We divide on both sides by 0.035.

$$28 \div 0.035 = p$$
$$800 = p$$

$$0.035_\wedge \overline{)28.000_\wedge}$$
$$\underline{28000}$$
$$0$$

$$800.$$

The purchase price is \$800.

13. The total tax rate is the city tax rate plus the state tax rate, or $1\% + 6\% = 7\%$.

We find the sales tax using an equation.

$$\text{Sales tax} = \text{Sales tax rate} \times \text{Purchase price}$$
$$t = 7\% \times 665$$

This tells us what to do. We multiply.

$$\begin{array}{r} 665 \\ \times\, 0.07 \\ \hline 46.55 \end{array}$$

The sales tax is \$46.55.

15. *Restate:*　Sales tax is what percent of purchase price?

Translate:　$1030.40 = r \times 18,400$

Solve:　We divide on both sides by 18,400.

$$1030.40 \div 18,400 = r$$
$$0.056 = r$$
$$5.6\% = r$$

The sales tax rate is 5.6%.

17.
$$2.3 \times y = 85.1$$
$$\frac{2.3 \times y}{2.3} = \frac{85.1}{2.3}$$
$$y = 37$$

19.
$$4.3 \times t = 34.4$$
$$\frac{4.3 \times t}{4.3} = \frac{34.4}{4.3}$$
$$t = 8$$

21. $\dfrac{13}{11} = 13 \div 11$

$$\begin{array}{r} 1.1818 \\ 11\overline{)13.0000} \\ \underline{11} \\ 20 \\ \underline{11} \\ 90 \\ \underline{88} \\ 20 \\ \underline{11} \\ 90 \\ \underline{88} \\ 2 \end{array}$$

We get a repeating decimal.

$$\frac{13}{11} = 1.1\overline{8}$$

23. We find the sales tax using an equation.

$$\text{Sales tax} = \text{Sales tax rate} \times \text{Purchase price}$$
$$t = 5.4\% \times 96,568.95$$

We multiply using a calculator.

$$0.054 \times 96,568.95 = 5214.7233 \approx 5214.72$$

The sales tax is \$5214.72.

Exercise Set 7.7

1.
$$\text{Commission} = \text{Commission rate} \times \text{Sales}$$
$$C = 6\% \times 45,000$$

This tells us what to do. We multiply.

$$\begin{array}{r} 45,000 \\ \times\quad 0.06 \\ \hline 2700.00 \end{array} \qquad (6\% = 0.06)$$

The commission is \$2700.

3.
$$\text{Commission} = \text{Commission rate} \times \text{Sales}$$
$$120 = r \times 2400$$

To solve this equation we divide on both sides by 2400:

$$120 \div 2400 = r$$

We can divide, but this time we simplify by removing a factor of 1:

$$r = \frac{120}{2400} = \frac{1}{20} \cdot \frac{120}{120} = \frac{1}{20} = 0.05 = 5\%$$

The commission rate is 5%.

5.
$$\text{Commission} = \text{Commission rate} \times \text{Sales}$$
$$392 = 40\% \times S$$

To solve this equation we divide on both sides by 40%:

$$392 \div 40\% = S$$
$$392 \div 0.4 = S$$
$$980 = S$$

$$0.4_\wedge \overline{)392.0_\wedge}$$
$$\underline{3600}$$
$$320$$
$$\underline{320}$$
$$0$$
$$\underline{0}$$
$$0$$

$$980.$$

There were $980 worth of dishwashers sold.

7. Commission = Commission rate × Sales
$$C \quad = \quad 7\% \quad \times 98,000$$

This tells us what to do. We multiply.

$$
\begin{array}{r}
9\,8,0\,0\,0 \\
\times \quad 0.0\,7 \\
\hline
6\,8\,6\,0.0\,0
\end{array}
\qquad (7\% = 0.07)
$$

The commission is $6860.

9. First we find the commission.

Commission = Commission rate × Sales
$$C \quad = \quad 2\% \quad \times 990$$

This tells us what to do. We multiply.

$$
\begin{array}{r}
9\,9\,0 \\
\times 0.0\,2 \\
\hline
1\,9.8\,0
\end{array}
\qquad (2\% = 0.02)
$$

The commission is $19.80.

The total wages were salary plus commission, or

$$
\begin{array}{r}
5\,0\,0.0\,0 \\
+ \quad 1\,9.8\,0 \\
\hline
5\,1\,9.8\,0
\end{array}
$$

The total wages for that month were $519.80.

11. Discount = Rate of discount × Marked price
$$D \quad = \quad 10\% \quad \times \quad \$300$$

Convert 10% to decimal notation and multiply.

$$
\begin{array}{r}
3\,0\,0 \\
\times \quad 0.1 \\
\hline
3\,0.0
\end{array}
\qquad (10\% = 0.10 = 0.1)
$$

The discount is $30.

Sale price = Marked price − Discount
$$S \quad = \quad 300 \quad - \quad 30$$

We subtract:
$$
\begin{array}{r}
3\,0\,0 \\
- \quad 3\,0 \\
\hline
2\,7\,0
\end{array}
$$

The sale price is $270.

13. Discount = Rate of discount × Marked price
$$D \quad = \quad 15\% \quad \times \quad \$350$$

Convert 15% to decimal notation and multiply.

$$
\begin{array}{r}
3\,5\,0 \\
\times \; 0.1\,5 \\
\hline
1\,7\,5\,0 \\
3\,5\,0\,0 \\
\hline
5\,2.5\,0
\end{array}
\qquad (15\% = 0.15)
$$

The discount is $52.50.

Sale price = Marked price − Discount
$$S \quad = \quad 350 \quad - \quad 52.50$$

We subtract:
$$
\begin{array}{r}
3\,5\,0.0\,0 \\
- \quad 5\,2.5\,0 \\
\hline
2\,9\,7.5\,0
\end{array}
$$

The sale price is $297.50.

15. Discount = Rate of discount × Marked price
$$12.50 \quad = \quad 10\% \quad \times \quad M$$

To solve the equation we divide on both sides by 10%.
$$12.50 \div 10\% = M$$
$$12.50 \div 0.1 = M$$
$$125 = M$$

The marked price is $125.

Sale price = Marked price − Discount
$$S \quad = \quad 125.00 \quad - \quad 12.50$$

We subtract:
$$
\begin{array}{r}
1\,2\,5.0\,0 \\
- \quad 1\,2.5\,0 \\
\hline
1\,1\,2.5\,0
\end{array}
$$

The sale price is $112.50.

17. Discount = Rate of discount × Marked price
$$240 \quad = \quad R \quad \times \quad 600$$

To solve the equation we divide on both sides by 600.
$$240 \div 600 = R$$

We can simplify by removing a factor of 1:
$$R = \frac{240}{600} = \frac{2}{5} \cdot \frac{120}{120} = \frac{2}{5} = 0.4 = 40\%$$

The rate of discount is 40%.

Sale price = Marked price − Discount
$$S \quad = \quad 600 \quad - \quad 240$$

We subtract:
$$
\begin{array}{r}
6\,0\,0 \\
- \quad 2\,4\,0 \\
\hline
3\,6\,0
\end{array}
$$

The sale price is $360.

19. Discount = Marked price − Sale price
$$D \quad = \quad 1275 \quad - \quad 888$$

We subtract:
$$
\begin{array}{r}
1\,2\,7\,5 \\
- \quad 8\,8\,8 \\
\hline
3\,8\,7
\end{array}
$$

The discount is $387.

Discount = Rate of discount × Marked price
$$387 \quad = \quad R \quad \times \quad 1275$$

To solve the equation we divide on both sides by 1275.
$$387 \div 1275 = R$$
$$0.30353 \approx R$$
$$30.353\% \approx R, \text{ or}$$
$$30\frac{6}{17}\% = R$$

The commission rate is $30\frac{6}{17}\%$, or about 30.353%.

21. $\frac{5}{9} = 5 \div 9$

$$
\begin{array}{r}
0.5\,5 \\
9\overline{)5.0\,0} \\
4\,5 \\
\hline
5\,0 \\
4\,5 \\
\hline
5
\end{array}
$$

We get a repeating decimal.
$$\frac{5}{9} = 0.\overline{5}$$

23. $\dfrac{11}{12} = 11 \div 12$

$$
\begin{array}{r}
0.9166 \\
12\overline{)11.0000} \\
\underline{108} \\
20 \\
\underline{12} \\
80 \\
\underline{72} \\
80 \\
\underline{72} \\
8
\end{array}
$$

We get a repeating decimal.

$\dfrac{11}{12} = 0.91\overline{6}$

25. $\dfrac{15}{7} = 15 \div 7$

$$
\begin{array}{r}
2.142857 \\
7\overline{)15.000000} \\
\underline{14} \\
10 \\
\underline{7} \\
30 \\
\underline{28} \\
20 \\
\underline{14} \\
60 \\
\underline{56} \\
40 \\
\underline{35} \\
50 \\
\underline{49} \\
1
\end{array}
$$

We get a repeating decimal.

$\dfrac{15}{7} = 2.\overline{142857}$

27. Commission = Commission rate × Sales

$\quad\quad\; C \quad = \quad\quad 7.5\% \quad\quad \times\ 78,990$

We multiply.

$$
\begin{array}{r}
78,990 \\
\times\ 0.075 \\
\hline
394950 \\
5529300 \\
\hline
5924.250
\end{array}
$$
$\quad\quad (7.5\% = 0.075)$

The commission is $5924.25.

We subtract to find how much the seller gets for the house after paying the commission.

$\quad \$78,990 - \$5924.25 = \$73,065.75$

Exercise Set 7.8

1. We take 13% of $200:

$13\% \times 200 = 0.13 \times 200$
$\quad\quad\quad\quad\quad = 26$

$$
\begin{array}{r}
200 \\
\times 0.13 \\
\hline
600 \\
2000 \\
\hline
26.00
\end{array}
$$

The interest is $26.

3. We take 12.4% of $2000:

$12.4\% \times 2000 = 0.124 \times 2000$
$\quad\quad\quad\quad\quad\quad = 248$

$$
\begin{array}{r}
2000 \\
\times 0.124 \\
\hline
8000 \\
40000 \\
200000 \\
\hline
248.000
\end{array}
$$

The interest is $248.

5. Interest = (Interest for 1 year) × $\dfrac{1}{4}$

$\quad\quad = (14\% \times 4300) \times \dfrac{1}{4}$

$\quad\quad = 0.14 \times 4300 \times \dfrac{1}{4}$

$\quad\quad = 602 \times \dfrac{1}{4}$

$$
\begin{array}{r}
4300 \\
\times\ 0.14 \\
\hline
17200 \\
43000 \\
\hline
602.00
\end{array}
$$

$\quad\quad = \dfrac{602}{4}$

$\quad\quad = 150.5$

$$
\begin{array}{r}
150.5 \\
4\overline{)602.0} \\
\underline{400} \\
202 \\
\underline{200} \\
20 \\
\underline{20} \\
0
\end{array}
$$

The interest is $150.50.

7. Interest = (Interest for 1 year) × $\dfrac{60}{360}$

$\quad\quad = (14.5\% \times 5000) \times \dfrac{60}{360}$

$\quad\quad = 0.145 \times 5000 \times \dfrac{1}{6} \quad\quad \left(\dfrac{60}{360} = \dfrac{1}{6}\right)$

$\quad\quad = 725 \times \dfrac{1}{6}$

$\quad\quad = \dfrac{725}{6}$

$$
\begin{array}{r}
120.833 \\
6\overline{)725.000} \\
\underline{600} \\
125 \\
\underline{120} \\
50 \\
\underline{48} \\
20 \\
\underline{18} \\
20 \\
\underline{18} \\
2
\end{array}
$$

$\quad\quad = 120.83\overline{3}$

$\quad\quad = 120.83 \quad$ (Rounded to the nearest hundredth)

The interest is $120.83.

9. We take 9.4% of $4400:
$$9.4\% \times 4400 = 0.094 \times 4400$$
$$= 413.6$$

$$\begin{array}{r} 4\ 4\ 0\ 0 \\ \times\ 0.0\ 9\ 4 \\ \hline 1\ 7\ 6\ 0\ 0 \\ 3\ 9\ 6\ 0\ 0\ 0 \\ \hline 4\ 1\ 3.6\ 0\ 0 \end{array}$$

The interest is $413.60.

11. a) Find the interest at the end of 1 year.
$$I = (r \times P) \times t$$
$$= (10\% \times \$400) \times 1$$
$$= 0.1 \times \$400$$
$$= \$40$$

b) Find the principal after 1 year.
$$\$400 + \$40 = \$440$$

c) Going into the second year the principal is $440. Find the interest for 1 year after that.
$$I = (r \times P) \times t$$
$$= (10\% \times \$440) \times 1$$
$$= 0.1 \times \$440$$
$$= \$44$$

d) Find the new principal after 2 years.
$$\$440 + \$44 = \$484$$

The amount in the account after 2 years is $484.

13. a) Find the interest at the end of 1 year.
$$I = (r \times P) \times t$$
$$= (8.8\% \times \$200) \times 1$$
$$= 0.088 \times \$200$$

$$\begin{array}{r} 2\ 0\ 0 \\ \times\ 0.0\ 8\ 8 \\ \hline 1\ 6\ 0\ 0 \\ 1\ 6\ 0\ 0\ 0 \\ \hline 1\ 7.6\ 0\ 0 \end{array}$$

$$= \$17.60$$

b) Find the principal after 1 year.
$$\$200.00 + \$17.60 = \$217.60$$

c) Going into the second year the principal is $217.60. Find the interest for 1 year after that.
$$I = (r \times P) \times t$$
$$= (8.8\% \times \$217.60) \times 1$$
$$= 0.088 \times \$217.60$$

$$\begin{array}{r} 2\ 1\ 7.6 \\ \times\ 0.0\ 8\ 8 \\ \hline 1\ 7\ 4\ 0\ 8 \\ 1\ 7\ 4\ 0\ 8\ 0 \\ \hline 1\ 9.1\ 4\ 8\ 8 \end{array}$$

$$= \$19.1488$$
$$= \$19.15 \quad \text{(Rounded to the nearest hundredth)}$$

d) Find the new principal after 2 years.
$$\$217.60 + \$19.15 = \$236.75$$

The amount in the account after 2 years is $236.75.

15. a) Find the interest at the end of $\frac{1}{2}$ year.
$$I = (r \times P) \times t$$
$$= (7\% \times \$400) \times \frac{1}{2}$$
$$= (0.07 \times \$400) \times \frac{1}{2}$$
$$= \$28 \times \frac{1}{2}$$
$$= \$14$$

b) Find the principal after $\frac{1}{2}$ year.
$$\$400 + \$14 = \$414$$

c) Going into the second half of the year the principal is $414. Find the interest for $\frac{1}{2}$ year after that.
$$I = (r \times P) \times t$$
$$= (7\% \times \$414) \times \frac{1}{2}$$
$$= (0.07 \times \$414) \times \frac{1}{2}$$
$$= \$28.98 \times \frac{1}{2}$$
$$= \$14.49$$

d) Find the new principal after 1 year.
$$\$414.00 + \$14.49 = \$428.49$$

The amount in the account after 1 year is $428.49.

17. a) Find the interest at the end of $\frac{1}{2}$ year.
$$I = (r \times P) \times t$$
$$= (9\% \times \$2000) \times \frac{1}{2}$$
$$= (0.09 \times \$2000) \times \frac{1}{2}$$
$$= \$180 \times \frac{1}{2}$$
$$= \$90$$

b) Find the principal after $\frac{1}{2}$ year.
$$\$2000 + \$90 = \$2090$$

c) Going into the second half of the year the principal is $2090. Find the interest for $\frac{1}{2}$ year after that.
$$I = (r \times P) \times t$$
$$= (9\% \times \$2090) \times \frac{1}{2}$$
$$= (0.09 \times \$2090) \times \frac{1}{2}$$
$$= \$188.10 \times \frac{1}{2}$$
$$= \$94.05$$

d) Find the new principal after 1 year.
$$\$2090.00 + \$94.05 = \$2184.05$$

The amount in the account after 1 year is $2184.05.

19.
$$\frac{9}{10} = \frac{x}{5}$$
$$9 \cdot 5 = 10 \cdot x$$
$$\frac{9 \cdot 5}{10} = x$$
$$\frac{45}{10} = x$$
$$\frac{9}{2} = x, \text{ or }$$
$$4.5 = x$$

21.
$$\frac{3}{4} = \frac{6}{x}$$
$$3 \cdot x = 4 \cdot 6$$
$$x = \frac{4 \cdot 6}{3}$$
$$x = \frac{24}{3}$$
$$x = 8$$

23. $\frac{100}{3} = 33\frac{1}{3}$

$$\begin{array}{r} 33 \\ 3\overline{\smash)100} \\ \underline{90} \\ 10 \\ \underline{9} \\ 1 \end{array}$$

25. $\frac{38}{3} = 12\frac{2}{3}$

$$\begin{array}{r} 12 \\ 3\overline{\smash)38} \\ \underline{30} \\ 8 \\ \underline{6} \\ 2 \end{array}$$

27. Interest = (Interest for 1 year) $\times \frac{3}{4}$

$$= (7.75\% \times \$24,680) \times \frac{3}{4}$$
$$= (0.0775 \times \$24,680) \times \frac{3}{4}$$
$$= \$1434.525 \quad \text{Using a calculator}$$
$$= \$1434.53 \quad \text{Rounded to the nearest hundredth}$$

The interest is \$1434.53.

29. a) We find the interest at the end of $\frac{1}{2}$ year.
$$I = (r \times P) \times t$$
$$= (6.4\% \times \$24,800) \times \frac{1}{2}$$
$$= \$793.60$$

b) Find the principal after $\frac{1}{2}$ year.
$$\$24,800 + \$793.60 = \$25,593.60$$

c) Going into the second half of the first year, the principal is \$25,593.60. Now we find the interest for $\frac{1}{2}$ year after that.

$$I = (r \times P) \times t$$
$$= (6.4\% \times \$25,593.60) \times \frac{1}{2}$$
$$= \$818.9952$$
$$\approx \$819.00$$

d) We find the new principal after 1 year.
$$\$25,593.60 + \$819.00 = \$26,412.60$$

e) Going into the second year, the principal is \$26,412.60. Now we find the interest for $\frac{1}{2}$ year after that.

$$I = (r \times P) \times t$$
$$= (6.4\% \times \$26,412.60) \times \frac{1}{2}$$
$$= \$845.2032$$
$$\approx \$845.20$$

f) We find the new principal after $1\frac{1}{2}$ years.
$$\$26,412.60 + \$845.20 = \$27,257.80$$

g) Going into the second half of the second year, the principal is \$27,257.80. Now we find the interest for $\frac{1}{2}$ year after that.

$$I = (r \times P) \times t$$
$$= (6.4\% \times \$27,257.80) \times \frac{1}{2}$$
$$= \$872.2496$$
$$\approx \$872.25$$

h) We find the new principal after 2 years.
$$\$27,257.80 + \$872.25 = \$28,130.05$$

The amount after 2 years is \$28,130.05.

Chapter 8

Descriptive Statistics

Exercise Set 8.1

1. To find the average, add the numbers. Then divide by the number of addends.

$$\frac{16 + 18 + 29 + 14 + 29 + 19 + 15}{7} = \frac{140}{7} = 20$$

The average is 20.

To find the median, first list the numbers in order from smallest to largest. Then locate the middle number.

$$14, 15, 16, 18, 19, 29, 29$$
$$\uparrow$$
$$\text{Middle number}$$

The median is 18.

Find the mode:

The number that occurs most often is 29. The mode is 29.

3. To find the average, add the numbers. Then divide by the number of addends.

$$\frac{5 + 30 + 20 + 20 + 35 + 5 + 25}{7} = \frac{140}{7} = 20$$

The average is 20.

To find the median, first list the numbers in order from smallest to largest. Then locate the middle number.

$$5, 5, 20, 20, 25, 30, 35$$
$$\uparrow$$
$$\text{Middle number}$$

The median is 20.

Find the mode:

There are two numbers that occur most often, 5 and 20. Thus the modes are 5 and 20.

5. Find the average:

$$\frac{1.2 + 4.3 + 5.7 + 7.4 + 7.4}{5} = \frac{26}{5} = 5.2$$

The average is 5.2.

Find the median:

$$1.2, 4.3, 5.7, 7.4, 7.4$$
$$\uparrow$$
$$\text{Middle number}$$

The median is 5.7.

Find the mode:

The number that occurs most often is 7.4. The mode is 7.4.

7. Find the average:

$$\frac{234 + 228 + 234 + 229 + 234 + 278}{6} = \frac{1437}{6} = 239.5$$

The average is 239.5.

Find the median:

$$228, 229, 234, 234, 234, 278$$
$$\uparrow$$
$$\text{Middle number}$$

The median is halfway between 234 and 234. Although it seems clear that this is 234, we can compute it as follows:

$$\frac{234 + 234}{2} = \frac{468}{2} = 234$$

The median is 234.

Find the mode:

The number that occurs most often is 234. The mode is 234.

9. Find the average weight:

$$\frac{250 + 255 + 260 + 260}{4} = \frac{1025}{4} = 256.25$$

The average weight is 256.25 lb.

Find the median:

$$250, 255, 260, 260$$
$$\uparrow$$
$$\text{Middle number}$$

The median is halfway between 255 and 260. We find it as follows:

$$\frac{255 + 260}{2} = \frac{515}{2} = 257.5$$

The median is 257.5 lb.

Find the mode:

The number that occurs most often is 260. The mode is 260 lb.

11. Find the average:

$$\frac{43° + 40° + 23° + 38° + 54° + 35° + 47°}{7} = \frac{280°}{7} = 40°$$

The average temperature was 40°.

Find the median:

$$23°, 35°, 38°, 40°, 43°, 47°, 54°$$
$$\uparrow$$
$$\text{Middle number}$$

The median is 40°.

Find the mode:

No number repeats, so each number in the set is a mode. The modes are 43°, 40°, 23°, 38°, 54°, 35°, and 47°.

13. We divide the total number of miles, 522, by the number of gallons, 18.

$$\frac{522}{18} = 29$$

The average was 29 miles per gallon.

15. To find the GPA we first add the grade point values for each hour taken. This is done by first multiplying the grade point value by the number of hours in the course and then adding as follows:

$$
\begin{array}{llr}
\text{B} & 3.00 \cdot 4 = & 12 \\
\text{B} & 3.00 \cdot 5 = & 15 \\
\text{B} & 3.00 \cdot 3 = & 9 \\
\text{C} & 2.00 \cdot 4 = & 8 \\
\hline
& & 44 \ \text{(Total)}
\end{array}
$$

The total number of hours taken is

$$4 + 5 + 3 + 4, \text{ or } 16.$$

We divide 44 by 16.

$$\frac{44}{16} = 2.75$$

The student's grade point average is 2.75.

17. Find the average price per pound:

$$\frac{\$9.79 + \$9.59 + \$9.69 + \$9.79 + \$9.89}{5} = \frac{\$48.75}{5} = \$9.75$$

The average price per pound of steak was $9.75.

Find the median price per pound:

List the prices in order:

$$\$9.59, \$9.69, \$9.79, \$9.79, \$9.89$$
$$\uparrow$$
$$\text{Middle number}$$

The median is $9.79.

Find the mode:

The number that occurs most often is $9.79. The mode is $9.79.

19. We can find the total of the five scores needed as follows:

$$80 + 80 + 80 + 80 + 80 = 400.$$

The total of the scores on the first four tests is

$$80 + 74 + 81 + 75 = 310.$$

Thus the student needs to get at least

$$400 - 310, \text{ or } 90$$

to get a B. We can check this as follows:

$$\frac{80 + 74 + 81 + 75 + 90}{5} = \frac{400}{5} = 80.$$

21. To find the average we first add all the salaries. We do this by multiplying each salary by the number of employees earning that salary as follows:

$$
\begin{array}{lll}
\text{Owner} & \$29,200 \cdot 1 = & \$\ \ 29,200 \\
\text{Salesperson} & \$19,600 \cdot 5 = & \$\ \ 98,000 \\
\text{Secretary} & \$14,800 \cdot 3 = & \$\ \ 44,400 \\
\text{Custodian} & \$13,000 \cdot 1 = & \$\ \ 13,000 \\
\hline
& & \$184,600 \ \text{(Total)}
\end{array}
$$

The total number of employees is $1 + 5 + 3 + 1$, or 10. We divide $184,600 by 10:

$$\frac{\$184,600}{10} = \$18,460$$

The average salary is $18,460.

23. $\dfrac{2}{3} \cdot \dfrac{2}{3} = \dfrac{2 \cdot 2}{3 \cdot 3} = \dfrac{4}{9}$

25.
$$
\begin{array}{r}
1.414 \\
\times\ 1.414 \\
\hline
5\ 656 \\
14\ 140 \\
565\ 600 \\
1\ 414\ 000 \\
\hline
1.999\ 396
\end{array}
$$

27. Divide the total by the number of games. Use a calculator.

$$\frac{4621}{27} \approx 171.15$$

Drop the amount to the right of the decimal point.

$$\underline{171} \cdot \boxed{15}$$

This is the $\underline{\ \ }\!\uparrow$ $\uparrow\!\underline{\ \ }$ Drop this
average.   amount.

The bowler's average is 171.

Exercise Set 8.2

1. Go down the left column (headed "Penny Number") to 40. Then go across to the column headed "Length" (immediately to the right of the Penny Number column) and read the entry, 5. The Length column gives the length in inches, so a 40-penny nail is 5 inches long.

3. Go down the left column (headed "Penny Number") to 10. Then go across to the columns headed "Approx. Number Per Pound," and find the column headed "Box Nails." The entry is 94, so there are approximately 94 10-penny nails in a pound.

5. First, go down the "Penny Number" column to 16, and then go across to the "Finishing Nails" column to find how many 16-penny finishing nails there are in a pound. There are approximately 90. Then go down the "Penny Number" column to 10 and across to the "Common Nails" column to find how many 10-penny nails there are in a pound. There are approximately 69. The difference, 90 − 69, or 21, tells us that there are approximately 21 more 16-penny finishing nails in a pound.

7. Go down the "Penny Number" column to 20, and then go across to the "Box Nails" column. The entry is 52, so there are approximately 52 box nails in a pound.

9. Go down the "Penny Number" column to 4, and then go across to the "Finishing Nails" column. The entry is 548, so there are approximately 548 4-penny finishing nails in one pound. Then there are approximately 5×548, or 2740 nails in 5 pounds.

11. Go down the "Activity" column to Racquetball, and then go across to the "154 lb" column under the heading "Calories Burned in 30 Minutes." The entry is 294, so 294 calories are burned by a 154-pound person after 30 minutes of racquetball.

13. Go down the "110 lb" column to 216, and then go across to the "Activity" column. The entry is Calisthenics, so calisthenics burns 216 calories in 30 minutes for a 110-pound person.

15. Going down the "Activity" column to Aerobic Dance and then across to the "154 lb" column, we find the entry 282. Next, going down the "Activity" column to Tennis and across to the "154 lb" column, we find the entry 222. Since $282 > 222$, aerobic dance burns more calories in 30 minutes for a 154-pound person.

17. Going down the "Activity" column to Tennis and then across to the "110 lb" column, we see that 165 calories are burned after 30 minutes of tennis for a 110-pound person. Now, 2 hours, or 120 minutes, is 4×30 minutes, so we multiply to find the answer: 4×165 calories $= 660$ calories.

19. Go down the "132 lb" column and find the smallest entry. It is 132 in the bottom row. Go across the bottom row to the "Activity" column, and read the entry, Moderate Walking.

21. We observe that 120 pounds is approximately half-way between 110 pounds and 132 pounds (or is approximately the average of 110 pounds and 132 pounds). Therefore, we would expect the number of calories burned by a 120-pound person during 30 minutes of moderate walking to be approximately half-way between (or the average of) the numbers of calories burned by a 110-pound person and a 132-pound person during 30 minutes of moderate walking. Going down the "Activity" column to Moderate Walking and then across, first to the "110 lb" column, and then to the "132 lb" column, we find the entries 111 and 132, respectively. The number half-way between these two (or their average) is $\frac{111 + 132}{2}$, or 121.5, or approximately 121. (We round down since 120 is closer to 110 than to 132.) Therefore, you would expect to burn about 121 calories during 30 minutes of moderate walking.

23. There are more bottle symbols beside 1995 than any other year. Therefore, the year in which the greatest number of bottles was sold is 1995.

25. There was positive growth between every pair of consecutive years except 1993 and 1994. The pair of years for which the amount of positive growth was the least was 1991

and 1992 (represented by an increase of 1 bottle symbol as opposed to 2 or 3 symbols for each of the other pairs).

27. Sales for 1992 are represented by 5 bottles. Since each bottle symbol stands for 1000 bottles sold, we multiply to find 5×1000, or 5000, bottles were sold in 1992.

29. We look for a row of the chart containing fewer symbols than the one immediately below it. The only such row is the one showing sales in 1994. Therefore, in 1994 there was a decline in the number of bottles sold.

31. The key at the bottom of the pictograph tells us that one baseball symbol represents 3 times at bat.

33. Singles are represented by 5 whole symbols (5×3) and 1/3 of another symbol ($1/3 \times 3$) for a total of $15 + 1$, or 16.

35. Triples are represented by 1 whole symbol, so there were 1×3, or 3, triples. In Exercise 33 we calculated that there were 16 singles. Then there were $16 - 3$, or 13 fewer triples than singles

37. $12 \div 3 = 4$, so we are looking for a row containing exactly 4 whole symbols. Only the row representing outs contains exactly 4 whole symbols, so the player made an out exactly 12 times.

39. Since each flame symbol will represent 10 calories, we divide the number of calories for each tablespread by 10 to determine the number of symbols to use.

Jam: $54 \div 10 = 5.4$

Mayonnaise: $51 \div 10 = 5.1$

Peanut Butter: $94 \div 10 = 9.4$

Honey: $64 \div 10 = 6.4$

Syrup: $60 \div 10 = 6$

We fill in the pictograph with the appropriate number of symbols beside each tablespread.

41. **Method 1**: Solve using an equation.

Restate: What is 90% of 30,955?

Translate: $a = 90\% \times 30{,}955$

Solve: We convert to decimal notation and multiply.

$$\begin{array}{r} 3\,0,9\,5\,5 \\ \times\qquad 0.9 \\ \hline 2\,7,8\,5\,9\,.5 \end{array}$$

Thus, 27,859.5 sq mi of Maine are forest.

Method 2: Solve using a proportion.

Restate: What is   90%   of 30,955?
                 ↓   ↓   ↓
            amount   number   base

Translate: $\dfrac{90}{100} = \dfrac{a}{30,955}$

Solve : $90 \cdot 30,955 = 100 \cdot a$

$$\dfrac{90 \cdot 30,955}{100} = a$$

$$\dfrac{2,785,950}{100} = a$$

$$27,859.50 = a$$

Thus, 27,859.5 sq mi of Maine are forest.

Exercise Set 8.3

1. We look for the longest bar and find that it represents Los Angeles, CA.

3. Go to the right end of the bar representing Pittsburgh, PA, and then go down to the "Weeks of Growing Season" scale. We can read, fairly accurately, that the growing season is 22 weeks long.

5. Going to the right end of the bar representing Fresno, CA, and then down to the scale, we find that the growing season in Fresno is 40 weeks long. Doing the same for the bar representing Ogden, UT, we find that the growing season in Ogden is 20 weeks long. Since $40 \div 20 = 2$, the growing season in Fresno, CA, is 2 times as long as in Ogden, UT.

7. Going to the right end of the bar representing Los Angeles, CA, and then down to the scale, we find that the growing season in Los Angeles is 52 weeks long. Now $\frac{1}{2} \times 52 = 26$, so we go to 26 on the "Weeks of Growing Season" scale and then go up, looking for the bar that ends closest to 26 weeks. It is the top bar. We then go across to the left and read the name of the city, Syracuse, NY.

9. Go to the top of the bar representing New York. Then go across to the "Daily Expenses" scale. We can read, fairly accurately, that the average daily expenses are approximately $270.

11. We look for the shortest bar and find that it represents San Francisco.

13. Going to the top of the bar representing Washington and then across to the "Daily Expenses" scale, we find that the average daily expenses are approximately $240. Doing the same for the bar representing San Francisco, we find that the average daily expenses in San Francisco are approximately $190. Subtracting, we find that the average daily expenses in Washington are $240 - $190, or approximately $50, more than in San Francisco.

15. Label the marks on the vertical scale with the names of the activities and title it "Activity." Label the horizontal scale appropriately by 100's and title it "Calories burned per hour." Then draw horizontal bars to show the calories burned by each activity.

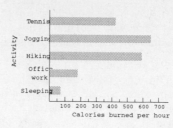

17. Disregarding jogging, the activity in which the most calories per hour are burned is hiking. Thus, if you were trying to lose weight by exercising and could not jog, hiking would be the most beneficial exercise.

19. Find the highest point on the line, and go down to the "Time" scale. The temperature was highest at 3 P.M.

21. Finding the highest point on the line and going straight across to the "Temperature" scale, we see that the highest temperature was about 51°F. Doing the same for the lowest point on the line, we see that the lowest temperature was about 36°F. The difference is 51°F − 36°F, or approximately 15°F.

23. Reading the graph from left to right, we see that the line went down the most between 5 P.M. and 6 P.M.

25. Find the highest point on the line, and go down to the "Year" scale. Estimated sales are the greatest in 1995.

27. Find 1993 on the bottom scale and go up to the point on the line directly above it. Then go straight across to the "Estimated Sales" scale. We are at about 17.0, so estimated sales in 1993 are approximately $17.0 million.

29. The graph shows estimated sales of about $19.5 million in 1995 and about $18.0 million in 1997. Now, $19.5 million − $18.0 million = $1.5 million, so estimated sales in 1995 are approximately $1.5 million greater than in 1997.

31. First, label the horizontal scale with the times, and title it "Time (P.M.)." Then label the vertical scale by 10's, and title it "Number of cars." Next, mark points above the times at appropriate levels, and then draw line segments connecting them.

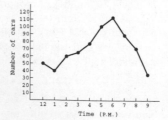

33. Reading the graph from left to right, we see that the line went down the most between 8 P.M. and 9 P.M.

35. We add the number of cars for each hour and divide by the number of hours:

$$\frac{50 + 40 + 60 + 65 + 77 + 100 + 112 + 88 + 70 + 35}{10}$$

$$= \frac{697}{10}$$

$$= 69.7$$

The average number of cars passing through the intersection in a 1-hr period was 69.7.

37. Method 1: Solve using an equation.

Restate: 9 is what percent of 50?

↓ ↓   ↓   ↓ ↓
Translate: 9 = n × 50

Solve: We divide on both sides by 50.

$$9 \div 50 = n$$
$$0.18 = n$$
$$18\% = n$$

Thus, 18% of the pizzas will be ordered with extra cheese.

Method 2: Solve using a proportion.

Restate: 9   is   what percent   of   50?
↓   ↓   ↓
amount   number   base

Translate: $\dfrac{n}{100} = \dfrac{9}{50}$

Solve : $50 \cdot n = 100 \cdot 9$

$$n = \frac{100 \cdot 9}{50}$$

$$n = \frac{900}{50}$$

$$n = 18$$

Thus, 18% of the pizzas will be ordered with extra cheese.

39. Method 1: Solve using an equation.

34 is what percent of 51?

↓ ↓   ↓   ↓ ↓
Translate: 34 = n × 51

Solve: We divide on both sides by 51.

$$34 \div 51 = n$$
$$0.66\overline{6} = n$$
$$66.\overline{6}\% = n, \text{or}$$
$$66\frac{2}{3}\% = n$$

34 is $66.\overline{6}\%$, or $66\frac{2}{3}\%$, of 51.

Method 2: Solve using a proportion.

34   is   what percent   of   51?
↓   ↓   ↓
amount   number   base

Translate: $\dfrac{n}{100} = \dfrac{34}{51}$

Solve : $51 \cdot n = 100 \cdot 34$

$$n = \frac{100 \cdot 34}{51}$$

$$n = \frac{100 \cdot 2 \cdot \cancel{17}}{3 \cdot \cancel{17}}$$

$$n = \frac{200}{6} = 66\frac{2}{3}, \text{or } 66.\overline{6}$$

34 is $66\frac{2}{3}\%$, or $66.\overline{6}\%$, of 51.

Exercise Set 8.4

---

1. We see from the graph that 3.7% of all records sold are jazz.

3. We see from the graph that 9% of all records sold are country. Find 9% of 3000: $0.09 \times 3000 = 270$. Then 270 of the records are country.

5. We see from the graph that 6.8% of all records sold are classical.

7. The largest section of the circle represents gas purchased. Therefore, the most is spent on gas purchased.

9. We see from the graph that 4¢ of each dollar is spent on dividends.

11. 4¢ of each dollar is spent on dividends, and 12¢ on wages, salaries, and employee benefits. Adding these amounts, we find that 4¢ + 12¢, or 16¢, of every dollar is spent on these items all together.

13. First find the number of degrees required to represent each type of expenditure. Transportation accounts for 15% of the expenditures. 15% of 360° ($0.15 \times 360°$) is 54° (54°/5° = 10.8 intervals). Meals account for 20% of the expenditures. 20% of 360° ($0.2 \times 360°$) is 72° more (72°/5° = 14.4 intervals). Lodging accounts for 32% of the expenditures. 32% of 360° ($0.32 \times 360°$) is another 115.2° (115.2°/5° = 23.04 intervals). Recreation accounts for 18% of the expenditures. 18% of 360° ($0.18 \times 360°$) is 64.8° more (64.8°/5° = 12.96 intervals). Other accounts for the remaining 15%. We found above that 15% of 360° is 54° (and 10.8 intervals). Note that this last section accounts exactly for the remainder of the circle. Using this information we can draw a circle graph and label it appropriately.

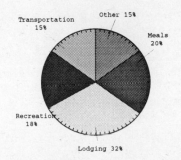

15. Method 1: Solve using an equation.

16 is what percent of 64?

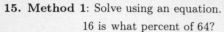

Translate: $16 = n \times 64$

Solve: We divide on both sides by 64.

$$16 \div 64 = n$$
$$0.25 = n$$
$$25\% = n$$

16 is 25% of 64.

Method 2: Solve using a proportion.

16   is   what percent   of   64?
↓   ↓   ↓
amount   number   base

Translate: $\dfrac{n}{100} = \dfrac{16}{64}$

Solve : $64 \cdot n = 100 \cdot 16$

$$n = \frac{100 \cdot 16}{64}$$

$$n = \frac{1600}{64}$$

$$n = 25$$

16 is 25% of 64.

17. $\dfrac{4}{7} \div \dfrac{16}{21} = \dfrac{4}{7} \cdot \dfrac{21}{16} = \dfrac{4 \cdot 21}{7 \cdot 16} = \dfrac{4 \cdot 3 \cdot 7}{7 \cdot 4 \cdot 4} = \dfrac{4 \cdot 7}{4 \cdot 7} \cdot \dfrac{3}{4} = \dfrac{3}{4}$

Chapter 9

Geometry and Measure: Length and Area

Exercise Set 9.1

1. 1 foot = 12 in.

This is the relation stated on page 419 of the text.

3. $1 \text{ in.} = 1 \text{ in.} \times \dfrac{1 \text{ ft}}{12 \text{ in.}}$ Multiplying by 1 using $\dfrac{1 \text{ ft}}{12 \text{ in.}}$ to eliminate in.

$= \dfrac{1 \text{ in.}}{12 \text{ in.}} \times 1 \text{ ft}$

$= \dfrac{1}{12} \times \dfrac{\text{in.}}{\text{in.}} \times 1 \text{ ft}$

$= \dfrac{1}{12} \times 1 \text{ ft}$ The $\dfrac{\text{in.}}{\text{in.}}$ acts like 1, so we can omit it.

$= \dfrac{1}{12} \text{ ft}$

5. 1 mi = 5280 ft

This is the relation stated on page 419 of the text.

7. 3 yd = 3 × 1 yd

= 3 × 36 in. Substituting 36 in. for 1 yd

= 108 in. Multiplying

9. $84 \text{ in.} = \dfrac{84 \text{ in.}}{1} \times \dfrac{1 \text{ ft}}{12 \text{ in.}}$ Multiplying by 1 using $\dfrac{1 \text{ ft}}{12 \text{ in.}}$

$= \dfrac{84 \text{ in.}}{12 \text{ in.}} \times 1 \text{ ft}$

$= \dfrac{84}{12} \times \dfrac{\text{in.}}{\text{in.}} \times 1 \text{ ft}$

$= 7 \times 1 \text{ ft}$ The $\dfrac{\text{in.}}{\text{in.}}$ acts like 1, so we can omit it.

$= 7 \text{ ft}$

11. $18 \text{ in.} = \dfrac{18 \text{ in.}}{1} \times \dfrac{1 \text{ ft}}{12 \text{ in.}}$ Multiplying by 1 using $\dfrac{1 \text{ ft}}{12 \text{ in.}}$

$= \dfrac{18 \text{ in.}}{12 \text{ in.}} \times 1 \text{ ft}$

$= \dfrac{18}{12} \times \dfrac{\text{in.}}{\text{in.}} \times 1 \text{ ft}$

$= \dfrac{3}{2} \times 1 \text{ ft}$ The $\dfrac{\text{in.}}{\text{in.}}$ acts like 1, so we can omit it.

$= \dfrac{3}{2} \text{ ft, or } 1\dfrac{1}{2} \text{ ft}$

13. 5 mi = 5 × 1 mi

= 5 × 5280 ft Substituting 5280 ft for 1 mi

= 26,400 ft Multiplying

15. $36 \text{ in.} = \dfrac{36 \text{ in.}}{1} \times \dfrac{1 \text{ ft}}{12 \text{ in.}}$ Multiplying by 1 using $\dfrac{1 \text{ ft}}{12 \text{ in.}}$

$= \dfrac{36 \text{ in.}}{12 \text{ in.}} \times 1 \text{ ft}$

$= \dfrac{36}{12} \times \dfrac{\text{in.}}{\text{in.}} \times 1 \text{ ft}$

$= 3 \times 1 \text{ ft}$ The $\dfrac{\text{in.}}{\text{in.}}$ acts like 1, so we can omit it.

$= 3 \text{ ft}$

17. $10 \text{ ft} = 10 \text{ ft} \times \dfrac{1 \text{ yd}}{3 \text{ ft}}$ Multiplying by 1 using $\dfrac{1 \text{ yd}}{3 \text{ ft}}$

$= \dfrac{10}{3} \times \dfrac{\text{ft}}{\text{ft}} \times 1 \text{ yd}$

$= \dfrac{10}{3} \times 1 \text{ yd}$

$= \dfrac{10}{3} \text{ yd, or } 3\dfrac{1}{3} \text{ yd}$

19. 10 mi = 10 × 1 mi

= 10 × 5280 ft Substituting 5280 ft for 1 mi

= 52,800 ft Multiplying

21. $4\dfrac{1}{2} \text{ ft} = 4\dfrac{1}{2} \text{ ft} \times \dfrac{1 \text{ yd}}{3 \text{ ft}}$

$= \dfrac{9}{2} \text{ ft} \times \dfrac{1 \text{ yd}}{3 \text{ ft}}$

$= \dfrac{9}{6} \times \dfrac{\text{ft}}{\text{ft}} \times 1 \text{ yd}$

$= \dfrac{3}{2} \times 1 \text{ yd}$

$= \dfrac{3}{2} \text{ yd, or } 1\dfrac{1}{2} \text{ yd}$

23. $36 \text{ in.} = 36 \text{ in.} \times \dfrac{1 \text{ ft}}{12 \text{ in.}} \times \dfrac{1 \text{ yd}}{3 \text{ ft}}$

$= \dfrac{36}{12 \cdot 3} \times \dfrac{\text{in.}}{\text{in.}} \times \dfrac{\text{ft}}{\text{ft}} \times 1 \text{ yd}$

$= \dfrac{36}{36} \times 1 \text{ yd}$

$= 1 \times 1 \text{ yd}$

$= 1 \text{ yd}$

25. $330 \text{ ft} = 330 \text{ ft} \times \dfrac{1 \text{ yd}}{3 \text{ ft}}$

$\phantom{330 \text{ ft}} = \dfrac{330}{3} \times \dfrac{\text{ft}}{\text{ft}} \times 1 \text{ yd}$

$\phantom{330 \text{ ft}} = 110 \times 1 \text{ yd}$
$\phantom{330 \text{ ft}} = 110 \text{ yd}$

27. $3520 \text{ yd} = 3520 \times 1 \text{ yd} \times \dfrac{3 \text{ ft}}{1 \text{ yd}} \times \dfrac{1 \text{ mi}}{5280 \text{ ft}}$

$\phantom{3520 \text{ yd}} = \dfrac{3520 \cdot 3}{5280} \times \dfrac{\text{yd}}{\text{yd}} \times \dfrac{\text{ft}}{\text{ft}} \times 1 \text{ mi}$

$\phantom{3520 \text{ yd}} = \dfrac{10,560}{5280} \times 1 \text{ mi}$

$\phantom{3520 \text{ yd}} = 2 \times 1 \text{ mi}$
$\phantom{3520 \text{ yd}} = 2 \text{ mi}$

29. $100 \text{ yd} = 100 \times 1 \text{ yd}$
$\phantom{100 \text{ yd}} = 100 \times 3 \text{ ft}$
$\phantom{100 \text{ yd}} = 300 \text{ ft}$

31. $360 \text{ in.} = 360 \text{ in.} \times \dfrac{1 \text{ ft}}{12 \text{ in.}}$

$\phantom{360 \text{ in.}} = \dfrac{360}{12} \times \dfrac{\text{in.}}{\text{in.}} \times 1 \text{ ft}$

$\phantom{360 \text{ in.}} = 30 \times 1 \text{ ft}$
$\phantom{360 \text{ in.}} = 30 \text{ ft}$

33. $1 \text{ in.} = 1 \text{ in.} \times \dfrac{1 \text{ ft}}{12 \text{ in.}} \times \dfrac{1 \text{ yd}}{3 \text{ ft}}$

$\phantom{1 \text{ in.}} = \dfrac{1}{12 \cdot 3} \times \dfrac{\text{in.}}{\text{in.}} \times \dfrac{\text{ft}}{\text{ft}} \times 1 \text{ yd}$

$\phantom{1 \text{ in.}} = \dfrac{1}{36} \times 1 \text{ yd}$

$\phantom{1 \text{ in.}} = \dfrac{1}{36} \text{ yd}$

35. $2 \text{ mi} = 2 \times 1 \text{ mi}$
$\phantom{2 \text{ mi}} = 2 \times 5280 \times 1 \text{ ft}$
$\phantom{2 \text{ mi}} = 2 \times 5280 \times 12 \text{ in.}$
$\phantom{2 \text{ mi}} = 126,720 \text{ in.}$

37. $9.25\% = \dfrac{9.25}{100}$

$ = \dfrac{9.25}{100} \cdot \dfrac{100}{100}$

$ = \dfrac{925}{10,000}$

$ = \dfrac{25 \cdot 37}{25 \cdot 400}$

$ = \dfrac{25}{25} \cdot \dfrac{37}{400}$

$ = \dfrac{37}{400}$

39. a) First find decimal notation by division.

$$8 \overline{\smash{\big)}\ 11.000} \quad \begin{array}{r} 1.375 \end{array}$$

```
         1 . 3 7 5
   8 ⟌ 1 1 . 0 0 0
       8
       ‾‾‾
       3 0
       2 4
       ‾‾‾
         6 0
         5 6
         ‾‾‾
           4 0
           4 0
           ‾‾‾
             0
```

$\dfrac{11}{8} = 1.375$

b) Convert the decimal notation to percent notation. Move the decimal point two places to the right and write a % symbol.

1.37.5

$\dfrac{11}{8} = 137.5\%$

41. $\dfrac{1}{4} = \dfrac{1}{4} \cdot \dfrac{25}{25} = \dfrac{25}{100} = 25\%$

43. First find the height, in inches, of the stack of $1 bills.

1.169 × the distance to the moon

$ = 1.169 \times 238,866 \text{ mi}$
$ = 279,234.354 \times 1 \text{ mi}$
$ = 279,234.354 \times 5280 \text{ ft}$
$ = 1,474,357,389 \times 1 \text{ ft}$
$ = 1,474,357,389 \times 12 \text{ in.}$
$ = 17,692,288,670 \text{ in.}$

We divide this measure by 4.265 trillion to find the thickness, in inches, of a $1 bill.

$17,692,288,670 \div 4,265,000,000,000 \approx 0.0041$

A $1 bill is approximately 0.0041 in. thick.

Exercise Set 9.2

1. a) 1 km = _____ m

Think: To go from km to m in the table is a move of 3 places to the right. Thus, we move the decimal point 3 places to the right.

1   1.000.

1 km = 1000 m

b) 1 m = _____ km

Think: To go from m to km in the table is a move of 3 places to the left. Thus, we move the decimal point 3 places to the left.

1   0.001.

1 m = 0.001 km

3. a) 1 dam = _____ m

Think: To go from dam to m in the table is a move of 1 place to the right. Thus, we move the decimal point 1 place to the right.

 1   1.0.
     ⌞↑

1 dam = 10 m

b) 1 m = _____ dam

Think: To go from m to dam in the table is a move of 1 place to the left. Thus, we move the decimal point 1 place to the left.

 1   0.1.
     ↑⌟

1 m = 0.1 dam

5. a) 1 cm = _____ m

Think: To go from cm to m in the table is a move of 2 places to the left. Thus, we move the decimal point 2 places to the left.

 1   0.01.
     ↑⌟

1 cm = 0.01 m

b) 1 m = _____ cm

Think: To go from m to cm in the table is a move of 2 places to the right. Thus, we move the decimal point 2 places to the right.

 1   1.00.
     ⌞↑

1 m = 100 cm

7. 6.7 km = _____ m

Think: To go from km to m in the table is a move of 3 places to the right. Thus, we move the decimal point 3 places to the right.

 6.7   6.700.
     ⌞___↑

6.7 km = 6700 m

9. 98 cm = _____ m

Think: To go from cm to m in the table is a move of 2 places to the left. Thus, we move the decimal point 2 places to the left.

 98   0.98.
     ↑⌟

98 cm = 0.98 m

11. 8921 m = _____ km

Think: To go from m to km in the table is a move of 3 places to the left. Thus, we move the decimal point 3 places to the left.

 8921   8.921.
     ↑⌟

8921 m = 8.921 km

13. 56.66 m = _____ km

Think: To go from m to km in the table is a move of 3 places to the left. Thus, we move the decimal point 3 places to the left.

 56.66   0.056.66
     ↑⌟

56.66 m = 0.05666 km

15. 5666 m = _____ cm

Think: To go from m to cm in the table is a move of 2 places to the right. Thus, we move the decimal point 2 places to the right.

 5666   5666.00.
     ⌞↑

5666 m = 566,600 cm

17. 477 cm = _____ m

Think: To go from cm to m in the table is a move of 2 places to the left. Thus, we move the decimal point 2 places to the left.

 477   4.77.
     ↑⌟

477 cm = 4.77 m

19. 6.88 m = _____ cm

Think: To go from m to cm in the table is a move of 2 places to the right. Thus, we move the decimal point 2 places to the right.

 6.88   6.88.
     ⌞↑

6.88 m = 688 cm

21. 1 mm = _____ cm

Think: To go from mm to cm in the table is a move of 1 place to the left. Thus, we move the decimal point 1 place to the left.

 1   0.1.
     ↑⌟

1 mm = 0.1 cm

23. 1 km = _____ cm

Think: To go from km to cm in the table is a move of 5 places to the right. Thus, we move the decimal point 5 places to the right.

 1   1.00000.
     ⌞____↑

1 km = 100,000 cm

25. 14.2 cm = _____ mm

Think: To go from cm to mm in the table is a move of 1 place to the right. Thus, we move the decimal point 1 place to the right.

 14.2   14.2.
     ⌞↑

14.2 cm = 142 mm

27. 8.2 mm = _____ cm

Think: To go from mm to cm in the table is a move of 1 place to the left. Thus, we move the decimal point 1 place to the left.

8.2   0.8.2

8.2 mm = 0.82 cm

29. 4500 mm = _____ cm

Think: To go from mm to cm in the table is a move of 1 place to the left. Thus, we move the decimal point 1 place to the left.

4500   450.0.

4500 mm = 450 cm

31. 0.024 mm = _____ m

Think: To go from mm to m in the table is a move of 3 places to the left. Thus, we move the decimal point 3 places to the left.

0.024   0.000.024

0.024 mm = 0.000024 m

33. 6.88 m = _____ dam

Think: To go from m to dam in the table is a move of 1 place to the left. Thus, we move the decimal point 1 place to the left.

6.88   0.6.88

6.88 m = 0.688 dam

35. 2.3 dam = _____ dm

Think: To go from dam to dm in the table is a move of 2 places to the right. Thus, we move the decimal point 2 places to the right.

2.3   2.30.

2.3 dam = 230 dm

37. 392 dam = _____ km

Think: To go from dam to km in the table is a move of 2 places to the left. Thus, we move the decimal point 2 places to the left.

392   3.92.

392 dam = 3.92 km

39. $330 \text{ ft} \approx 330 \text{ ft} \times \dfrac{1 \text{ m}}{3.3 \text{ ft}} = \dfrac{330}{3.3} \times \dfrac{\text{ft}}{\text{ft}} \times 1 \text{ m} = 100 \text{ m}$

41. 1171.352 km = 1171.352 × 1 km
      ≈ 1171.352 × 0.621 mi
      = 727.409592 mi

43. $65 \text{ mph} = 65 \dfrac{\text{mi}}{\text{hr}} = 65 \times \dfrac{1 \text{ mi}}{\text{hr}} \approx 65 \times \dfrac{1.609 \text{ km}}{\text{hr}} =$ 104.585 km/h

45.
```
        1 .7 5
   1 2 | 2 1 .0 0
         1 2
          9 0
          8 4
           6 0
           6 0
            0
```

47. To divide by 100, move the decimal point 2 places to the left.

23.4   0.23.4

23.4 ÷ 100 = 0.234

49.
```
      3 .1 4      (2 decimal places)
    × 4 .4 1      (2 decimal places)
      3 1 4
  1 2 5 6 0
  1 2 5 6 0 0
  1 3 .8 4 7 4    (4 decimal places)
```

Round to the nearest hundredth:

1 3 . 8 4 7 4   Thousandths digit is 5 or higher. Round up.

1 3 . 8 5

51. $48 \cdot \dfrac{1}{12} = \dfrac{48 \cdot 1}{12} = \dfrac{48}{12} = 4$

Exercise Set 9.3

1. Perimeter = 4 mm + 6 mm + 7 mm
          = (4 + 6 + 7) mm
          = 17 mm

3. Perimeter = 3.5 in. + 3.5 in. + 4.25 in.+
               0.5 in. + 3.5 in.
          = (3.5 + 3.5 + 4.25 + 0.5 + 3.5) in.
          = 15.25 in.

5. $P = 4 \cdot s$ Perimeter of a square
     $P = 4 \cdot 3.25 \text{ m}$
     $P = 13 \text{ m}$

7. $P = 2 \cdot (l + w)$ Perimeter of a rectangle
     $P = 2 \cdot (5 \text{ ft} + 10 \text{ ft})$
     $P = 2 \cdot (15 \text{ ft})$
     $P = 30 \text{ ft}$

9. $P = 2 \cdot (l + w)$ Perimeter of a rectangle
     $P = 2 \cdot (34.67 \text{ cm} + 4.9 \text{ cm})$
     $P = 2 \cdot (39.57 \text{ cm})$
     $P = 79.14 \text{ cm}$

11. $P = 4 \cdot s$ Perimeter of a square
      $P = 4 \cdot 22 \text{ ft}$
      $P = 88 \text{ ft}$

13. $P = 4 \cdot s$ Perimeter of a square
      $P = 4 \cdot 45.5 \text{ mm}$
      $P = 182 \text{ mm}$

15. *Familiarize.* First we find the perimeter of the field. Then we multiply to find the cost of the fence wire. We make a drawing.

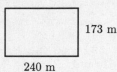

173 m

240 m

Translate. The perimeter of the field is given by

$$P = 2 \cdot (l + w) = 2 \cdot (240 \text{ m} + 173 \text{ m}).$$

Solve. We calculate the perimeter.

$$P = 2 \cdot (240 \text{ m} + 173 \text{ m}) = 2 \cdot (413 \text{ m}) = 826 \text{ m}$$

Then we multiply to find the cost of the fence wire.

$$\begin{aligned} \text{Cost} &= \$1.45/\text{m} \times \text{Perimeter} \\ &= \$1.45/\text{m} \times 826 \text{ m} \\ &= \$1197.70 \end{aligned}$$

Check. Repeat the calculations.

State. The perimeter of the field is 826 m. The fence will cost $1197.70.

17. *Familiarize.* We make a drawing and let P = the perimeter.

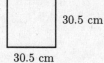

30.5 cm

30.5 cm

Translate. The perimeter of the square is given by

$$P = 4 \cdot s = 4 \cdot (30.5 \text{ cm}).$$

Solve. We do the calculation.

$$P = 4 \cdot (30.5 \text{ cm}) = 122 \text{ cm}.$$

Check. Repeat the calculation.

State. The perimeter of the tile is 122 cm.

19. *Familiarize.* We label the missing lengths on the drawing and let P = the perimeter.

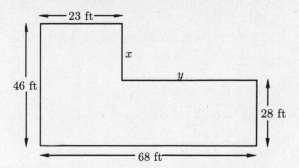

Translate. First we find the missing lengths x and y.

28 ft	plus	how many more ft	is	46 ft
↓	↓	↓	↓	↓
28	+	x	=	46

23 ft	plus	how many more ft	is	68 ft
↓	↓	↓	↓	↓
23	+	y	=	68

Solve. We solve for x and y.

$$\begin{aligned} 28 + x &= 46 & \qquad 23 + y &= 68 \\ x &= 46 - 28 & y &= 68 - 23 \\ x &= 18 & y &= 45 \end{aligned}$$

a) To find the perimeter we add the lengths of the sides of the house.

$$\begin{aligned} P &= 23 \text{ ft} + 18 \text{ ft} + 45 \text{ ft} + 28 \text{ ft} + 68 \text{ ft} + 46 \text{ ft} \\ &= (23 + 18 + 45 + 28 + 68 + 46) \text{ ft} \\ &= 228 \text{ ft} \end{aligned}$$

b) Next we find t, the total cost of the gutter.

Cost per foot	times	Number of feet	is	Total cost
↓	↓	↓	↓	↓
4.59	×	228	=	t

We carry out the multiplication.

$$\begin{array}{r} 2\,2\,8 \\ \times\ 4\,.5\,9 \\ \hline 2\,0\,5\,2 \\ 1\,1\,4\,0\,0 \\ 9\,1\,2\,0\,0 \\ \hline 1\,0\,4\,6\,.5\,2 \end{array}$$

Thus, $t = 1046.52$.

Check. We can repeat the calculations.

State. (a) The perimeter of the house is 228 ft. (b) The total cost of the gutter is $1046.52.

21. 56.1%

a) Replace the percent symbol with × 0.01.

56.1 × 0.01

b) Move the decimal point two places to the left.

0.56.1

Thus, 56.1% = 0.561.

23. a) First find decimal notation by division.

$$
\begin{array}{r}
1.125 \\
8\,\overline{)\,9.000} \\
\underline{8}\quad\;\; \\
1\,0\;\;\; \\
\underline{8}\;\;\; \\
2\,0\;\; \\
\underline{1\,6}\;\; \\
4\,0\; \\
\underline{4\,0}\; \\
0
\end{array}
$$

$\dfrac{9}{8} = 1.125$

b) Convert the decimal notation to percent notation. Move the decimal point two places to the right and write a % symbol.

1.12.5

$\dfrac{9}{8} = 112.5\%$, or $112\dfrac{1}{2}\%$

25. $10^2 = 10 \cdot 10 = 100$

Exercise Set 9.4

1. $A = l \cdot w$ Area of a rectangular region
$A = (5 \text{ km}) \cdot (3 \text{ km})$
$A = 5 \cdot 3 \cdot \text{ km} \cdot \text{ km}$
$A = 15 \text{ km}^2$

3. $A = l \cdot w$ Area of a rectangular region
$A = (2 \text{ in.}) \cdot (0.7 \text{ in.})$
$A = 2 \cdot 0.7 \cdot \text{ in.} \cdot \text{ in.}$
$A = 1.4 \text{ in}^2$

5. $A = s \cdot s$ Area of a square
$A = \left(2\dfrac{1}{2} \text{ yd}\right) \cdot \left(2\dfrac{1}{2} \text{ yd}\right)$
$A = \left(\dfrac{5}{2} \text{ yd}\right) \cdot \left(\dfrac{5}{2} \text{ yd}\right)$
$A = \dfrac{5}{2} \cdot \dfrac{5}{2} \cdot \text{ yd} \cdot \text{ yd}$
$A = \dfrac{25}{4} \text{ yd}^2$, or $6\dfrac{1}{4} \text{ yd}^2$

7. $A = s \cdot s$ Area of a square
$A = (90 \text{ ft}) \cdot (90 \text{ ft})$
$A = 90 \cdot 90 \cdot \text{ ft} \cdot \text{ ft}$
$A = 8100 \text{ ft}^2$

9. $A = l \cdot w$ Area of a rectangular region
$A = (10 \text{ ft}) \cdot (5 \text{ ft})$
$A = 10 \cdot 5 \cdot \text{ ft} \cdot \text{ ft}$
$A = 50 \text{ ft}^2$

11. $A = l \cdot w$ Area of a rectangular region
$A = (34.67 \text{ cm}) \cdot (4.9 \text{ cm})$
$A = 34.67 \cdot 4.9 \cdot \text{ cm} \cdot \text{ cm}$
$A = 169.883 \text{ cm}^2$

13. $A = l \cdot w$ Area of a rectangular region
$A = \left(4\dfrac{2}{3} \text{ in.}\right) \cdot \left(8\dfrac{5}{6} \text{ in.}\right)$
$A = \left(\dfrac{14}{3} \text{ in.}\right) \cdot \left(\dfrac{53}{6} \text{ in.}\right)$
$A = \dfrac{14}{3} \cdot \dfrac{53}{6} \cdot \text{ in.} \cdot \text{ in.}$
$A = \dfrac{2 \cdot 7 \cdot 53}{3 \cdot 2 \cdot 3} \text{ in}^2$
$A = \dfrac{2}{2} \cdot \dfrac{7 \cdot 53}{3 \cdot 3} \text{ in}^2$
$A = \dfrac{371}{9} \text{ in}^2$, or $41\dfrac{2}{9} \text{ in}^2$

15. $A = s \cdot s$ Area of a square
$A = (22 \text{ ft}) \cdot (22 \text{ ft})$
$A = 22 \cdot 22 \cdot \text{ ft} \cdot \text{ ft}$
$A = 484 \text{ ft}^2$

17. $A = s \cdot s$ Area of a square
$A = (56.9 \text{ km}) \cdot (56.9 \text{ km})$
$A = 56.9 \cdot 56.9 \cdot \text{ km} \cdot \text{ km}$
$A = 3237.61 \text{ km}^2$

19. $A = s \cdot s$ Area of a square
$A = \left(5\dfrac{3}{8} \text{ yd}\right) \cdot \left(5\dfrac{3}{8} \text{ yd}\right)$
$A = \left(\dfrac{43}{8} \text{ yd}\right) \cdot \left(\dfrac{43}{8} \text{ yd}\right)$
$A = \dfrac{43}{8} \cdot \dfrac{43}{8} \cdot \text{ yd} \cdot \text{ yd}$
$A = \dfrac{1849}{64} \text{ yd}^2$, or $28\dfrac{57}{64} \text{ yd}^2$

21. *Familiarize.* We draw a picture

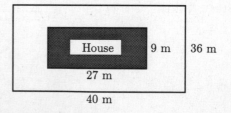

Translate. We let $A =$ the area left over.

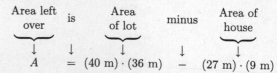

Area left over	is	Area of lot	minus	Area of house
↓	↓	↓	↓	↓
A	$=$	$(40 \text{ m}) \cdot (36 \text{ m})$	$-$	$(27 \text{ m}) \cdot (9 \text{ m})$

Solve. The area of the lot is

$(40 \text{ m}) \cdot (36 \text{ m}) = 40 \cdot 36 \cdot \text{ m} \cdot \text{ m} = 1440 \text{ m}^2.$

The area of the house is

$$(27 \text{ m}) \cdot (9 \text{ m}) = 27 \cdot 9 \cdot \text{ m} \cdot \text{ m} = 243 \text{ m}^2.$$

The area left over is

$$A = 1440 \text{ m}^2 - 243 \text{ m}^2 = 1197 \text{ m}^2.$$

Check. Repeat the calculations.

State. The area left over is 1197 m².

23. **Familiarize.** We use the drawing in the text.

Translate. We let A = the area of the sidewalk.

Area of sidewalk	is	Total area	minus	Area of building
↓	↓	↓	↓	↓
A	$=$	$(113.4 \text{ m}) \times (75.4 \text{ m})$	$-$	$(110 \text{ m}) \times (72 \text{ m})$

Solve. The total area is

$$(113.4 \text{ m}) \times (75.4 \text{ m}) = 113.4 \times 75.4 \times \text{ m} \times \text{ m} = 8550.36 \text{ m}^2.$$

The area of the building is

$$(110 \text{ m}) \times (72 \text{ m}) = 110 \times 72 \times \text{ m} \times \text{ m} = 7920 \text{ m}^2.$$

The area of the sidewalk is

$$A = 8550.36 \text{ m}^2 - 7920 \text{ m}^2 = 630.36 \text{ m}^2.$$

Check. Repeat the calculations.

State. The area of the sidewalk is 630.36 m².

25. **Familiarize.** The dimensions are as follows:

Two walls are 15 ft by 8 ft.

Two walls are 20 ft by 8 ft.

The ceiling is 15 ft by 20 ft.

The total area of the walls and ceiling is the total area of the rectangles described above less the area of the windows and the door.

Translate. a) We let A = the total area of the walls and ceiling. The total area of the two 15 ft by 8 ft walls is

$$2 \cdot (15 \text{ ft}) \cdot (8 \text{ ft}) = 2 \cdot 15 \cdot 8 \cdot \text{ ft} \cdot \text{ ft} = 240 \text{ ft}^2$$

The total area of the two 20 ft by 8 ft walls is

$$2 \cdot (20 \text{ ft}) \cdot (8 \text{ ft}) = 2 \cdot 20 \cdot 8 \cdot \text{ ft} \cdot \text{ ft} = 320 \text{ ft}^2$$

The area of the ceiling is

$$(15 \text{ ft}) \cdot (20 \text{ ft}) = 15 \cdot 20 \cdot \text{ ft} \cdot \text{ ft} = 300 \text{ ft}^2$$

The area of the two windows is

$$2 \cdot (3 \text{ ft}) \cdot (4 \text{ ft}) = 2 \cdot 3 \cdot 4 \cdot \text{ ft} \cdot \text{ ft} = 24 \text{ ft}^2$$

The area of the door is

$$\left(2\frac{1}{2} \text{ ft}\right) \cdot \left(6\frac{1}{2} \text{ ft}\right) = \left(\frac{5}{2} \text{ ft}\right) \cdot \left(\frac{13}{2} \text{ ft}\right)$$
$$= \frac{5}{2} \cdot \frac{13}{2} \cdot \text{ ft} \cdot \text{ ft}$$
$$= \frac{65}{4} \text{ ft}^2, \text{ or } 16\frac{1}{4} \text{ ft}^2$$

Thus

$$A = 240 \text{ ft}^2 + 320 \text{ ft}^2 + 300 \text{ ft}^2 - 24 \text{ ft}^2 - 16\frac{1}{4} \text{ ft}^2$$
$$= 819\frac{3}{4} \text{ ft}^2, \text{ or } 819.75 \text{ ft}^2$$

b) We divide to find how many gallons of paint are needed.

$$819.75 \div 86.625 \approx 9.46$$

It will be necessary to buy 10 gallons of paint in order to have the required 9.46 gallons.

c) We multiply to find the cost of the paint.

$$10 \times \$17.95 = \$179.50$$

Check. We repeat the calculations.

State. (a) The total area of the walls and ceiling is 819.75 ft². (b) 10 gallons of paint are needed. (c) It will cost $179.50 to paint the room.

27.

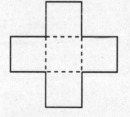

Each side is 4 cm.

The region is composed of 5 squares, each with sides of length 4 cm. The area is

$$A = 5 \cdot (s \cdot s) = 5 \cdot (4 \text{ cm} \cdot 4 \text{ cm}) = 5 \cdot 4 \cdot 4 \text{ cm} \cdot \text{ cm} = 80 \text{ cm}^2$$

29.   0.452   0.45.2   Move the decimal point
                         2 places to the right.

Write a % symbol: 45.2%

$0.452 = 45.2\%$

31. We multiply by 1 to get 100 in the denominator.

$$\frac{11}{20} = \frac{11}{20} \cdot \frac{5}{5} = \frac{55}{100} = 55\%$$

Exercise Set 9.5

1. $A = b \cdot h$ Area of a parallelogram

   $A = 8 \text{ cm} \cdot 4 \text{ cm}$ Substituting 8 cm for b and 4 cm for h

   $A = 32 \text{ cm}^2$

3. $A = \frac{1}{2} \cdot b \cdot h$ Area of a triangle

   $A = \frac{1}{2} \cdot 15 \text{ in.} \cdot 8 \text{ in.}$ Substituting 15 in. for b and 8 in. for h

   $A = 60 \text{ in}^2$

5. $A = \frac{1}{2} \cdot h \cdot (a + b)$ Area of a trapezoid

$A = \frac{1}{2} \cdot 8 \text{ ft} \cdot (6 + 20) \text{ ft}$ Substituting 8 ft for h, 6 ft for a, and 20 ft for b

$A = \frac{8 \cdot 26}{2} \text{ ft}^2$

$A = 104 \text{ ft}^2$

7. $A = b \cdot h$ Area of a parallelogram
$A = 8 \text{ m} \cdot 8 \text{ m}$
$A = 64 \text{ m}^2$

9. $A = \frac{1}{2} \cdot h \cdot (a + b)$ Area of a trapezoid

$A = \frac{1}{2} \cdot 7 \text{ in.} \cdot (4.5 + 8.5) \text{ in.}$ Substituting 7 in. for h, 4.5 in. for a, and 8.5 in. for b

$A = \frac{7 \cdot 13}{2} \text{ in}^2$

$A = \frac{91}{2} \text{ in}^2$

$A = 45.5 \text{ in}^2$

11. $A = b \cdot h$ Area of a parallelogram
$A = 2.3 \text{ cm} \cdot 3.5 \text{ cm}$ Substituting 2.3 cm for b and 3.5 cm for h

$A = 8.05 \text{ cm}^2$

13. $A = \frac{1}{2} \cdot h \cdot (a + b)$ Area of a trapezoid

$A = \frac{1}{2} \cdot 18 \text{ cm} \cdot (9 + 24) \text{ cm}$ Substituting 18 cm for h, 9 cm for a, and 24 cm for b

$A = \frac{18 \cdot 33}{2} \text{ cm}^2$

$A = 297 \text{ cm}^2$

15. $A = \frac{1}{2} \cdot b \cdot h$ Area of a triangle

$A = \frac{1}{2} \cdot 4 \text{ m} \cdot 3.5 \text{ m}$ Substituting 4 m for b and 3.5 m for h

$A = \frac{4 \cdot 3.5}{2} \text{ m}^2$

$A = 7 \text{ m}^2$

17. $A = b \cdot h$ Area of a parallelogram

$A = 12\frac{1}{4} \text{ ft} \cdot 4\frac{1}{2} \text{ ft}$ Substituting $12\frac{1}{4}$ ft for b and $4\frac{1}{2}$ ft for h

$A = \frac{49}{4} \cdot \frac{9}{2} \cdot \text{ ft}^2$

$A = \frac{441}{8} \text{ ft}^2$

$A = 55\frac{1}{8} \text{ ft}^2$

19. Familiarize. We look for the kinds of figures whose areas we can calculate using area formulas that we already know.

Translate. The shaded region consists of a square region with a triangular region removed from it. The sides of

the square are 30 cm, and the triangle has base 30 cm and height 15 cm. We find the area of the square using the formula $A = s \cdot s$, and the area of the triangle using $A = \frac{1}{2} \cdot b \cdot h$. Then we subtract.

Solve. Area of the square: $A = 30 \text{ cm} \cdot 30 \text{ cm} = 900 \text{ cm}^2$.

Area of the triangle: $A = \frac{1}{2} \cdot 30 \text{ cm} \cdot 15 \text{ cm} = 225 \text{ cm}^2$.

Area of the shaded region: $A = 900 \text{ cm}^2 - 225 \text{ cm}^2 = 675 \text{ cm}^2$.

Check. We repeat the calculations.

State. The area of the shaded region is 675 cm².

21. Familiarize. We look for the kinds of figures whose areas we can calculate using area formulas that we already know.

Translate. The shaded region consists of 8 triangles, each with base 43 in. and height 52 in. We will find the area of one triangle using the formula $A = \frac{1}{2} \cdot b \cdot h$. Then we will multiply by 8.

Solve. $A = \frac{1}{2} \cdot 43 \text{ in.} \cdot 52 \text{ in.} = 1118 \text{ in}^2$

Then we multiply by 8: $8 \cdot 1118 \text{ in}^2 = 8944 \text{ in}^2$

Check. We repeat the calculations.

State. The area of the shaded region is 8944 in².

23. Familiarize. We make a drawing, shading the area left over after the triangular piece is cut from the sailcloth.

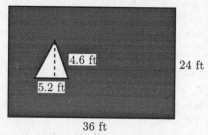

Translate. The shaded region consists of a rectangular region with a triangular region removed from it. The rectangular region has dimensions 36 ft by 24 ft, and the triangular region has base 5.2 ft and height 4.6 ft. We will find the area of the rectangular region using the formula $A = b \cdot h$, and the area of the triangle using $A = \frac{1}{2} \cdot b \cdot h$. Then we will subtract to find the area of the shaded region.

Solve. Area of the rectangle: $A = 36 \text{ ft} \cdot 24 \text{ ft} = 864 \text{ ft}^2$.

Area of the triangle: $A = \frac{1}{2} \cdot 5.2 \text{ ft} \cdot 4.6 \text{ ft} = 11.96 \text{ ft}^2$.

Area of the shaded region: $A = 864 \text{ ft}^2 - 11.96 \text{ ft}^2 = 852.04 \text{ ft}^2$.

Check. We repeat the calculation.

State. The area left over is 852.04 ft².

25. $P = 10 \text{ cm} + 8 \text{ cm} + 6 \text{ cm} = (10 + 8 + 6) \text{ cm} = 24 \text{ cm}$

$A = \frac{1}{2} \cdot b \cdot h$

$A = \frac{1}{2} \cdot 10 \text{ cm} \cdot 5 \text{ cm}$

$A = \frac{10 \cdot 5}{2} \text{ cm}^2$

$A = 25 \text{ cm}^2$

Exercise Set 9.6

1. $d = 2 \cdot r$

$d = 2 \cdot 7 \text{ cm} = 14 \text{ cm}$

3. $d = 2 \cdot r$

$d = 2 \cdot \frac{3}{4} \text{ in.} = \frac{6}{4} \text{ in.} = \frac{3}{2} \text{ in., or } 1\frac{1}{2} \text{ in.}$

5. $r = \frac{d}{2}$

$r = \frac{32 \text{ ft}}{2} = 16 \text{ ft}$

7. $r = \frac{d}{2}$

$r = \frac{1.4 \text{ cm}}{2} = 0.7 \text{ cm}$

9. $C = 2 \cdot \pi \cdot r$

$C \approx 2 \cdot \frac{22}{7} \cdot 7 \text{ cm} = \frac{2 \cdot 22 \cdot 7}{7} \text{ cm} = 44 \text{ cm}$

11. $C = 2 \cdot \pi \cdot r$

$C \approx 2 \cdot \frac{22}{7} \cdot \frac{3}{4} \text{ in.} = \frac{2 \cdot 22 \cdot 3}{7 \cdot 4} \text{ in.} = \frac{132}{28} \text{ in.} = \frac{33}{7} \text{ in.,}$

or $4\frac{5}{7}$ in.

13. $C = \pi \cdot d$

$C \approx 3.14 \cdot 32 \text{ ft} = 100.48 \text{ ft}$

15. $C = \pi \cdot d$

$C \approx 3.14 \cdot 1.4 \text{ cm} = 4.396 \text{ cm}$

17. $A = \pi \cdot r \cdot r$

$A \approx \frac{22}{7} \cdot 7 \text{ cm} \cdot 7 \text{ cm} = \frac{22}{7} \cdot 49 \text{ cm}^2 = 154 \text{ cm}^2$

19. $A = \pi \cdot r \cdot r$

$A \approx \frac{22}{7} \cdot \frac{3}{4} \text{ in.} \cdot \frac{3}{4} \text{ in.} = \frac{22 \cdot 3 \cdot 3}{7 \cdot 4 \cdot 4} \text{ in}^2 = \frac{99}{56} \text{ in}^2, \text{ or } 1\frac{43}{56} \text{ in}^2$

21. $A = \pi \cdot r \cdot r$

$A \approx 3.14 \cdot 16 \text{ ft} \cdot 16 \text{ ft}$ $\quad (r = \frac{d}{2}; r = \frac{32 \text{ ft}}{2} = 16 \text{ ft})$

$A = 3.14 \cdot 256 \text{ ft}^2$

$A = 803.84 \text{ ft}^2$

23. $A = \pi \cdot r \cdot r$

$A \approx 3.14 \cdot 0.7 \text{ cm} \cdot 0.7 \text{ cm}$

$\quad (r = \frac{d}{2}; r = \frac{1.4 \text{ cm}}{2} = 0.7 \text{ cm})$

$A = 3.14 \cdot 0.49 \text{ cm}^2 = 1.5386 \text{ cm}^2$

25. $r = \frac{d}{2}$

$r = \frac{6 \text{ cm}}{2} = 3 \text{ cm}$

The radius is 3 cm.

$C = \pi \cdot d$

$C \approx 3.14 \cdot 6 \text{ cm} = 18.84 \text{ cm}$

The circumference is about 18.84 cm.

$A = \pi \cdot r \cdot r$

$A \approx 3.14 \cdot 3 \text{ cm} \cdot 3 \text{ cm} = 28.26 \text{ cm}^2$

The area is about 28.26 cm^2.

27. $A = \pi \cdot r \cdot r$

$A \approx 3.14 \cdot 220 \text{ mi} \cdot 220 \text{ mi} = 151,976 \text{ mi}^2$

The broadcast area is about 151,976 mi^2.

29. $C = \pi \cdot d$

$C \approx 3.14 \cdot 1.1 \text{ ft} = 3.454 \text{ ft}$

The circumference of the elm tree is about 3.454 ft.

31. $C = \pi \cdot d$

$7.85 \text{ cm} \approx 3.14 \cdot d$ Substituting 7.85 cm for C and 3.14 for π

$\frac{7.85 \text{ cm}}{3.14} = d$ Dividing on both sides by 3.14

$2.5 \text{ cm} = d$

The diameter is about 2.5 cm.

$r = \frac{d}{2}$

$r = \frac{2.5 \text{ cm}}{2} = 1.25 \text{ cm}$

The radius is about 1.25 cm.

$A = \pi \cdot r \cdot r$

$A \approx 3.14 \cdot 1.25 \text{ cm} \cdot 1.25 \text{ cm} = 4.90625 \text{ cm}^2$

The area is about 4.90625 cm^2.

33. Find the area of the larger circle (pool plus walk). Its diameter is 1 yd + 20 yd + 1 yd, or 22 yd. Thus its radius is $\frac{22}{2}$ yd, or 11 yd.

$A = \pi \cdot r \cdot r$

$A \approx 3.14 \cdot 11 \text{ yd} \cdot 11 \text{ yd} = 379.94 \text{ yd}^2$

Find the area of the pool. Its diameter is 20 yd. Thus its radius is $\frac{20}{2}$ yd, or 10 yd.

$A = \pi \cdot r \cdot r$

$A \approx 3.14 \cdot 10 \text{ yd} \cdot 10 \text{ yd} = 314 \text{ yd}^2$

We subtract to find the area of the walk:

$A = 379.94 \text{ yd}^2 - 314 \text{ yd}^2$

$A = 65.94 \text{ yd}^2$

The area of the walk is 65.94 yd^2.

35. The perimeter consists of the circumferences of three semi-circles, each with diameter 8 ft, and one side of a square of length 8 ft. We first find the circumference of one semi-circle. This is one-half the circumference of a circle with diameter 8 ft:

$$\frac{1}{2} \cdot \pi \cdot d \approx \frac{1}{2} \cdot 3.14 \cdot 8 \text{ ft} = 12.56 \text{ ft}$$

Then we multiply by 3:

$$3 \cdot (12.56 \text{ ft}) = 37.68 \text{ ft}$$

Finally we add the circumferences of the semicircles and the length of the side of the square:

$$37.68 \text{ ft} + 8 \text{ ft} = 45.68 \text{ ft}$$

The perimeter is 45.68 ft.

37. The perimeter consists of three-fourths of the circumference of a circle with radius 4 yd and two sides of a square with sides of length 4 yd. We first find three-fourths of the circumference of the circle:

$$\frac{3}{4} \cdot 2 \cdot \pi \cdot r \approx 0.75 \cdot 2 \cdot 3.14 \cdot 4 \text{ yd} = 18.84 \text{ yd}$$

Then we add this length to the lengths of two sides of the square:

$$18.84 \text{ yd} + 4 \text{ yd} + 4 \text{ yd} = 26.84 \text{ yd}$$

The perimeter is 26.84 yd.

39. The perimeter consists of three-fourths of the perimeter of a square with side of length 10 yd and the circumference of a semicircle with diameter 10 yd. First we find three-fourths of the perimeter of the square:

$$\frac{3}{4} \cdot 4 \cdot s = \frac{3}{4} \cdot 4 \cdot 10 \text{ yd} = 30 \text{ yd}$$

Then we find one-half of the circumference of a circle with diameter 10 yd:

$$\frac{1}{2} \cdot \pi \cdot d \approx \frac{1}{2} \cdot 3.14 \cdot 10 \text{ yd} = 15.7 \text{ yd}$$

Then we add:

$$30 \text{ yd} + 15.7 \text{ yd} = 45.7 \text{ yd}$$

The perimeter is 45.7 yd.

41. The shaded region consists of a circle of radius 8 m, with two circles each of diameter 8 m, removed. First we find the area of the large circle:

$$A = \pi \cdot r \cdot r \approx 3.14 \cdot 8 \text{ m} \cdot 8 \text{ m} = 200.96 \text{ m}^2$$

Then we find the area of one of the small circles: The radius is $\frac{8 \text{ m}}{2} = 4$ m.

$$A = \pi \cdot r \cdot r \approx 3.14 \cdot 4 \text{ m} \cdot 4 \text{ m} = 50.24 \text{ m}^2$$

We multiply this area by 2 to find the area of the two small circles:

$$2 \cdot 50.24 \text{ m}^2 = 100.48 \text{ m}^2$$

Finally we subtract to find the area of the shaded region:

$$200.96 \text{ m}^2 - 100.48 \text{ m}^2 = 100.48 \text{ m}^2$$

The area of the shaded region is 100.48 m².

43. The shaded region consists of one-half of a circle with diameter 2.8 cm and a triangle with base 2.8 cm and height 2.8 cm. First we find the area of the semicircle. The radius is $\frac{2.8 \text{ cm}}{2} = 1.4$ cm.

$$A = \frac{1}{2} \cdot \pi \cdot r \cdot r \approx \frac{1}{2} \cdot 3.14 \cdot 1.4 \text{ cm} \cdot 1.4 \text{ cm} = 3.0772 \text{ cm}^2$$

Then we find the area of the triangle:

$$A = \frac{1}{2} \cdot b \cdot h = \frac{1}{2} \cdot 2.8 \text{ cm} \cdot 2.8 \text{ cm} = 3.92 \text{ cm}^2$$

Finally we add to find the area of the shaded region:

$$3.0772 \text{ cm}^2 + 3.92 \text{ cm}^2 = 6.9972 \text{ cm}^2$$

The area of the shaded region is 6.9972 cm².

45. The shaded area consists of a rectangle of dimensions 10.2 in. by 12.8 in., with the area of two semicircles, each of diameter 10.2 in., removed. This is equivalent to removing one circle with diameter 10.2 in. from the rectangle. First we find the area of the rectangle:

$$l \cdot w = (10.2 \text{ in.}) \cdot (12.8 \text{ in.}) = 130.56 \text{ in}^2$$

Then we find the area of the circle. The radius is $\frac{10.2 \text{ in.}}{2} = 5.1$ in..

$$\pi \cdot r \cdot r \approx 3.14 \cdot 5.1 \text{ in.} \cdot 5.1 \text{ in.} = 81.6714 \text{ in}^2$$

Finally we subtract to find the area of the shaded region:

$$130.56 \text{ in}^2 - 81.6714 \text{ in}^2 = 48.8886 \text{ in}^2$$

47. 0.875

a) Move the decimal point 2 places to the right.   0.87.5
                                                          ⌞↑

b) Add a percent symbol   87.5%
$0.875 = 87.5\%$

49. $0.\overline{6}$

a) Move the decimal point 2 places to the right. $0.66.\overline{6}$
                                                          ⌞↑

b) Add a percent symbol $66.\overline{6}\%$
$0.\overline{6} = 66.\overline{6}\%$

51. a) Find decimal notation using long division.

$$
\begin{array}{r}
0.375 \\
8\ \overline{\smash{)}3.000} \\
\underline{2\ 4} \\
6\ 0 \\
\underline{5\ 6} \\
4\ 0 \\
\underline{4\ 0} \\
0
\end{array}
$$

$\frac{3}{8} = 0.375$

b) Convert the decimal notation to percent notation. Move the decimal point two places to the right, and write a % symbol.

0.37.5
└───↑

$\frac{3}{8} = 37.5\%$

53. a) Find decimal notation using long division.

$$
\begin{array}{r}
0.66 \\
3\ \overline{\smash{)}2.00} \\
\underline{1\ 8} \\
2\ 0 \\
\underline{1\ 8} \\
2
\end{array}
$$

$\frac{2}{3} = 0.\overline{6}$

b) Convert the decimal notation to percent notation. Move the decimal point two places to the right, and write a % symbol.

0.66.$\overline{6}$
└───↑

$\frac{2}{3} = 00.\overline{0}\%$

55. Find $3927 \div 1250$ using a calculator.

$$\frac{3927}{1250} = 3.1416 \approx 3.142 \qquad \text{Rounding}$$

57. The height of the stack of tennis balls is three times the diameter of one ball, or $3 \cdot d$.

The circumference of one ball is given by $\pi \cdot d$.

The circumference of one ball is greater than the height of the stack of balls, because $\pi > 3$.

Exercise Set 9.7

1. $\sqrt{100} = 10$

The square root of 100 is 10 because $10^2 = 100$.

3. $\sqrt{441} = 21$

The square root of 441 is 21 because $21^2 = 441$.

5. $\sqrt{625} = 25$

The square root of 625 is 25 because $25^2 = 625$.

7. $\sqrt{361} = 19$

The square root of 361 is 19 because $19^2 = 361$.

9. $\sqrt{529} = 23$

The square root of 529 is 23 because $23^2 = 529$.

11. $\sqrt{10,000} = 100$

The square root of 10,000 is 100 because $100^2 = 10,000$.

13. $\sqrt{48} \approx 6.928$

15. $\sqrt{8} \approx 2.828$

17. $\sqrt{18} \approx 4.243$

19. $\sqrt{6} \approx 2.449$

21. $\sqrt{10} \approx 3.162$

23. $\sqrt{75} \approx 8.660$

25. $\sqrt{196} = 14$

27. $\sqrt{183} \approx 13.528$

29.
$$
\begin{aligned}
a^2 + b^2 &= c^2 &&\text{Pythagorean equation} \\
3^2 + 5^2 &= c^2 &&\text{Substituting} \\
9 + 25 &= c^2 \\
34 &= c^2 \\
\sqrt{34} &= c &&\text{Exact answer} \\
5.831 &\approx c &&\text{Approximation}
\end{aligned}
$$

31.
$$
\begin{aligned}
a^2 + b^2 &= c^2 &&\text{Pythagorean equation} \\
7^2 + 7^2 &= c^2 &&\text{Substituting} \\
49 + 49 &= c^2 \\
98 &= c^2 \\
\sqrt{98} &= c &&\text{Exact answer} \\
9.899 &\approx c &&\text{Approximation}
\end{aligned}
$$

33.
$$
\begin{aligned}
a^2 + b^2 &= c^2 \\
a^2 + 12^2 &= 13^2 \\
a^2 + 144 &= 169 \\
a^2 &= 169 - 144 = 25 \\
a &= 5
\end{aligned}
$$

35.
$$
\begin{aligned}
a^2 + b^2 &= c^2 \\
6^2 + b^2 &= 10^2 \\
36 + b^2 &= 100 \\
b^2 &= 100 - 36 = 64 \\
b &= 8
\end{aligned}
$$

37.
$$
\begin{aligned}
a^2 + b^2 &= c^2 \\
5^2 + 12^2 &= c^2 \\
25 + 144 &= c^2 \\
169 &= c^2 \\
13 &= c
\end{aligned}
$$

39.
$$
\begin{aligned}
a^2 + b^2 &= c^2 \\
18^2 + b^2 &= 30^2 \\
324 + b^2 &= 900 \\
b^2 &= 900 - 324 = 576 \\
b &= 24
\end{aligned}
$$

41. $a^2 + b^2 = c^2$
$a^2 + 1^2 = 20^2$
$a^2 + 1 = 400$
$a^2 = 400 - 1 = 399$
$a = \sqrt{399}$
$a \approx 19.975$

43. $a^2 + b^2 = c^2$
$1^2 + b^2 = 15^2$
$1 + b^2 = 225$
$b^2 = 225 - 1 = 224$
$b = \sqrt{224}$
$b \approx 14.967$

45. *Familiarize.* We first make a drawing. In it we see a right triangle. We let w = the length of the wire.

9 m

Translate. We substitute 9 for a, 13 for b, and w for c in the Pythagorean equation.

$$a^2 + b^2 = c^2$$
$$9^2 + 13^2 = w^2$$

Solve. We solve the equation for w.

$$81 + 169 = w^2$$
$$250 = w^2$$
$$\sqrt{250} = w \qquad \text{Exact answer}$$
$$15.811 \approx w \qquad \text{Approximation}$$

Check. $9^2 + 13^2 = 81 + 169 = 250 = (\sqrt{250})^2$

State. The length of the wire is $\sqrt{250}$m, or about 15.811 m.

47. *Familiarize.* We refer to the drawing in the text. We let d = the distance from home to second base.

Translate. We substitute 65 for a, 65 for b, and d for c in the Pythagorean equation.

$$a^2 + b^2 = c^2$$
$$65^2 + 65^2 = d^2$$

Solve. We solve the equation for d.

$$4225 + 4225 = d^2$$
$$8450 = d^2$$
$$\sqrt{8450} = d$$
$$91.924 \approx d$$

Check. $65^2 + 65^2 = 4225 + 4225 = 8450 = (\sqrt{8450})^2$

State. The distance from home to second base is $\sqrt{8450}$ ft, or about 91.924 ft.

49. *Familiarize.* We refer to the drawing in the text.

Translate. We substitute in the Pythagorean equation.

$$a^2 + b^2 = c^2$$
$$20^2 + h^2 = 30^2$$

Solve. We solve the equation for h.

$$400 + h^2 = 900$$
$$h^2 = 900 - 400$$
$$h^2 = 500$$
$$h = \sqrt{500}$$
$$h \approx 22.361$$

Check. $20^2 + (\sqrt{500})^2 = 400 + 500 = 900 = 30^2$

State. The height of the tree is $\sqrt{500}$ ft, or about 22.361 ft.

51. *Familiarize.* We refer to the drawing in the text. We let h = the plane's horizontal distance from the airport.

Translate. We substitute 4100 for a, h for b, and 15,100 for c in the Pythagorean equation.

$$a^2 + b^2 = c^2$$
$$4100^2 + h^2 = 15,100^2$$

Solve. We solve the equation for h.

$$16,810,000 + h^2 = 228,010,000$$
$$h^2 = 228,010,000 - 16,810,000$$
$$h^2 = 211,200,000$$
$$h = \sqrt{211,200,000}$$
$$h \approx 14,532.7$$

Check. $4100^2 + (\sqrt{211,200,000})^2 = 16,810,000 + 211,200,000 = 228,010,000 = 15,100^2$

State. The plane's horizontal distance from the airport is about 14,532.7 ft.

53. a) Replace the percent symbol with $\times 0.01$.

$$45.6 \times 0.01$$

b) Move the decimal point two places to the left.

$$0.45.6$$

Thus, $45.6\% = 0.456$.

55. a) Replace the percent symbol with $\times 0.01$.

$$123 \times 0.01$$

b) Move the decimal point two places to the left.

$$1.23.$$

Thus, $123\% = 1.23$.

57. a) Replace the percent symbol with $\times 0.01$.

$$0.41 \times 0.01$$

b) Move the decimal point two places to the left.

$$0.00.41$$

Thus, $0.41\% = 0.0041$.

59. To find the areas we must first use the Pythagorean equation to find the height of each triangle and then use the formula for the area of a triangle.

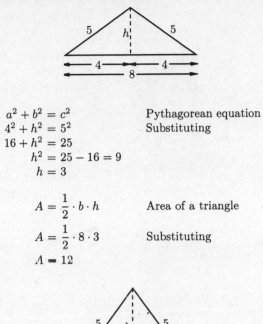

$$a^2 + b^2 = c^2 \qquad \text{Pythagorean equation}$$
$$4^2 + h^2 = 5^2 \qquad \text{Substituting}$$
$$16 + h^2 = 25$$
$$h^2 = 25 - 16 = 9$$
$$h = 3$$

$$A = \frac{1}{2} \cdot b \cdot h \qquad \text{Area of a triangle}$$

$$A = \frac{1}{2} \cdot 8 \cdot 3 \qquad \text{Substituting}$$

$$A = 12$$

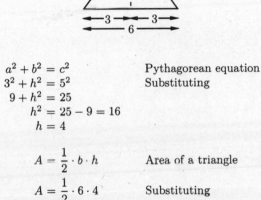

$$a^2 + b^2 = c^2 \qquad \text{Pythagorean equation}$$
$$3^2 + h^2 = 5^2 \qquad \text{Substituting}$$
$$9 + h^2 = 25$$
$$h^2 = 25 - 9 = 16$$
$$h = 4$$

$$A = \frac{1}{2} \cdot b \cdot h \qquad \text{Area of a triangle}$$

$$A = \frac{1}{2} \cdot 6 \cdot 4 \qquad \text{Substituting}$$

$$A = 12$$

The areas of the triangles are the same (12 square units).

Chapter 10

More on Measures

Exercise Set 10.1

1. $V = l \cdot w \cdot h$
   $V = 12 \text{ cm} \cdot 8 \text{ cm} \cdot 8 \text{ cm}$
   $V = 12 \cdot 64 \text{ cm}^3$
   $V = 768 \text{ cm}^3$

3. $V = l \cdot w \cdot h$
   $V = 7.5 \text{ in.} \cdot 2 \text{ in.} \cdot 3 \text{ in.}$
   $V = 7.5 \cdot 6 \text{ in}^3$
   $V = 45 \text{ in}^3$

5. $V = l \cdot w \cdot h$
   $V = 10 \text{ m} \cdot 5 \text{ m} \cdot 1.5 \text{ m}$
   $V = 10 \cdot 7.5 \text{ m}^3$
   $V = 75 \text{ m}^3$

7. $V = l \cdot w \cdot h$
   $V = 6\frac{1}{2} \text{ yd} \cdot 5\frac{1}{2} \text{ yd} \cdot 10 \text{ yd}$
   $V = \frac{13}{2} \cdot \frac{11}{2} \cdot 10 \text{ yd}^3$
   $V = \frac{715}{2} \text{ yd}^3$
   $V = 357\frac{1}{2} \text{ yd}^3$

9. $1 \text{ L} = 1000 \text{ mL} = 1000 \text{ cm}^3$

   These conversion relations appear in the text on page 477.

11. $87 \text{ L} = 87 \times (1 \text{ L})$
    $= 87 \times (1000 \text{ mL})$
    $= 87,000 \text{ mL}$

13. $49 \text{ mL} = 49 \times (1 \text{ mL})$
    $= 49 \times (0.001 \text{ L})$
    $= 0.049 \text{ L}$

15. $0.401 \text{ mL} = 0.401 \times (1 \text{ mL})$
    $= 0.401 \times (0.001 \text{ L})$
    $= 0.000401 \text{ L}$

17. $78.1 \text{ L} = 78.1 \times (1 \text{ L})$
    $= 78.1 \times (1000 \text{ cm}^3)$
    $= 78,100 \text{ cm}^3$

19. $10 \text{ qt} = 10 \times 1 \text{ qt}$
    $= 10 \times 2 \text{ pt}$
    $= 10 \times 2 \times 1 \text{ pt}$
    $= 10 \times 2 \times 16 \text{ oz}$
    $= 320 \text{ oz}$

21. $20 \text{ cups} = 20 \text{ cups} \cdot \frac{1 \text{ pt}}{2 \text{ cups}} = \frac{20}{2} \cdot 1 \text{ pt} = 10 \text{ pt}$

23. $8 \text{ gal} = 8 \times 1 \text{ gal}$
    $= 8 \times 4 \text{ qt}$
    $= 32 \text{ qt}$

25. We convert 0.5 L to milliliters:
    $0.5 \text{ L} = 0.5 \times (1 \text{ L})$
    $= 0.5 \times (1000 \text{ mL})$
    $= 500 \text{ mL}$

27. First convert 3 L to milliliters:
    $3 \text{ L} = 3 \times (1 \text{ L})$
    $= 3 \times (1000 \text{ mL})$
    $= 3000 \text{ mL}$

    Then translate to an equation.

    $$\underbrace{\text{Amount of saline solution, in mL}} \div \underbrace{\begin{array}{c}\text{Number} \\ \text{of hours}\end{array}} = \underbrace{\begin{array}{c}\text{Amount of saline solution} \\ \text{per hour}\end{array}}$$
    $$\downarrow \qquad \downarrow \quad \downarrow \quad \downarrow \qquad \downarrow$$
    $$3000 \quad \div \quad 24 \quad = \quad s$$
    $$\qquad\qquad 125 \quad = s \quad \text{Carrying out the division}$$

    Thus, 125 mL per hour must be administered.

29. First convert 32 oz to gallons:

    $$32 \text{ oz} = 32 \text{ oz} \cdot \frac{1 \text{ pt}}{16 \text{ oz}} \cdot \frac{1 \text{ qt}}{2 \text{ pt}} \cdot \frac{1 \text{ gal}}{4 \text{ qt}}$$
    $$= \frac{32}{16 \cdot 2 \cdot 4} \times 1 \text{ gal} = 0.25 \text{ gal}$$

    Thus 0.25 gal per day are wasted by one person. We multiply to find how many gallons are wasted in a week (7 days) by one person:

    $$7 \times 0.25 \text{ gal} = 1.75 \text{ gal}$$

    Next we multiply to find how many gallons are wasted in a month (30 days) by one person:

    $$30 \times 0.25 \text{ gal} = 7.5 \text{ gal}$$

    We multiply again to find how many gallons are wasted in a year (365 days) by one person:

    $$365 \times 0.25 \text{ gal} = 91.25 \text{ gal}$$

    To find how much water is wasted in this country in a day, we multiply 0.25 gal by 261 million:

    $$261,000,000 \times 0.25 \text{ gal} = 62,250,000 \text{ gal}$$

    To find how much water is wasted in this country in a year, we multiply 91.25 gal by 261 million:

    $$261,000,000 \times 91.25 \text{ gal} = 23,816,250,000 \text{ gal}$$

31. Interest = (Interest for 1 year) $\times \frac{1}{2}$

$$= (6.4\% \times \$600) \times \frac{1}{2}$$

$$= 0.064 \times \$600 \times \frac{1}{2}$$

$$= \$38.40 \times \frac{1}{2}$$

$$= \$19.20$$

The interest is \$19.20.

33. $10^3 = 10 \cdot 10 \cdot 10 = 1000$

35. $7^2 = 7 \cdot 7 = 49$

37. First find the volume of one one-dollar bill in cubic inches:

$$V = l \cdot w \cdot h$$
$$V = 6.0625 \text{ in.} \times 2.3125 \text{ in.} \times 0.0041 \text{ in.}$$
$$V = 0.05748 \text{ in}^3 \qquad \text{Rounding}$$

Then multiply to find the volume of one million one-dollar bills in cubic inches:

$$1,000,000 \times 0.05748 \text{ in}^3 = 57,480 \text{ in}^3$$

Thus the volume of one million one-dollar bills is about $57,480 \text{ in}^3$.

Next we convert to cubic feet:

$$57,480 \text{ in}^3 = 57,480 \times \text{ in.} \times \text{ in.} \times \text{ in.}$$
$$= 57,480 \times \text{ in.} \times \text{ in.} \times \text{ in.} \times$$
$$\frac{1 \text{ ft}}{12 \text{ in.}} \times \frac{1 \text{ ft}}{12 \text{ in.}} \times \frac{1 \text{ ft}}{12 \text{ in.}}$$
$$= \frac{57,480}{12 \cdot 12 \cdot 12} \times \frac{\text{in.}}{\text{in.}} \times \frac{\text{in.}}{\text{in.}} \times \frac{\text{in.}}{\text{in.}} \times$$
$$\text{ft} \times \text{ft} \times \text{ft}$$
$$\approx 33.3 \text{ ft}^3$$

Exercise Set 10.2

1. $V = Bh = \pi \cdot r^2 \cdot h$
$\approx 3.14 \times 8 \text{ in.} \times 8 \text{ in.} \times 4 \text{ in.}$
$\approx 803.84 \text{ in}^3$

3. $V = Bh = \pi \cdot r^2 \cdot h$
$\approx 3.14 \times 5 \text{ cm} \times 5 \text{ cm} \times 4.5 \text{ cm}$
$\approx 353.25 \text{ cm}^3$

5. $V = Bh = \pi \cdot r^2 \cdot h$
$\approx \frac{22}{7} \times 210 \text{ yd} \times 210 \text{ yd} \times 300 \text{ yd}$
$\approx 41,580,000 \text{ yd}^3$

7. $V = \frac{4}{3} \cdot \pi \cdot r^3$
$\approx \frac{4}{3} \times 3.14 \times (100 \text{ in.})^3$
$\approx \frac{4 \times 3.14 \times 1,000,000 \text{ in}^3}{3}$
$\approx 4,186,666\frac{2}{3} \text{ in}^3$

9. $V = \frac{4}{3} \cdot \pi \cdot r^3$
$\approx \frac{4}{3} \times 3.14 \times (3.1 \text{ m})^3$
$\approx \frac{4 \times 3.14 \times 29.791 \text{ m}^3}{3}$
$\approx 124.72 \text{ m}^3$

11. $V = \frac{4}{3} \cdot \pi \cdot r^3$
$\approx \frac{4}{3} \times \frac{22}{7} \times (7 \text{ km})^3$
$\approx \frac{4 \times 22 \times 343 \text{ km}^3}{3 \times 7}$
$\approx 1437\frac{1}{3} \text{ km}^3$

13. $V = \frac{1}{3} \cdot \pi \cdot r^2 \cdot h$
$\approx \frac{1}{3} \times 3.14 \times 33 \text{ ft} \times 33 \text{ ft} \times 100 \text{ ft}$
$\approx 113,982 \text{ ft}^3$

15. $V = \frac{1}{3} \cdot \pi \cdot r^2 \cdot h$
$\approx \frac{1}{3} \times \frac{22}{7} \times 1.4 \text{ cm} \times 1.4 \text{ cm} \times 12 \text{ cm}$
$\approx 24.64 \text{ cm}^3$

17. We must find the radius of the base in order to use the formula for the volume of a circular cylinder.

$$r = \frac{d}{2} = \frac{14 \text{ yd}}{2} = 7 \text{ yd}$$
$$V = Bh = \pi \cdot r^2 \cdot h$$
$$\approx \frac{22}{7} \times 7 \text{ yd} \times 7 \text{ yd} \times 220 \text{ yd}$$
$$\approx 33,880 \text{ yd}^3$$

19. We must find the radius of the silo in order to use the formula for the volume of a circular cylinder.

$$r = \frac{d}{2} = \frac{6 \text{ m}}{2} = 3 \text{ m}$$
$$V = Bh = \pi \cdot r^2 \cdot h$$
$$\approx 3.14 \times 3 \text{ m} \times 3 \text{ m} \times 13 \text{ m}$$
$$\approx 367.38 \text{ m}^3$$

21. First we find the radius of the ball:

$$r = \frac{d}{2} = \frac{6.5 \text{ cm}}{2} = 3.25 \text{ cm}$$

Then we find the volume, using the formula for the volume of a sphere.

$$V = \frac{4}{3} \cdot \pi \cdot r^3$$
$$\approx \frac{4}{3} \cdot 3.14 \cdot (3.25 \text{ cm})^3$$
$$\approx 143.72 \text{ cm}^3$$

23. First we find the radius of the earth:

$$\frac{3980 \text{ mi}}{2} = 1990 \text{ mi}$$

Then we find the volume, using the formula for the volume of a sphere.

$$V = \frac{4}{3} \cdot \pi \cdot r^3$$
$$\approx \frac{4}{3} \cdot 3.14 \cdot (1990 \text{ mi})^3$$
$$\approx 32,993,441,150 \text{ mi}^3$$

25. First find the radius of the tank in inches:

$$r = \frac{d}{2} = \frac{16 \text{ in.}}{2} = 8 \text{ in.}$$

Then convert 8 in. to feet:

$$8 \text{ in.} = 8 \text{ in.} \times \frac{1 \text{ ft}}{12 \text{ in.}} = \frac{8}{12} \times \frac{\text{in.}}{\text{in.}} \times \text{ft} = \frac{2}{3} \text{ ft}$$

Find the volume of the tank:

$$V = Bh = \pi \cdot r^2 \cdot h$$
$$\approx 3.14 \times \frac{2}{3} \text{ ft} \times \frac{2}{3} \text{ ft} \times 5 \text{ ft} \approx 6.97 \text{ ft}^3$$

The volume of the tank is $6.9\overline{7}$ ft^3.

We multiply to find the number of gallons the tank will hold:

$$6.9\overline{7} \times 7.5 \text{ gal} = 52.\overline{3} \text{ gal}$$

The tank will hold $52.\overline{3}$ gallons of water.

Exercise Set 10.3

1. 1 T = 2000 lb

This conversion relation is given in the text on page 485.

3. $6000 \text{ lb} = 6000 \text{ lb} \times \frac{1 \text{ T}}{2000 \text{ lb}}$ Multiplying by 1 using $\frac{1 \text{ T}}{2000 \text{ lb}}$

$$= \frac{6000}{2000} \times \frac{\text{lb}}{\text{lb}} \times 1 \text{ T}$$
$$= 3 \times 1 \text{ T}$$ The $\frac{\text{lb}}{\text{lb}}$ acts like 1, so we can omit it.
$$= 3 \text{ T}$$

5. $4 \text{ lb} = 4 \times 1 \text{ lb}$
$$= 4 \times 16 \text{ oz}$$ Substituting 16 oz for 1 lb
$$= 64 \text{ oz}$$

7. $6.32 \text{ T} = 6.32 \times 1 \text{ T}$
$$= 6.32 \times 2000 \text{ lb}$$ Substituting 2000 lb for 1 T
$$= 12,640 \text{ lb}$$

9. $3200 \text{ oz} = 3200 \text{ oz} \times \frac{1 \text{ lb}}{16 \text{ oz}} \times \frac{1 \text{ T}}{2000 \text{ lb}}$
$$= \frac{3200}{16 \times 2000} \text{ T}$$
$$= \frac{1}{10} \text{ T, or } 0.1 \text{ T}$$

11. $80 \text{ oz} = 80 \text{ oz} \times \frac{1 \text{ lb}}{16 \text{ oz}}$
$$= \frac{80}{16} \text{ lb}$$
$$= 5 \text{ lb}$$

13. 1 kg = _____ g

Think: To go from kg to g in the table is a move of 3 places to the right. Thus, we move the decimal point 3 places to the right.

1 1.000.

1 kg = 1000 g

15. 1 dag = _____ g

Think: To go from dag to g in the table is a move of 1 place to the right. Thus, we move the decimal point 1 place to the right.

1 1.0.

1 dag = 10 g

17. 1 cg = _____ g

Think: To go from cg to g in the table is a move of 2 places to the left. Thus, we move the decimal point 2 places to the left.

1 0.01.

1 cg = 0.01 g

19. 1 g = _____ mg

Think: To go from g to mg in the table is a move of 3 places to the right. Thus, we move the decimal point 3 places to the right.

1 1.000.

1 g = 1000 mg

21. 1 g = _____ dg

Think: To go from g to dg in the table is a move of 1 place to the right. Thus, we move the decimal point 1 place to the right.

1 1.0.

1 g = 10 dg

23. Complete: 234 kg = _____ g

Think: To go from kg to g in the table is a move of 3 places to the right. Thus, we move the decimal point 3 places to the right.

234 234.000.

234 kg = 234,000 g

25. Complete: 5200 g = _____ kg

Think: To go from g to kg in the table is a move of 3 places to the left. Thus, we move the decimal point 3 places to the left.

5200   5.200.

5200 g = 5.2 kg

27. Complete: 67 hg = _____ kg

Think: To go from hg to kg in the table is a move of 1 place to the left. Thus, we move the decimal point 1 place to the left.

67   6.7.

67 hg = 6.7 kg

29. Complete: 0.502 dg = _____ g

Think: To go from dg to g in the table is a move of 1 place to the left. Thus, we move the decimal point 1 place to the left.

0.502   0.0.502

0.502 dg = 0.0502 g

31. Complete: 8492 g = _____ kg

Think: To go from g to kg in the table is a move of 3 places to the left. Thus, we move the decimal point 3 places to the left.

8492   8.492.

8492 g = 8.492 kg

33. Complete: 585 mg = _____ cg

Think: To go from mg to cg in the table is a move of 1 place to the left. Thus, we move the decimal point 1 place to the left.

585   58.5.

585 mg = 58.5 cg

35. Complete: 8 kg = _____ cg

Think: To go from kg to cg in the table is a move of 5 places to the right. Thus, we move the decimal point 5 places to the right.

8   8.00000.

8 kg = 800,000 cg

37. 1 t = 1000 kg

This conversion relation is given in the text on page 486.

39. Complete: 3.4 cg = _____ dag

Think: To go from cg to dag in the table is a move of 3 places to the left. Thus, we move the decimal point 3 places to the left.

3.4   0.003.4

3.4 cg = 0.0034 dag

41. 1 day = 24 hr

This conversion relation is given in the text on page 488.

43. 1 min = 60 sec

This conversion relation is given in the text on page 488.

45. 1 yr = $365\frac{1}{4}$ days

This conversion relation is given in the text on page 488.

47. $180 \text{ sec} = 180 \text{ sec} \cdot \frac{1 \text{ min}}{60 \text{ sec}} \cdot \frac{1 \text{ hr}}{60 \text{ min}}$

$= \frac{180}{60 \cdot 60} \text{ hr}$

$= 0.05 \text{ hr}$

49. $492 \text{ sec} = 492 \text{ sec} \times \frac{1 \text{ min}}{60 \text{ sec}}$

$= \frac{492}{60} \text{ min}$

$= 8.2 \text{ min}$

51. $156 \text{ hr} = 156 \text{ hr} \cdot \frac{1 \text{ day}}{24 \text{ hr}}$

$= \frac{156}{24} \text{ days}$

$= 6.5 \text{ days}$

53. $645 \text{ min} = 645 \text{ min} \cdot \frac{1 \text{ hr}}{60 \text{ min}}$

$= \frac{645}{60} \text{ hr}$

$= 10.75 \text{ hr}$

55. 2 wk = 2 × 1 wk

= 2 × 7 days   Substituting 7 days for 1 wk

= 14 days

= 14 × 1 day

= 14 × 24 hr   Substituting 24 hr for 1 day

= 336 hr

57. $756 \text{ hr} = 756 \text{ hr} \cdot \frac{1 \text{ day}}{24 \text{ hr}} \cdot \frac{1 \text{ wk}}{7 \text{ days}}$

$= \frac{756}{24 \cdot 7} \text{ wk}$

$= 4.5 \text{ wk}$

59. $2922 \text{ wk} = 2922 \text{ wk} \cdot \frac{7 \text{ days}}{1 \text{ wk}} \cdot \frac{1 \text{ yr}}{365\frac{1}{4} \text{ days}}$

$= \frac{2922 \cdot 7}{365\frac{1}{4}} \text{ yr}$

$= 56 \text{ yr}$

61. $2^4 = 2 \cdot 2 \cdot 2 \cdot 2 = 16$

63. $5^3 = 5 \cdot 5 \cdot 5 = 125$

65. 5.43 m = _____ cm

Think: To go from m to cm in the table is a move of 2 places to the right. Thus, we move the decimal point 2 places to the right.

5.43   5.43.

5.43 m = 543 cm

67. 1 lb = 453.5 g   Using the conversion given

453.5 g = _____ kg

Think: To go from g to kg in the table is a move of 3 places to the left. Thus, we move the decimal point 3 places to the left.

453.5   0.453.5

1 lb = 0.4535 kg

69. First, use a proportion to determine how many eggs are in one pound.

$$\begin{matrix} \text{eggs} \rightarrow \\ \text{price} \rightarrow \end{matrix} \quad \frac{12}{0.90} = \frac{x}{0.60} \quad \begin{matrix} \leftarrow \text{eggs} \\ \leftarrow \text{price} \end{matrix}$$

$$12 \cdot 0.60 = 0.90 \cdot x$$

$$\frac{12 \cdot 0.60}{0.90} = x$$

$$8 = x$$

There are 8 eggs in one pound.

Then use a proportion to find the weight of one egg.

$$\begin{matrix} \text{weight} \rightarrow \\ \text{eggs} \rightarrow \end{matrix} \quad \frac{16}{8} = \frac{w}{1} \quad \begin{matrix} \leftarrow \text{weight} \\ \leftarrow \text{eggs} \end{matrix}$$

$$16 \cdot 1 = 8 \cdot w$$

$$\frac{16 \cdot 1}{8} = w$$

$$2 = w$$

One egg weighs 2 oz.

71. Use a calculator.

1,000,000,000 sec

$$= 1,000,000,000 \text{ sec} \times \frac{1 \text{ min}}{60 \text{ sec}} \times \frac{1 \text{ hr}}{60 \text{ min}} \times$$

$$\frac{1 \text{ day}}{24 \text{ hr}} \times \frac{1 \text{ yr}}{365\frac{1}{4} \text{ days}}$$

$$\approx 31.7 \text{ yr} \quad \text{Rounded to the nearest tenth}$$

One billion seconds is approximately 31.7 years.

73. $1 \text{ mg} = \frac{1}{1000} \text{ g}$ Using the conversion on page 486 of the text

$$\frac{1}{1000} \text{ g} = 1000 \times \frac{1}{1,000,000} \text{ g}$$

$$= 1000 \times 1 \text{ } \mu\text{g} \quad \text{Using the definition of } \mu\text{g}$$

$$= 1000 \text{ } \mu\text{g}$$

$$1 \text{ mg} = 1000 \text{ } \mu\text{g}$$

75. Convert 125 μg to milligrams.

$$125 \text{ } \mu\text{g} = 125 \times 1 \text{ } \mu\text{g}$$

$$= 125 \times \frac{1}{1,000,000} \text{ g}$$

$$= 125 \times \frac{1}{1000} \times \frac{1}{1000} \text{ g}$$

$$= \frac{125}{1000} \times 1 \text{ mg}$$

$$= 0.125 \text{ mg, or } \frac{1}{8} \text{ mg}$$

$$125 \text{ } \mu\text{g} = 0.125 \text{ mg}$$

77. First convert 500 mg to grams.

Think: To go from mg to g in the table is a move of 3 places to the left. Thus, we move the decimal point 3 places to the left.

500   0.500.

500 mg = 0.5 g

Now translate to an equation.

$$\underbrace{\begin{matrix} \text{Dosage per} \\ \text{tablet} \\ \text{in g} \end{matrix}}_{} \quad \text{times} \quad \underbrace{\begin{matrix} \text{Number of} \\ \text{tablets} \end{matrix}}_{} \quad \text{is} \quad \underbrace{\begin{matrix} \text{Total} \\ \text{dosage} \\ \text{in g} \end{matrix}}_{}$$

$$0.5 \quad \times \quad n \quad = \quad 2$$

To solve we divide on both sides by 0.5.

$$n = 2 \div 0.5$$

$$n = 4$$

Thus, 4 tablets would have to be taken.

79. We solve using a proportion.

$$\begin{matrix} \text{cephalexin} \rightarrow \\ \text{liquid} \rightarrow \end{matrix} \quad \frac{250}{5} = \frac{400}{a} \quad \begin{matrix} \leftarrow \text{cephalexin} \\ \leftarrow \text{liquid} \end{matrix}$$

Solve: $250 \cdot a = 5 \cdot 400$

$$a = \frac{5 \cdot 400}{250}$$

$$a = 8$$

Thus, 8 mL of liquid would be required.

Exercise Set 10.4

1. By laying a ruler or piece of paper horizontally between the scales on page 493, we see that 178°F ≈ 80°C.

3. By laying a ruler or piece of paper horizontally between the scales on page 493, we see that 140°F ≈ 60°C.

5. By laying a ruler or piece of paper horizontally between the scales on page 493, we see that 68°F ≈ 20°C.

7. By laying a ruler or piece of paper horizontally between the scales on page 493, we see that 10°F ≈ −10°C.

9. By laying a ruler or piece of paper horizontally between the scales on page 493, we see that 86°C ≈ 190°F.

11. By laying a ruler or piece of paper horizontally between the scales on page 493, we see that 58°C ≈ 140°F.

13. By laying a ruler or piece of paper horizontally between the scales on page 493, we see that −10°C ≈ 10°F.

15. By laying a ruler or piece of paper horizontally between the scales on page 493, we see that 5°C ≈ 40°F.

17. $F = \dfrac{9}{5} \cdot C + 32$

     $F = \dfrac{9}{5} \cdot 25 + 32$

     $= 45 + 32$

     $= 77$

Thus, $25°C = 77°F$.

19. $F = \dfrac{9}{5} \cdot C + 32$

     $F = \dfrac{9}{5} \cdot 40 + 32$

     $= 72 + 32$

     $= 104$

Thus, $40°C = 104°F$.

21. $F = \dfrac{9}{5} \cdot C + 32$

     $F = \dfrac{9}{5} \cdot 3000 + 32$

     $= 5400 + 32$

     $= 5432$

Thus, $3000°C = 5432°F$.

23. $C = \dfrac{5}{9} \cdot (F - 32)$

     $C = \dfrac{5}{9} \cdot (86 - 32)$

     $= \dfrac{5}{9} \cdot 54$

     $= 30$

Thus, $86°F = 30°C$.

25. $C = \dfrac{5}{9} \cdot (F - 32)$

     $C = \dfrac{5}{9} \cdot (131 - 32)$

     $= \dfrac{5}{9} \cdot 99$

     $= 55$

Thus, $131°F = 55°C$.

27. $C = \dfrac{5}{9} \cdot (F - 32)$

     $C = \dfrac{5}{9} \cdot (98.6 - 32)$

     $= \dfrac{5}{9} \cdot 66.6$

     $= 37$

Thus, $98.6°F = 37°C$.

29. 23.4 cm = _____ mm

Think: To go from cm to mm in the table is a move of 1 place to the right. Thus, we move the decimal point 1 place to the right.

     23.4   23.4.

          $\llcorner\uparrow$

23.4 cm = 234 mm

31. 28 ft $= 28 \times 1$ ft $= 28 \times 12$ in. $= 336$ in.

33. 72.4 m = _____ cm

Think: To go from m to cm in the table is a move of 2 places to the right. Thus, we move the decimal point 2 places to the right.

     72.4   72.40.

            $\llcorner\uparrow$

72.4 m = 7240 cm

35. We first convert from Kelvin to Celsius temperature and then from Celsius to Fahrenheit temperature.

     $K = C + 273$

   $400 = C + 273$ Substituting

   $127 = C$ Subtracting 273 on both sides

Thus, $400°\text{Kelvin} = 127°C$.

     $F = \dfrac{9}{5} \cdot C + 32$

     $F = \dfrac{9}{5} \cdot 127 + 32$

     $= 228.6 + 32$

     $= 260.6$

Thus, $127°C = 260.6°F$, so the reaction will take place at $260.6°F$.

Exercise Set 10.5

1. 1 ft$^2 = 144$ in^2

This conversion relation is given in the text on page 497.

3. 1 mi$^2 = 640$ acres

This conversion relation is given in the text on page 497.

5. 1 in$^2 = 1$ in$^2 \times \dfrac{1 \text{ ft}^2}{144 \text{ in}^2}$ Multiplying by 1

                                  using $\dfrac{1 \text{ ft}^2}{144 \text{ in}^2}$

     $= \dfrac{1}{144} \times \dfrac{\text{in}^2}{\text{in}^2} \times 1 \text{ ft}^2$

     $= \dfrac{1}{144} \text{ ft}^2$

7. 22 yd$^2 = 22 \times 1$ yd^2

           $= 22 \times 9$ ft^2 Substituting 9 ft^2
                                for 1 yd^2

           $= 198$ ft^2

9. 44 yd$^2 = 44 \cdot 1$ yd^2

           $= 44 \cdot 9$ ft^2 Substituting 9 ft^2
                                for 1 yd^2

           $= 396$ ft^2

11. 20 mi$^2 = 20 \times 1$ mi^2

           $= 20 \cdot 640$ acres   Substituting 640 acres
                                   for 1 mi^2

           $= 12,800$ acres

13. 1 mi$^2 = 1 \cdot (1 \text{ mi})^2$

           $= 1 \cdot (5280 \text{ ft})^2$ Substituting 5280 ft
                                 for 1 mi

           $= 5280 \text{ ft} \cdot 5280 \text{ ft}$

           $= 27,878,400 \text{ ft}^2$

15. $720 \text{ in}^2 = 720 \text{ in}^2 \times \dfrac{1 \text{ ft}^2}{144 \text{ in}^2}$ Multiplying by 1

$\qquad\qquad\qquad\qquad\qquad$ using $\dfrac{1 \text{ ft}^2}{144 \text{ in}^2}$

$\qquad\quad = \dfrac{720}{144} \times \dfrac{\text{in}^2}{\text{in}^2} \times 1 \text{ ft}^2$

$\qquad\quad = 5 \text{ ft}^2$

17. $144 \text{ in}^2 = 1 \text{ ft}^2$

This conversion relation is given in the text on page 497.

19. $1 \text{ acre} = 1 \text{ acre} \cdot \dfrac{1 \text{ mi}^2}{640 \text{ acres}}$

$\qquad\quad = \dfrac{1}{640} \cdot \dfrac{\text{acres}}{\text{acres}} \cdot 1 \text{ mi}^2$

$\qquad\quad = \dfrac{1}{640} \text{ mi}^2$, or 0.0015625 mi^2

21. $5.21 \text{ km}^2 = \underline{\qquad} \text{ m}^2$

Think: To go from km to m in the table is a move of 3 places to the right. So we move the decimal point $2 \cdot 3$, or 6 places to the right.

5.21   5.210000.

$5.21 \text{ km}^2 = 5{,}210{,}000 \text{ m}^2$

23. $0.014 \text{ m}^2 = \underline{\qquad} \text{ cm}^2$

Think: To go from m to cm in the table is a move of 2 places to the right. So we move the decimal point $2 \cdot 2$, or 4 places to the right.

0.014   0.0140.

$0.014 \text{ m}^2 = 140 \text{ cm}^2$

25. $2345.6 \text{ mm}^2 = \underline{\qquad} \text{ cm}^2$

Think: To go from mm to cm in the table is a move of 1 place to the left. So we move the decimal point $2 \cdot 1$, or 2 places to the left.

2345.6   23.45.6

$2345.6 \text{ mm}^2 = 23.456 \text{ cm}^2$

27. $852.14 \text{ cm}^2 = \underline{\qquad} \text{ m}^2$

Think: To go from cm to m in the table is a move of 2 places to the left. So we move the decimal point $2 \cdot 2$, or 4 places to the left.

852.14   0.0852.14

$852.14 \text{ cm}^2 = 0.085214 \text{ m}^2$

29. $250{,}000 \text{ mm}^2 = \underline{\qquad} \text{ cm}^2$

Think: To go from mm to cm in the table is a move of 1 place to the left. So we move the decimal point $2 \cdot 1$, or 2 places to the left.

250,000   2500.00.

$250{,}000 \text{ mm}^2 = 2500 \text{ cm}^2$

31. $472{,}800 \text{ m}^2 = \underline{\qquad} \text{ km}^2$

Think: To go from m to km in the table is a move of 3 places to the left. So we move the decimal point $2 \cdot 3$, or 6 places to the left.

472,800   0.472800.

$472{,}800 \text{ m}^2 = 0.4728 \text{ km}^2$

33. Interest $= (\text{Interest for 1 year}) \times 1.5$

$\qquad\quad = (8\% \times 2000) \times 1.5$

$\qquad\quad = 0.08 \times 2000 \times 1.5$

$\qquad\quad = 160 \times 1.5$

$\qquad\quad = 240$

The interest is \$240.

35. $22{,}176 \text{ ft} = 22{,}176 \text{ ft} \times \dfrac{1 \text{ mi}}{5280 \text{ ft}}$ Multiplying by 1

$\qquad\qquad\qquad\qquad\qquad\qquad$ using $\dfrac{1 \text{ mi}}{5280 \text{ ft}}$

$\qquad\quad = \dfrac{22{,}176}{5280} \times \dfrac{\text{ft}}{\text{ft}} \times 1 \text{ mi}$

$\qquad\quad = 4.2 \text{ mi}$

37. $1 \text{ m}^2 = 1 \times 1 \text{ m} \times 1 \text{ m}$

$\qquad\quad = 1 \times 3.3 \text{ ft} \times 3.3 \text{ ft}$

$\qquad\quad = 1 \times 3.3 \times 3.3 \times \text{ ft} \times \text{ ft}$

$\qquad\quad = 10.89 \text{ ft}^2$

39. $2 \text{ yd}^2 = 2 \times 1 \text{ yd} \times 1 \text{ yd}$

$\qquad\quad = 2 \times 3 \text{ ft} \times 3 \text{ ft} \times \dfrac{1 \text{ m}}{3.3 \text{ ft}} \times \dfrac{1 \text{ m}}{3.3 \text{ ft}}$

$\qquad\quad = \dfrac{2 \times 3 \times 3}{3.3 \times 3.3} \times \text{ m} \times \text{ m}$

$\qquad\quad \approx 1.65 \text{ m}^2$

41. $20{,}175 \text{ ft}^2 = 20{,}175 \times 1 \text{ ft} \times 1 \text{ ft}$

$\qquad\quad = 20{,}175 \times 1 \text{ ft} \times 1 \text{ ft} \times$

$\qquad\qquad\quad \dfrac{1 \text{ m}}{3.3 \text{ ft}} \times \dfrac{1 \text{ m}}{3.3 \text{ ft}}$

$\qquad\quad = \dfrac{20{,}175}{3.3 \times 3.3} \times \text{ m} \times \text{ m}$

$\qquad\quad \approx 1852.6 \text{ m}^2$

Chapter 11

The Real-Number System

Exercise Set 11.1

1. The integer -200 corresponds to losing \$200, and the integer 600 corresponds to winning \$600.

3. The integer -34 corresponds to team A being 34 pins behind team B; the integer 15 corresponds to team B being 15 pins ahead of team C.

5. The integer 750 corresponds to a \$750 deposit, and the integer -125 corresponds to a \$125 withdrawal.

7. The integers 20, -150, and 300 correspond to the interception of the missile, the loss of the starship, and the capture of the base, respectively.

9. The number $\frac{10}{3}$ can be named $3\frac{1}{3}$ or $3.3\overline{3}$. The graph is $\frac{1}{3}$ of the way from 3 to 4.

$$\frac{10}{3}$$

-5 -4 -3 -2 -1 0 1 2 3 4 5

11. The graph of -4.3 is $\frac{3}{10}$ of the way from -4 to -5.

-4.3

-5 -4 -3 -2 -1 0 1 2 3 4 5

13. See Example 7 in the text.

15. We first find decimal notation for $\frac{5}{3}$. Since $\frac{5}{3}$ means $5 \div 3$, we divide.

$$
\begin{array}{r}
1.66\ldots \\
3\overline{)5.00} \\
\underline{3} \\
2\,0 \\
\underline{1\,8} \\
2\,0 \\
\underline{1\,8} \\
2
\end{array}
$$

Thus $\frac{5}{3} = 1.\overline{6}$ so $-\frac{5}{3} = -1.\overline{6}$.

17. We first find decimal notation for $\frac{7}{6}$. Since $\frac{7}{6}$ means $7 \div 6$, we divide.

$$
\begin{array}{r}
1.166\ldots \\
6\overline{)7.000} \\
\underline{6} \\
1\,0 \\
\underline{6} \\
4\,0 \\
\underline{3\,6} \\
4\,0 \\
\underline{3\,6} \\
4
\end{array}
$$

Thus $\frac{7}{6} = 1.1\overline{6}$ so $-\frac{7}{6} = -1.1\overline{6}$.

19. We first find decimal notation for $\frac{7}{8}$. Since $\frac{7}{8}$ means $7 \div 8$, we divide.

$$
\begin{array}{r}
0.875 \\
8\overline{)7.000} \\
\underline{6\,4} \\
6\,0 \\
\underline{5\,6} \\
4\,0 \\
\underline{4\,0} \\
0
\end{array}
$$

Thus $\frac{7}{8} = 0.875$ so $-\frac{7}{8} = -0.875$.

21. We first find decimal notation for $\frac{7}{20}$. Since $\frac{7}{20}$ means $7 \div 20$, we divide.

$$
\begin{array}{r}
0.35 \\
20\overline{)7.00} \\
\underline{6\,0} \\
1\,0\,0 \\
\underline{1\,0\,0} \\
0
\end{array}
$$

Thus $\frac{7}{20} = 0.35$ so $-\frac{7}{20} = -0.35$.

23. Since 9 is to the right of 0, we have $9 > 0$.

25. Since 8 is to the right of -8, we have $8 > -8$.

27. Since 0 is to the right of -7, we have $0 > -7$.

29. Since -4 is to the left of -3, we have $-4 < -3$.

31. Since -3 is to the right of -4, we have $-3 > -4$.

33. Since -10 is to the right of -14, we have $-10 > -14$.

35. Since -3.3 is to the left of -2.2, we have $-3.3 < -2.2$.

37. Since 17.2 is to the right of -1.67, we have $17.2 > -1.67$.

39. Since -14.34 is to the right of -17.88, we have $-14.34 > -17.88$.

41. We convert to decimal notation: $-\frac{14}{17} \approx -0.8235$ and $-\frac{27}{35} \approx -0.7714$. Since -0.8235 is to the left of -0.7714, we have $-\frac{14}{17} < -\frac{27}{35}$.

43. The distance of -7 from 0 is 7, so $|-7| = 7$.

45. The distance of 0 from 0 is 0, so $|0| = 0$.

47. The distance of -4 from 0 is 4, so $|-4| = 4$.

49. The distance of 325 from 0 is 325, so $|325| = 325$.

51. The distance of $-\dfrac{10}{7}$ from 0 is $\dfrac{10}{7}$, so $\left|-\dfrac{10}{7}\right| = \dfrac{10}{7}$.

53. The distance of 14.8 from 0 is 14.8, so $|14.8| = 14.8$.

55.

$$192 = 2 \cdot 2 \cdot 2 \cdot 2 \cdot 2 \cdot 2 \cdot 3$$

57.

$$260 = 2 \cdot 2 \cdot 5 \cdot 13$$

59. $18 = 2 \cdot 3 \cdot 3$

$24 = 2 \cdot 2 \cdot 2 \cdot 3$

The LCM is $2 \cdot 2 \cdot 2 \cdot 3 \cdot 3$, or 72.

61. $12 = 2 \cdot 2 \cdot 3$

$24 = 2 \cdot 2 \cdot 2 \cdot 3$

$36 = 2 \cdot 2 \cdot 3 \cdot 3$

The LCM is $2 \cdot 2 \cdot 2 \cdot 3 \cdot 3$, or 72.

63. $|4| = 4$, and $|-7| = 7$. Since 4 is to the left of 7 we have $|4| < |-7|$.

65. $-\dfrac{2}{3}, \dfrac{1}{2}, -\dfrac{3}{4}, -\dfrac{5}{6}, \dfrac{3}{8}, \dfrac{1}{6}$ can be written in decimal notation as $-0.\overline{6}, 0.5, -0.75, -0.8\overline{3}, 0.375, 0.1\overline{6}$, respectively. Listing from least to greatest, we have $-\dfrac{5}{6}, -\dfrac{3}{4}, -\dfrac{2}{3}, \dfrac{1}{6}, \dfrac{3}{8}, \dfrac{1}{2}$.

Exercise Set 11.2

1. $-9 + 2$ The absolute values are 9 and 2. The difference is $9 - 2$, or 7. The negative number has the larger absolute value, so the answer is negative. $-9 + 2 = -7$

3. $-10 + 6$ The absolute values are 10 and 6. The difference is $10 - 6$, or 4. The negative number has the larger absolute value, so the answer is negative. $-10 + 6 = -4$

5. $-8 + 8$ A positive and a negative number. The numbers have the same absolute value. The sum is 0. $-8 + 8 = 0$

7. $-3 + (-5)$ Two negatives. Add the absolute values, 3 and 5, getting 8. Make the answer negative. $-3 + (-5) = -8$

9. $-7 + 0$ One number is 0. The answer is the other number. $-7 + 0 = -7$

11. $0 + (-27)$ One number is 0. The answer is the other number. $0 + (-27) = -27$

13. $17 + (-17)$ A positive and a negative number. The numbers have the same absolute value. The sum is 0. $17 + (-17) = 0$

15. $-17 + (-25)$ Two negatives. Add the absolute values, 17 and 25, getting 42. Make the answer negative. $-17 + (-25) = -42$

17. $18 + (-18)$ A positive and a negative number. The numbers have the same absolute value. The sum is 0. $18 + (-18) = 0$

19. $-18 + 18$ A positive and a negative number. The numbers have the same absolute value. The sum is 0. $-18 + 18 = 0$

21. $8 + (-5)$ The absolute values are 8 and 5. The difference is $8 - 5$, or 3. The positive number has the larger absolute value, so the answer is positive. $8 + (-5) = 3$

23. $-4 + (-5)$ Two negatives. Add the absolute values, 4 and 5, getting 9. Make the answer negative. $-4 + (-5) = -9$

25. $13 + (-6)$ The absolute values are 13 and 6. The difference is $13 - 6$, or 7. The positive number has the larger absolute value, so the answer is positive. $13 + (-6) = 7$

27. $-25 + 25$ A positive and a negative number. The numbers have the same absolute value. The sum is 0. $-25 + 25 = 0$

29. $63 + (-18)$ The absolute values are 63 and 18. The difference is $63 - 18$, or 45. The positive number has the larger absolute value, so the answer is positive. $63 + (-18) = 45$

31. $-6.5 + 4.7$ The absolute values are 6.5 and 4.7. The difference is $6.5 - 4.7$, or 1.8. The negative number has the larger absolute value, so the answer is negative. $-6.5 + 4.7 = -1.8$

33. $-2.8 + (-5.3)$ Two negatives. Add the absolute values, 2.8 and 5.3, getting 8.1. Make the answer negative. $-2.8 + (-5.3) = -8.1$

35. $-\dfrac{3}{5} + \dfrac{2}{5}$ The absolute values are $\dfrac{3}{5}$ and $\dfrac{2}{5}$. The difference is $\dfrac{3}{5} - \dfrac{2}{5}$, or $\dfrac{1}{5}$. The negative number has the larger absolute value, so the answer is negative. $-\dfrac{3}{5} + \dfrac{2}{5} = -\dfrac{1}{5}$

37. $-\dfrac{3}{7} + \left(-\dfrac{5}{7}\right)$ Two negatives. Add the absolute values, $\dfrac{3}{7}$ and $\dfrac{5}{7}$, getting $\dfrac{8}{7}$. Make the answer negative. $-\dfrac{3}{7} + \left(-\dfrac{5}{7}\right) = -\dfrac{8}{7}$

39. $-\dfrac{5}{8} + \dfrac{1}{4}$ The absolute values are $\dfrac{5}{8}$ and $\dfrac{1}{4}$. The difference is $\dfrac{5}{8} - \dfrac{2}{8}$, or $\dfrac{3}{8}$. The negative number has the larger absolute value, so the answer is negative. $-\dfrac{5}{8} + \dfrac{1}{4} = -\dfrac{3}{8}$

41. $-\dfrac{3}{7} + \left(-\dfrac{2}{5}\right)$ Two negatives. Add the absolute values, $\dfrac{3}{7}$ and $\dfrac{2}{5}$, getting $\dfrac{15}{35} + \dfrac{14}{35}$, or $\dfrac{29}{35}$. Make the answer negative. $-\dfrac{3}{7} + \left(-\dfrac{2}{5}\right) = -\dfrac{29}{35}$

43. $-\frac{3}{5} + \left(-\frac{2}{15}\right)$ Two negatives. Add the absolute values, $\frac{3}{5}$ and $\frac{2}{15}$, getting $\frac{9}{15} + \frac{2}{15}$, or $\frac{11}{15}$. Make the answer negative. $-\frac{3}{5} + \left(-\frac{2}{15}\right) = -\frac{11}{15}$

45. $-5.7 + (-7.2) + 6.6 = -12.9 + 6.6$ Adding the negative numbers
$$= -6.3 \qquad \text{Adding the results}$$

47. $-\frac{7}{16} + \frac{7}{8}$ The absolute values are $\frac{7}{16}$ and $\frac{7}{8}$. The difference is $\frac{14}{16} - \frac{7}{16}$, or $\frac{7}{16}$. The positive number has the larger absolute value, so the answer is positive. $-\frac{7}{16} + \frac{7}{8} = \frac{7}{16}$

49. $75 + (-14) + (-17) + (-5)$
  a) $-14 + (-17) + (-5) = -36$ Adding the negative numbers
  b) $75 + (-36) = 39$ Adding the results

51. $-44 + \left(-\frac{3}{8}\right) + 95 + \left(-\frac{5}{8}\right)$
  a) $-44 + \left(-\frac{3}{8}\right) + \left(-\frac{5}{8}\right) = -45$ Adding the negative numbers
  b) $-45 + 95 = 50$ Adding the results

53. $98 + (-54) + 113 + (-998) + 44 + (-612) + (-18) + 334$
  a) $98 + 113 + 44 + 334 = 589$ Adding the positive numbers
  b) $-54 + (-998) + (-612) + (-18) = -1682$ Adding the negative numbers
  c) $589 + (-1682) = -1093$ Adding the results

55. The additive inverse of 24 is -24 because $24 + (-24) = 0$.

57. The additive inverse of -26.9 is 26.9 because $-26.9 + 26.9 = 0$.

59. If $x = 9$ then $-x = -(9) = -9$. (The additive inverse of 9 is -9.)

61. If $x = -\frac{14}{3}$, then $-x = -\left(-\frac{14}{3}\right) = \frac{14}{3}$. (The additive inverse of $-\frac{14}{3}$ is $\frac{14}{3}$.)

63. If $x = -65$ then $-(-x) = -[-(-65)] = -65$. (The opposite of the opposite of -65 is -65.)

65. If $x = \frac{5}{3}$, then $-(-x) = -\left(-\frac{5}{3}\right) = \frac{5}{3}$. (The opposite of the opposite of $\frac{5}{3}$ is $\frac{5}{3}$.)

67. $-(-14) = 14$

69. $-(10) = -10$

71. When x is positive, the opposite of x, $-x$ is negative.

73. Positive (The sum of two positive numbers is positive.)

75. If a is positive, $-a$ is negative. Thus $-a + b$, the sum of two negatives, is negative.

Exercise Set 11.3

1. $3 - 7 = 3 + (-7) = -4$

3. $0 - 7 = 0 + (-7) = -7$

5. $-8 - (-2) = -8 + 2 = -6$

7. $-10 - (-10) = -10 + 10 = 0$

9. $12 - 16 = 12 + (-16) = -4$

11. $20 - 27 = 20 + (-27) = -7$

13. $-9 - (-3) = -9 + 3 = -6$

15. $-11 - (-11) = -11 + 11 = 0$

17. $8 - (-3) = 8 + 3 = 11$

19. $-6 - 8 = -6 + (-8) = -14$

21. $-4 - (-9) = -4 + 9 = 5$

23. $2 - 9 = 2 + (-9) = -7$

25. $0 - 5 = 0 + (-5) = -5$

27. $-5 - (-2) = -5 + 2 = -3$

29. $2 - 25 = 2 + (-25) = -23$

31. $-42 - 26 = -42 + (-26) = -68$

33. $-71 - 2 = -71 + (-2) = -73$

35. $24 - (-92) = 24 + 92 = 116$

37. $-2.8 - 0 = -2.8 + 0 = -2.8$

39. $\frac{3}{8} - \frac{5}{8} = \frac{3}{8} + \left(-\frac{5}{8}\right) = -\frac{2}{8} = -\frac{1}{4}$

41. $\frac{3}{4} - \frac{2}{3} = \frac{9}{12} - \frac{8}{12} = \frac{9}{12} + \left(-\frac{8}{12}\right) = \frac{1}{12}$

43. $-\frac{3}{4} - \frac{2}{3} = -\frac{9}{12} - \frac{8}{12} = -\frac{9}{12} + \left(-\frac{8}{12}\right) = -\frac{17}{12}$

45. $-\frac{5}{8} - \left(-\frac{3}{4}\right) = -\frac{5}{8} - \left(-\frac{6}{8}\right) = -\frac{5}{8} + \frac{6}{8} = \frac{1}{8}$

47. $6.1 - (-13.8) = 6.1 + 13.8 = 19.9$

49. $-3.2 - 5.8 = -3.2 + (-5.8) = -9$

51. $0.99 - 1 = 0.99 + (-1) = -0.01$

53. $3 - 5.7 = 3 + (-5.7) = -2.7$

55. $7 - 10.53 = 7 + (-10.53) = -3.53$

57. $\frac{1}{6} - \frac{2}{3} = \frac{1}{6} - \frac{4}{6} = \frac{1}{6} + \left(-\frac{4}{6}\right) = -\frac{3}{6} = -\frac{1}{2}$

59. $-\frac{4}{7} - \left(-\frac{10}{7}\right) = -\frac{4}{7} + \frac{10}{7} = \frac{6}{7}$

61. $-\frac{7}{10} - \frac{10}{15} = -\frac{21}{30} - \frac{20}{30} = -\frac{21}{30} + \left(-\frac{20}{30}\right) = -\frac{41}{30}$

63. $\frac{1}{13} - \frac{1}{12} = \frac{12}{156} - \frac{13}{156} = \frac{12}{156} + \left(-\frac{13}{156}\right) = -\frac{1}{156}$

65. $18 - (-15) - 3 - (-5) + 2 = 18 + 15 + (-3) + 5 + 2 = 37$

67. $-31+(-28)-(-14)-17=(-31)+(-28)+14+(-17)=$
-62

69. $-93-(-84)-41-(-56)=(-93)+84+(-41)+56=6$

71. $-5-(-30)+30+40-(-12)=(-5)+30+30+40+12=107$

73. $132-(-21)+45-(-21)=132+21+45+21=219$

75. $A=l\cdot w$
$A=(8.4\text{ cm})\times(11.5\text{ cm})$
$A=(8.4)\times(11.5)\times(\text{cm})\times(\text{cm})$
$A=96.6\text{ cm}^2$

77. $36=2\cdot2\cdot3\cdot3$
$54=2\cdot3\cdot3\cdot3$
The LCM is $2\cdot2\cdot3\cdot3\cdot3$, or 108.

79. The integer $-10,415$ corresponds to the depth of the Marianas Trench, and the integer -8648 corresponds to the depth of the Puerto Rico Trench. We subtract the smaller number from the larger to find the difference in their depths: $-8648-(-10,415)=-8648+10,415=1767$. The Puerto Rico Trench is 1767 m higher than the Marianas Trench.

Exercise Set 11.4

1. -16

3. -24

5. -72

7. 16

9. 42

11. -120

13. -238

15. 1200

17. 98

19. -12.4

21. 24

23. 21.7

25. $\dfrac{2}{3}\cdot\left(-\dfrac{3}{5}\right)=-\left(\dfrac{2\cdot3}{3\cdot5}\right)=-\left(\dfrac{2}{5}\cdot\dfrac{3}{3}\right)=-\dfrac{2}{5}$

27. $-\dfrac{3}{8}\cdot\left(-\dfrac{2}{9}\right)=\dfrac{3\cdot2\cdot1}{4\cdot2\cdot3\cdot3}=\dfrac{3\cdot2}{3\cdot2}\cdot\dfrac{1}{4\cdot3}=\dfrac{1}{12}$

29. -17.01

31. $-\dfrac{5}{9}\cdot\dfrac{3}{4}=-\dfrac{5\cdot3}{3\cdot3\cdot4}=-\dfrac{3}{3}\cdot\dfrac{5}{3\cdot4}=-\dfrac{5}{12}$

33. $7\cdot(-4)\cdot(-3)\cdot5=7\cdot12\cdot5=7\cdot60=420$

35. $-\dfrac{2}{3}\cdot\dfrac{1}{2}\cdot\left(-\dfrac{6}{7}\right)=-\dfrac{2}{6}\cdot\left(-\dfrac{6}{7}\right)=\dfrac{2\cdot6}{7\cdot6}=\dfrac{2}{7}\cdot\dfrac{6}{6}=\dfrac{2}{7}$

37. $-3\cdot(-4)\cdot(-5)=12\cdot(-5)=-60$

39. $-2\cdot(-5)\cdot(-3)\cdot(-5)=10\cdot15=150$

41. $-\dfrac{2}{45}$

43. $-7\cdot(-21)\cdot13=147\cdot13=1911$

45. $-4\cdot(-1.8)\cdot7=(7.2)\cdot7=50.4$

47. $-\dfrac{1}{9}\cdot\left(-\dfrac{2}{3}\right)\cdot\left(\dfrac{5}{7}\right)=\dfrac{2}{27}\cdot\dfrac{5}{7}=\dfrac{10}{189}$

49. $4\cdot(-4)\cdot(-5)\cdot(-12)=-16\cdot(60)=-960$

51. $0.07\cdot(-7)\cdot6\cdot(-6)=0.07\cdot6\cdot(-7)\cdot(-6)=0.42\cdot(42)=17.64$

53. $\left(-\dfrac{5}{6}\right)\left(\dfrac{1}{8}\right)\left(-\dfrac{3}{7}\right)\left(-\dfrac{1}{7}\right)=\left(-\dfrac{5}{48}\right)\left(\dfrac{3}{49}\right)=$
$-\dfrac{5\cdot3}{16\cdot3\cdot49}=-\dfrac{3}{3}\cdot\dfrac{5}{16\cdot49}=-\dfrac{5}{784}$

55. $(-14)\cdot(-27)\cdot(-2)=378\cdot(-2)=-756$

57. $(-8)(-9)(-10)=72(-10)=-720$

59. $(-6)(-7)(-8)(-9)(-10)=42\cdot72\cdot(-10)=3024\cdot(-10)=-30,240$

61.

$$
\begin{array}{r|r}
 & 3 \\
3 & 9 \\
2 & 18 \\
2 & 36 \\
2 & 72 \\
2 & 144 \\
2 & 288 \\
2 & 576 \\
2 & 1152 \\
2 & 2304 \\
2 & 4608 \\
\end{array}
$$

$4608=2\cdot2\cdot2\cdot2\cdot2\cdot2\cdot2\cdot2\cdot2\cdot3\cdot3$

63. Method 1: Solve using an equation.

Translate. 23 is what percent of 69

$$23=\quad n\quad\times 69$$

Solve. We divide on both sides by 69 and convert to percent notation.

$$n\cdot69=23$$
$$\dfrac{n\cdot69}{69}=\dfrac{23}{69}$$
$$n=0.33\tfrac{1}{3},\text{ or }0.33\overline{3}=33\tfrac{1}{3}\%,\text{ or }33.\overline{3}\%$$

Thus, 23 is $33\tfrac{1}{3}\%$, or $33.\overline{3}\%$, of 69.

Method 2: Solve using a proportion.

$$\underset{\text{amount}}{23}\quad\text{is what percent of }\underset{\text{base}}{69}?$$
number

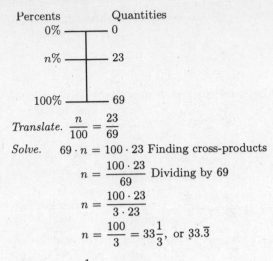

Percents   Quantities

0% ———————— 0

n% ——————— 23

100% ——————— 69

Translate. $\dfrac{n}{100} = \dfrac{23}{69}$

Solve. $69 \cdot n = 100 \cdot 23$ Finding cross-products

$n = \dfrac{100 \cdot 23}{69}$ Dividing by 69

$n = \dfrac{100 \cdot 23}{3 \cdot 23}$

$n = \dfrac{100}{3} = 33\dfrac{1}{3}$, or $33.\overline{3}$

Thus 23 is $33\dfrac{1}{3}\%$, or $33.\overline{3}\%$, of 69.

65. Method 1: Solve using an equation.

Translate. 40% of what number is 28.8

$\downarrow\downarrow\downarrow\downarrow\downarrow$

$40\%\timesb=28.8$

Solve. We convert 40% to decimal notation and divide.

$$0.4 \times b = 28.8$$
$$b = \dfrac{28.8}{0.4}$$
$$b = 72$$

Thus 40% of 72 is 28.8.

Method 2: Solve using a proportion.

40% of what number is 28.8?

$\downarrow\downarrow\downarrow$

number   base   amount

Percents   Quantities

0% ———————— 0

40% ——————— 28.8

100% ——————— b

Translate. $\dfrac{40}{100} = \dfrac{28.8}{b}$

Solve. $40 \cdot b = 100 \cdot 28.8$

$b = \dfrac{100 \cdot 28.8}{40}$

$b = \dfrac{2880}{40}$

$b = 72$

Thus 40% of 72 is 28.8.

67. $-6[(-5) + (-7)] = -6[-12] = 72$

69. $-(3^5) \cdot [-(2^3)] = -243[-8] = 1944$

71. $|(-2)^3 + 4^2| - (2 - 7)^2 = |(-2)^3 + 4^2| - (-5)^2 = |-8 + 16| - 25 = |8| - 25 = 8 - 25 = -17$

73. a) a and b have different signs;

b) either a or b is zero;

c) a and b have the same sign

Exercise Set 11.5

1. $36 \div (-6) = -6$ Check: $-6 \cdot (-6) = 36$

3. $\dfrac{26}{-2} = -13$ Check: $-13 \cdot (-2) = 26$

5. $\dfrac{-16}{8} = -2$ Check: $-2 \cdot 8 = -16$

7. $\dfrac{-48}{-12} = 4$ Check: $4(-12) = -48$

9. $\dfrac{-72}{9} = -8$ Check: $-8 \cdot 9 = -72$

11. $-100 \div (-50) = 2$ Check: $2(-50) = -100$

13. $-108 \div 9 = -12$ Check: $9(-12) = -108$

15. $\dfrac{200}{-25} = -8$ Check: $-8(-25) = 200$

17. Undefined

19. $\dfrac{81}{-9} = -9$ Check: $-9 \cdot (-9) = 81$

21. The reciprocal of $-\dfrac{15}{7}$ is $-\dfrac{7}{15}$ because $-\dfrac{15}{7} \cdot \left(-\dfrac{7}{15}\right) = 1$.

23. The reciprocal of 13 is $\dfrac{1}{13}$ because $13 \cdot \left(\dfrac{1}{13}\right) = 1$.

25. $\dfrac{3}{4} \div \left(-\dfrac{2}{3}\right) = \dfrac{3}{4} \cdot \left(-\dfrac{3}{2}\right) - -\dfrac{9}{8}$

27. $-\dfrac{5}{4} \div \left(-\dfrac{3}{4}\right) = -\dfrac{5}{4} \cdot \left(-\dfrac{4}{3}\right) = \dfrac{20}{12} = \dfrac{5 \cdot 4}{3 \cdot 4} = \dfrac{5}{3} \cdot \dfrac{4}{4} = \dfrac{5}{3}$

29. $-\dfrac{2}{7} \div \left(-\dfrac{4}{9}\right) = -\dfrac{2}{7} \cdot \left(-\dfrac{9}{4}\right) = \dfrac{18}{28} = \dfrac{9 \cdot 2}{14 \cdot 2} = \dfrac{9}{14} \cdot \dfrac{2}{2} = \dfrac{9}{14}$

31. $-\dfrac{3}{8} \div \left(-\dfrac{8}{3}\right) = -\dfrac{3}{8} \cdot \left(-\dfrac{3}{8}\right) = \dfrac{9}{64}$

33. $-6.6 \div 3.3 = -2$ Do the long division. Make the answer negative.

35. $\dfrac{-11}{-13} = \dfrac{11}{13}$

37. $\dfrac{48.6}{-3} = -16.2$ Do the long division. Make the answer negative.

39. $\dfrac{-9}{17 - 17} = \dfrac{-9}{0}$ Division by zero is undefined.

41. $8 - 2 \cdot 3 - 9 = 8 - 6 - 9$ Multiplying
$\qquad\qquad\quad = 2 - 9$ Doing all additions and
$\qquad\qquad\qquad\qquad\qquad\quad$ subtractions in order
$\qquad\qquad\quad = -7$ from left to right

43. $(8 - 2 \cdot 3) - 9 = (8 - 6) - 9$ Multiplying inside
$\qquad\qquad\qquad\qquad\qquad\qquad$ parentheses
$\qquad\qquad\quad = 2 - 9$ Subtracting inside
$\qquad\qquad\qquad\qquad\qquad$ parentheses
$\qquad\qquad\quad = -7$ Subtracting

45. $16 \cdot (-24) + 50 = -384 + 50$ Multiplying
$\qquad\qquad\qquad\quad = -334$ Adding

47. $2^4 + 2^3 - 10 = 16 + 8 - 10$ Evaluating exponential
$\qquad\qquad\qquad\qquad\qquad\qquad$ expressions
$\qquad\qquad\quad = 24 - 10$ Adding and subtract-
$\qquad\qquad\qquad\qquad\qquad\quad$ ing in order
$\qquad\qquad\quad = 14$ from left to right

49. $5^3 + 26 \cdot 71 - (16 + 25 \cdot 3)$
$\quad = 5^3 + 26 \cdot 71 - (16 + 75)$ Multiplying inside par-
$\quad =$ entheses
$\quad = 5^3 + 26 \cdot 71 - 91$ Adding inside paren-
$\quad =$ theses
$\quad = 125 + 26 \cdot 71 - 91$ Evaluating the expo-
$\quad =$ nential expression
$\quad = 125 + 1846 - 91$ Multiplying
$\quad = 1971 - 91$ Adding and subtract-
$\qquad\qquad\qquad\qquad$ ing in order from left
$\quad = 1880$ to right

51. $4 \cdot 5 - 2 \cdot 6 + 4 = 20 - 12 + 4$ Multiplying
$\qquad\qquad\qquad\quad = 8 + 4$
$\qquad\qquad\qquad\quad = 12$

53. $\dfrac{4^3}{8} = \dfrac{64}{8}$ Evaluating the exponential
$\qquad\qquad\quad$ expression
$\quad\ = 8$ Dividing

55. $8(-7) + 6(-5) = -56 - 30$ Multiplying
$\qquad\qquad\qquad = -86$

57. $19 - 5(-3) + 3 = 19 + 15 + 3$ Multiplying
$\qquad\qquad\qquad = 34 + 3$
$\qquad\qquad\qquad = 37$

59. $9 \div (-3) + 16 \div 8 = -3 + 2$ Dividing
$\qquad\qquad\qquad\qquad = -1$

61. $6 - 4^2 = 6 - 16$
$\qquad\qquad = -10$

63. $(3 - 8)^2 = (-5)^2$ Subtracting inside par-
$\qquad\qquad\qquad\quad$ entheses
$\qquad\ = 25$

65. $12 - 20^3 = 12 - 8000$
$\qquad\qquad\quad = -7988$

67. $2 \times 10^3 - 5000 = 2 \times 1000 - 5000$
$\qquad\qquad\qquad = 2000 - 5000$
$\qquad\qquad\qquad = -3000$

69. $6[9 - (3 - 4)] = 6[9 - (-1)]$ Subtracting inside the
$\qquad\qquad\qquad\qquad\qquad\qquad$ innermost parentheses
$\qquad\qquad\quad = 6[9 + 1]$
$\qquad\qquad\quad = 6[10]$
$\qquad\qquad\quad = 60$

71. $-1000 \div (-100) \div 10 = 10 \div 10$ Doing the divi-
$\qquad\qquad\qquad\qquad\qquad\qquad$ sions in order
$\qquad\qquad\qquad\quad = 1$ from left to right

73. $8 - (7 - 9) = 8 - (-2)$
$\qquad\qquad\quad = 8 + 2$
$\qquad\qquad\quad = 10$

75. $\dfrac{10 - 6^2}{9^2 + 3^2} = \dfrac{10 - 36}{81 + 9}$ Evaluating the exponential
$\qquad\qquad\qquad\qquad$ expressions
$\qquad\quad = \dfrac{-26}{90}$ Subtracting in the numera-
$\qquad\qquad\qquad\qquad$ tor, adding in the de-
$\qquad\qquad\qquad\qquad$ nominator
$\qquad\quad = -\dfrac{13}{45}$ Simplifying

77. $\dfrac{20(8 - 3) - 4(10 - 3)}{10(2 - 6) - 2(5 + 2)}$
$\qquad = \dfrac{20 \cdot 5 - 4 \cdot 7}{10(-4) - 2 \cdot 7}$ Doing the operations inside
$\qquad\qquad\qquad\qquad$ the parentheses
$\qquad = \dfrac{100 - 28}{-40 - 14}$ Multiplying
$\qquad = \dfrac{72}{-54}$ Subtracting
$\qquad = -\dfrac{4}{3}$ Simplifying

Chapter 12

Algebra: Solving Equations and Problems

1. $6x = 6 \cdot 7 = 42$

3. $\dfrac{x}{y} = \dfrac{9}{3} = 3$

5. $\dfrac{3p}{q} = \dfrac{3 \cdot 2}{6} = \dfrac{6}{6} = 1$

7. $\dfrac{x+y}{5} = \dfrac{10+20}{5} = \dfrac{30}{5} = 6$

9. $10(x+y) = 10(20+4) = 10 \cdot 24 = 240$
$10x + 10y = 10 \cdot 20 + 10 \cdot 4 = 200 + 40 = 240$

11. $10(x-y) = 10(20-4) = 10 \cdot 16 = 160$
$10x - 10y = 10 \cdot 20 - 10 \cdot 4 = 200 - 40 = 160$

13. $2(b+5) = 2 \cdot b + 2 \cdot 5 = 2b + 10$

15. $7(1-t) = 7 \cdot 1 - 7 \cdot t = 7 - 7t$

17. $6(5x+2) = 6 \cdot 5x + 6 \cdot 2 = 30x + 12$

19. $7(x+4+6y) = 7 \cdot x + 7 \cdot 4 + 7 \cdot 6y = 7x + 28 + 42y$

21. $-7(y-2) = -7 \cdot y - (-7) \cdot 2 = -7y - (-14) = -7y + 14$

23. $-9(-5x - 6y + 8) = -9(-5x) - (-9)6y + (-9)8 =$
$45x - (-54y) + (-72) = 45x + 54y - 72$

25. $-4(x - 3y - 2z) = -4 \cdot x - (-4)3y - (-4)2z =$
$-4x - (-12y) - (-8z) = -4x + 12y + 8z$

27. $3.1(-1.2x + 3.2y - 1.1) = 3.1(-1.2x) + (3.1)3.2y - 3.1(1.1) =$
$-3.72x + 9.92y - 3.41$

29. $2x + 4 = 2 \cdot x + 2 \cdot 2 = 2(x+2)$

31. $30 + 5y = 5 \cdot 6 + 5 \cdot y = 5(6+y)$

33. $14x + 21y = 7 \cdot 2x + 7 \cdot 3y = 7(2x + 3y)$

35. $5x + 10 + 15y = 5 \cdot x + 5 \cdot 2 + 5 \cdot 3y = 5(x + 2 + 3y)$

37. $8x - 24 = 8 \cdot x - 8 \cdot 3 = 8(x-3)$

39. $32 - 4y = 4 \cdot 8 - 4 \cdot y = 4(8-y)$

41. $8x + 10y - 22 = 2 \cdot 4x + 2 \cdot 5y - 2 \cdot 11 = 2(4x + 5y - 11)$

43. $-18x - 12y + 6 = -6 \cdot 3x - 6 \cdot 2y - 6 \cdot (-1) = -6(3x + 2y - 1)$,
or $-18x - 12y + 6 = 6 \cdot (-3x) + 6 \cdot (-2y) + 6 \cdot 1 =$
$6(-3x - 2y + 1)$

45. $9a + 10a = (9+10)a = 19a$

47. $10a - a = 10a - 1 \cdot a = (10-1)a = 9a$

49. $2x + 9z + 6x = 2x + 6x + 9z$
$= (2+6)x + 9z$
$= 8x + 9z$

51. $41a + 90 - 60a - 2 = 41a - 60a + 90 - 2$
$= (41 - 60)a + (90 - 2)$
$= -19a + 88$

53. $23 + 5t + 7y - t - y - 27$
$= 23 - 27 + 5t - 1 \cdot t + 7y - 1 \cdot y$
$= (23 - 27) + (5-1)t + (7-1)y$
$= -4 + 4t + 6y$, or $4t + 6y - 4$

55. $11x - 3x = (11-3)x = 8x$

57. $6n - n = (6-1)n = 5n$

59. $y - 17y = (1-17)y = -16y$

61. $-8 + 11a - 5b + 6a - 7b + 7$
$= 11a + 6a - 5b - 7b - 8 + 7$
$= (11+6)a + (-5-7)b + (-8+7)$
$= 17a - 12b - 1$

63. $9x + 2y - 5x = (9-5)x + 2y = 4x + 2y$

65. $11x + 2y - 4x - y = (11-4)x + (2-1)y = 7x + y$

67. $2.7x + 2.3y - 1.9x - 1.8y = (2.7 - 1.9)x + (2.3 - 1.8)y = 0.8x + 0.5y$

Exercise Set 12.2

1.
$$x + 5 = 12$$
$$x + 5 - 5 = 12 - 5 \quad \text{Subtracting 5 on both sides}$$
$$x + 0 = 7 \quad \text{Simplifying}$$
$$x = 7 \quad \text{Identity property of zero}$$

Check: $\quad x + 5 = 12$

$$\begin{array}{c|c} 7 + 5 & 12 \\ \hline 12 & \end{array} \quad \text{TRUE}$$

The solution is 7.

3.
$$x + 15 = -5$$
$$x + 15 - 15 = -5 - 15 \quad \text{Subtracting 15 on both sides}$$
$$x = -20$$

Check: $\quad x + 15 = -5$

$$\begin{array}{c|c} -20 + 15 & -5 \\ \hline -5 & \end{array} \quad \text{TRUE}$$

The solution is -20.

5.
$$x + 6 = -8$$
$$x + 6 - 6 = -8 - 6$$
$$x = -14$$

Check: $\quad x + 6 = -8$

$$\begin{array}{c|c} -14 + 6 & -8 \\ \hline -8 & \end{array} \quad \text{TRUE}$$

The solution is -14.

7.
$$x + 16 = -2$$
$$x + 16 - 16 = -2 - 16$$
$$x = -18$$

Check:
$$\begin{array}{c|c} x + 16 = -2 \\ \hline -18 + 16 & -2 \\ -2 & \end{array} \quad \text{TRUE}$$

The solution is -18.

9.
$$x - 9 = 6$$
$$x - 9 + 9 = 6 + 9$$
$$x = 15$$

Check:
$$\begin{array}{c|c} x - 9 = 6 \\ \hline 15 - 9 & 6 \\ 6 & \end{array} \quad \text{TRUE}$$

The solution is 15.

11.
$$x - 7 = -21$$
$$x - 7 + 7 = -21 + 7$$
$$x = -14$$

Check:
$$\begin{array}{c|c} x - 7 = -21 \\ \hline -14 - 7 & -21 \\ -21 & \end{array} \quad \text{TRUE}$$

The solution is -14.

13.
$$5 + t = 7$$
$$-5 + 5 + t = -5 + 7$$
$$t = 2$$

Check:
$$\begin{array}{c|c} 5 + t = 7 \\ \hline 5 + 2 & 7 \\ 7 & \end{array} \quad \text{TRUE}$$

The solution is 2.

15.
$$-7 + y = 13$$
$$7 + (-7) + y = 7 + 13$$
$$y = 20$$

Check:
$$\begin{array}{c|c} -7 + y = 13 \\ \hline -7 + 20 & 13 \\ 13 & \end{array} \quad \text{TRUE}$$

The solution is 20.

17.
$$-3 + t = -9$$
$$3 + (-3) + t = 3 + (-9)$$
$$t = -6$$

Check:
$$\begin{array}{c|c} -3 + t = -9 \\ \hline -3 + (-6) & -9 \\ -9 & \end{array} \quad \text{TRUE}$$

The solution is -6.

19.
$$r + \frac{1}{3} = \frac{8}{3}$$
$$r + \frac{1}{3} - \frac{1}{3} = \frac{8}{3} - \frac{1}{3}$$
$$r = \frac{7}{3}$$

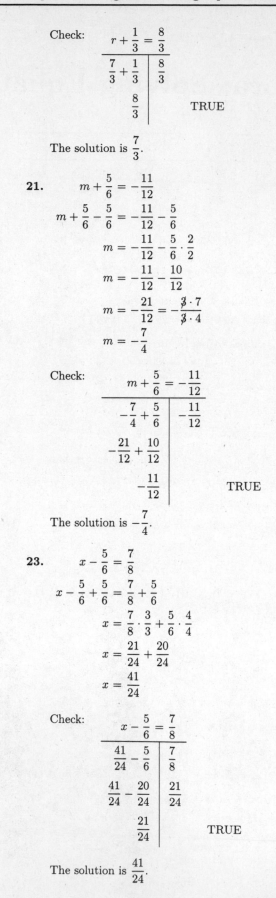

Check:
$$\begin{array}{c|c} r + \dfrac{1}{3} = \dfrac{8}{3} \\ \hline \dfrac{7}{3} + \dfrac{1}{3} & \dfrac{8}{3} \\[2mm] \dfrac{8}{3} & \end{array} \quad \text{TRUE}$$

The solution is $\frac{7}{3}$.

21.
$$m + \frac{5}{6} = -\frac{11}{12}$$
$$m + \frac{5}{6} - \frac{5}{6} = -\frac{11}{12} - \frac{5}{6}$$
$$m = -\frac{11}{12} - \frac{5}{6} \cdot \frac{2}{2}$$
$$m = -\frac{11}{12} - \frac{10}{12}$$
$$m = -\frac{21}{12} = -\frac{\cancel{3} \cdot 7}{\cancel{3} \cdot 4}$$
$$m = -\frac{7}{4}$$

Check:
$$\begin{array}{c|c} m + \dfrac{5}{6} = -\dfrac{11}{12} \\ \hline -\dfrac{7}{4} + \dfrac{5}{6} & -\dfrac{11}{12} \\[2mm] -\dfrac{21}{12} + \dfrac{10}{12} & \\[2mm] -\dfrac{11}{12} & \end{array} \quad \text{TRUE}$$

The solution is $-\frac{7}{4}$.

23.
$$x - \frac{5}{6} = \frac{7}{8}$$
$$x - \frac{5}{6} + \frac{5}{6} = \frac{7}{8} + \frac{5}{6}$$
$$x = \frac{7}{8} \cdot \frac{3}{3} + \frac{5}{6} \cdot \frac{4}{4}$$
$$x = \frac{21}{24} + \frac{20}{24}$$
$$x = \frac{41}{24}$$

Check:
$$\begin{array}{c|c} x - \dfrac{5}{6} = \dfrac{7}{8} \\ \hline \dfrac{41}{24} - \dfrac{5}{6} & \dfrac{7}{8} \\[2mm] \dfrac{41}{24} - \dfrac{20}{24} & \dfrac{21}{24} \\[2mm] \dfrac{21}{24} & \end{array} \quad \text{TRUE}$$

The solution is $\frac{41}{24}$.

25.
$$-\frac{1}{5} + z = -\frac{1}{4}$$
$$\frac{1}{5} - \frac{1}{5} + z = \frac{1}{5} - \frac{1}{4}$$
$$z = \frac{1}{5} \cdot \frac{4}{4} - \frac{1}{4} \cdot \frac{5}{5}$$
$$z = \frac{4}{20} - \frac{5}{20}$$
$$z = -\frac{1}{20}$$

Check:
$$\begin{array}{c|c} -\frac{1}{5} + z = -\frac{1}{4} & \\ \hline -\frac{1}{5} + \left(-\frac{1}{20}\right) & -\frac{1}{4} \\ -\frac{4}{20} + \left(-\frac{1}{20}\right) & -\frac{5}{20} \\ & -\frac{5}{20} \end{array}$$ TRUE

The solution is $-\frac{1}{20}$.

27.
$$x + 2.3 = 7.4$$
$$x + 2.3 - 2.3 = 7.4 - 2.3$$
$$x = 5.1$$

Check:
$$\begin{array}{c|c} x + 2.3 = 7.4 & \\ \hline 5.1 + 2.3 & 7.4 \\ 7.4 & \end{array}$$ TRUE

The solution is 5.1.

29.
$$7.6 = x - 4.8$$
$$7.6 + 4.8 = x - 4.8 + 4.8$$
$$12.4 = x$$

Check:
$$\begin{array}{c|c} 7.6 = x - 4.8 & \\ \hline 7.6 & 12.4 - 4.8 \\ & 7.6 \end{array}$$ TRUE

The solution is 12.4.

31.
$$-9.7 = -4.7 + y$$
$$4.7 + (-9.7) = 4.7 + (-4.7) + y$$
$$-5 = y$$

Check:
$$\begin{array}{c|c} -9.7 = -4.7 + y & \\ \hline -9.7 & -4.7 + (-5) \\ & -9.7 \end{array}$$ TRUE

The solution is -5.

33.
$$5\frac{1}{6} + x = 7$$
$$-5\frac{1}{6} + 5\frac{1}{6} + x = -5\frac{1}{6} + 7$$
$$x = -5\frac{1}{6} + 6\frac{6}{6}$$
$$x = 1\frac{5}{6}$$

Check:
$$\begin{array}{c|c} 5\frac{1}{6} + x = 7 & \\ \hline 5\frac{1}{6} + 1\frac{5}{6} & 7 \\ 7 & \end{array}$$ TRUE

The solution is $1\frac{5}{6}$.

35.
$$q + \frac{1}{3} = -\frac{1}{7}$$
$$q + \frac{1}{3} - \frac{1}{3} = -\frac{1}{7} - \frac{1}{3}$$
$$q = -\frac{1}{7} \cdot \frac{3}{3} - \frac{1}{3} \cdot \frac{7}{7}$$
$$q = -\frac{3}{21} - \frac{7}{21}$$
$$q = -\frac{10}{21}$$

Check:
$$\begin{array}{c|c} q + \frac{1}{3} = -\frac{1}{7} & \\ \hline -\frac{10}{21} + \frac{1}{3} & -\frac{1}{7} \\ -\frac{10}{21} + \frac{7}{21} & -\frac{3}{21} \\ -\frac{3}{21} & \end{array}$$ TRUE

The solution is $-\frac{10}{21}$.

37. $-3 + (-8)$ Two negative numbers. We add the absolute values, getting 11, and make the answer negative.
$$-3 + (-8) = -11$$

39. $-\frac{2}{3} \cdot \frac{5}{8} = -\frac{2 \cdot 5}{3 \cdot 8} = -\frac{2 \cdot 5}{3 \cdot 2 \cdot 4} = -\frac{2}{2} \cdot \frac{5}{3 \cdot 4} = -\frac{5}{12}$

41. The product of two negative numbers is positive:
$$(-3.2) \cdot (-4.1) = 13.12$$

43.
$$-356.788 = -699.034 + t$$
$$699.034 + (-356.788) = 699.034 + (-699.034) + t$$
$$342.246 = t$$

The solution is 342.246.

45.
$$x + \frac{4}{5} = -\frac{2}{3} - \frac{4}{15}$$
$$x + \frac{4}{5} - \frac{4}{5} = -\frac{2}{3} - \frac{4}{15} - \frac{4}{5}$$
$$x = -\frac{2}{3} \cdot \frac{5}{5} - \frac{4}{15} - \frac{4}{5} \cdot \frac{3}{3}$$
$$x = -\frac{10}{15} - \frac{4}{15} - \frac{12}{15}$$
$$x = -\frac{26}{15}$$

The solution is $-\frac{26}{15}$.

47. $16 + x - 22 = -16$
$\qquad x - 6 = -16$ Adding on the left side
$\qquad x - 6 + 6 = -16 + 6$
$\qquad\qquad x = -10$

The solution is -10.

49. $\qquad x + 3 = 3 + x$
$\qquad x + 3 - 3 = 3 + x - 3$
$\qquad\qquad x = x$

$x = x$ is true for all real numbers. Thus the solution is all real numbers.

51. $-\dfrac{3}{2} + x = -\dfrac{5}{17} - \dfrac{3}{2}$

$\dfrac{3}{2} - \dfrac{3}{2} + x = \dfrac{3}{2} - \dfrac{5}{17} - \dfrac{3}{2}$

$\qquad\qquad x = \left(\dfrac{3}{2} - \dfrac{3}{2}\right) - \dfrac{5}{17}$

$\qquad\qquad x = -\dfrac{5}{17}$

The solution is $-\dfrac{5}{17}$.

Exercise Set 12.3

1. $6x = 36$

$\dfrac{6x}{6} = \dfrac{36}{6}$ Dividing by 6 on both sides

$1 \cdot x = 6$ Simplifying
$\qquad x = 6$ Identity property of 1

Check: $\dfrac{6x = 36}{\underset{\displaystyle 36}{6 \cdot 6 \,\big|\, 36}}$ TRUE

The solution is 6.

3. $5x = 45$

$\dfrac{5x}{5} = \dfrac{45}{5}$ Dividing by 5 on both sides

$1 \cdot x = 9$ Simplifying
$\qquad x = 9$ Identity property of 1

Check: $\dfrac{5x = 45}{\underset{\displaystyle 45}{5 \cdot 9 \,\big|\, 45}}$ TRUE

The solution is 9.

5. $84 = 7x$

$\dfrac{84}{7} = \dfrac{7x}{7}$

$12 = 1 \cdot x$
$12 = x$

Check: $\dfrac{84 = 7x}{84 \,\big|\, \underset{\displaystyle 84}{7 \cdot 12}}$ TRUE

The solution is 12.

7. $\qquad\qquad -x = 40$
$\qquad\qquad -1 \cdot x = 40$
$\qquad -1 \cdot (-1 \cdot x) = -1 \cdot 40$
$\qquad\qquad 1 \cdot x = -40$
$\qquad\qquad\quad x = -40$

Check: $\dfrac{-x = 40}{\underset{\displaystyle 40}{-(-40) \,\big|\, 40}}$ TRUE

The solution is -40.

9. $\quad -2x = -10$

$\dfrac{-2x}{-2} = \dfrac{-10}{-2}$

$1 \cdot x = 5$
$\quad x = 5$

Check: $\dfrac{-2x = -10}{-2 \cdot 5 \,\big|\, \underset{\displaystyle -10}{-10}}$ TRUE

The solution is 5.

11. $\quad 7x = -49$

$\dfrac{7x}{7} = \dfrac{-49}{7}$

$1 \cdot x = -7$
$\quad x = -7$

Check: $\dfrac{7x = -49}{\underset{\displaystyle -49}{7(-7) \,\big|\, -49}}$ TRUE

The solution is -7.

13. $\quad -12x = 72$

$\dfrac{-12x}{-12} = \dfrac{72}{-12}$

$1 \cdot x = -6$
$\quad x = -6$

Check: $\dfrac{-12x = 72}{-12(-6) \,\big|\, \underset{\displaystyle 72}{72}}$ TRUE

The solution is -6.

15. $\quad -21x = -126$

$\dfrac{-21x}{-21} = \dfrac{-126}{-21}$

$1 \cdot x = 6$
$\quad x = 6$

Check: $\dfrac{-21x = -126}{-21 \cdot 6 \,\big|\, \underset{\displaystyle -126}{-126}}$ TRUE

The solution is 6.

17. $\quad \dfrac{1}{7}t = -9$

$7 \cdot \dfrac{1}{7}t = 7 \cdot (-9)$

$1 \cdot t = -63$
$\quad t = -63$

Check: $\dfrac{\dfrac{1}{7}t = -9}{\dfrac{1}{7} \cdot (-63) \,\big|\, \underset{\displaystyle -9}{-9}}$ TRUE

The solution is -63.

19. $\dfrac{3}{4}x = 27$

$\dfrac{4}{3} \cdot \dfrac{3}{4}x = \dfrac{4}{3} \cdot 27$

$1 \cdot x = \dfrac{4 \cdot \cancel{3} \cdot 3 \cdot 3}{\cancel{3} \cdot 1}$

$x = 36$

Check: $\dfrac{3}{4}x = 27$

$\begin{array}{c|c} \dfrac{3}{4} \cdot 36 & 27 \\ \hline 27 & 27 \end{array}$ TRUE

The solution is 36.

21. $-\dfrac{1}{3}t = 7$

$-3 \cdot \left(-\dfrac{1}{3}\right) \cdot t = -3 \cdot 7$

$1 \cdot t = -21$

$t = -21$

Check: $-\dfrac{1}{3}t = 7$

$\begin{array}{c|c} -\dfrac{1}{3} \cdot (-21) & 7 \\ \hline & 7 \end{array}$ TRUE

The solution is -21.

23. $-\dfrac{1}{3}m = \dfrac{1}{5}$

$-3 \cdot \left(-\dfrac{1}{3}m\right) = -3 \cdot \dfrac{1}{5}$

$1 \cdot m = -\dfrac{3}{5}$

$m = -\dfrac{3}{5}$

Check: $-\dfrac{1}{3}m = \dfrac{1}{5}$

$\begin{array}{c|c} -\dfrac{1}{3} \cdot \left(-\dfrac{3}{5}\right) & \dfrac{1}{5} \\ \hline \dfrac{1}{5} & \end{array}$ TRUE

The solution is $-\dfrac{3}{5}$.

25. $-\dfrac{3}{5}r = \dfrac{9}{10}$

$-\dfrac{5}{3} \cdot \left(-\dfrac{3}{5}r\right) = -\dfrac{5}{3} \cdot \dfrac{9}{10}$

$1 \cdot r = -\dfrac{\cancel{5} \cdot \cancel{3} \cdot 3}{\cancel{3} \cdot \cancel{5} \cdot 2}$

$r = -\dfrac{3}{2}$

Check: $-\dfrac{3}{5}r = \dfrac{9}{10}$

$\begin{array}{c|c} -\dfrac{3}{5} \cdot \left(-\dfrac{3}{2}\right) & \dfrac{9}{10} \\ \hline \dfrac{9}{10} & \end{array}$ TRUE

The solution is $-\dfrac{3}{2}$.

27. $-\dfrac{3}{2}r = -\dfrac{27}{4}$

$-\dfrac{2}{3} \cdot \left(-\dfrac{3}{2}r\right) = -\dfrac{2}{3} \cdot \left(-\dfrac{27}{4}\right)$

$1 \cdot r = \dfrac{\cancel{2} \cdot \cancel{3} \cdot 3 \cdot 3}{\cancel{3} \cdot \cancel{2} \cdot 2}$

$r = \dfrac{9}{2}$

Check: $\dfrac{3}{2}r - -\dfrac{27}{4}$

$\begin{array}{c|c} -\dfrac{3}{2} \cdot \dfrac{9}{2} & -\dfrac{27}{4} \\ \hline -\dfrac{27}{4} & \end{array}$ TRUE

The solution is $\dfrac{9}{2}$.

29. $6.3x = 44.1$

$\dfrac{6.3x}{6.3} = \dfrac{44.1}{6.3}$

$1 \cdot x = 7$

$x = 7$

Check: $6.3x = 44.1$

$\begin{array}{c|c} 6.3 \cdot 7 & 44.1 \\ \hline 44.1 & \end{array}$ TRUE

The solution is 7.

31. $-3.1y = 21.7$

$\dfrac{-3.1y}{-3.1} = \dfrac{21.7}{-3.1}$

$1 \cdot y = -7$

$y = -7$

Check: $3.1y = 21.7$

$\begin{array}{c|c} -3.1(-7) & 21.7 \\ \hline 21.7 & \end{array}$ TRUE

The solution is -7.

33. $38.7m = 309.6$

$\dfrac{38.7m}{38.7} = \dfrac{309.6}{38.7}$

$1 \cdot m = 8$

$m = 8$

Check: $38.7m = 309.6$

$\begin{array}{c|c} 38.7 \cdot 8 & 309.6 \\ \hline 309.6 & \end{array}$ TRUE

The solution is 8.

35.
$$-\frac{2}{3}y = -10.6$$
$$-\frac{3}{2}\cdot\left(-\frac{2}{3}y\right) = -\frac{3}{2}\cdot(-10.6)$$
$$1\cdot y = \frac{31.8}{2}$$
$$y = 15.9$$

Check:
$$-\frac{2}{3}y = -10.6$$

$-\frac{2}{3}\cdot(15.9)$	-10.6
$-\dfrac{31.8}{3}$	
-10.6	TRUE

The solution is 15.9.

37. $C = 2\cdot\pi\cdot r$
$C \approx 2\cdot 3.14\cdot 10\text{ ft} = 62.8\text{ ft}$

$d = 2\cdot r$
$d = 2\cdot 10\text{ ft} = 20\text{ ft}$

$A = \pi\cdot r\cdot r$
$A \approx 3.14\cdot 10\text{ ft}\cdot 10\text{ ft} = 314\text{ ft}^2$

39. $V = \ell\cdot w\cdot h$
$V = 25\text{ ft}\cdot 10\text{ ft}\cdot 32\text{ ft} = 8000\text{ ft}^3$

41.
$$-0.2344m = 2028.732$$
$$\frac{-0.2344m}{-0.2344} = \frac{2028.732}{-0.2344}$$
$$1\cdot m = -8655$$
$$m = -8655$$

43. For all x, $0\cdot x = 0$. There is no solution to $0\cdot x = 9$.

45.
$$2|x| = -12$$
$$\frac{2|x|}{2} = \frac{-12}{2}$$
$$1\cdot|x| = -6$$
$$|x| = -6$$

Absolute value cannot be negative. The equation has no solution.

Exercise Set 12.4

1.
$$5x + 6 = 31$$
$5x + 6 - 6 = 31 - 6$ Subtracting 6 on both sides
$5x = 25$ Simplifying
$\dfrac{5x}{5} = \dfrac{25}{5}$ Dividing by 5 on both sides
$x = 5$ Simplifying

Check:

$5x + 6 = 31$	
$5\cdot 5 + 6$	31
$25 + 6$	
31	TRUE

The solution is 5.

3.
$$8x + 4 = 68$$
$8x + 4 - 4 = 68 - 4$ Subtracting 4 on both sides
$8x = 64$ Simplifying
$\dfrac{8x}{8} = \dfrac{64}{8}$ Dividing by 8 on both sides
$x = 8$ Simplifying

Check:

$8x + 4 = 68$	
$8\cdot 8 + 4$	68
$64 + 4$	
68	TRUE

The solution is 8.

5.
$$4x - 6 = 34$$
$4x - 6 + 6 = 34 + 6$ Adding 6 on both sides
$4x = 40$
$\dfrac{4x}{4} = \dfrac{40}{4}$ Dividing by 4 on both sides
$x = 10$

Check:

$4x - 6 = 34$	
$4\cdot 10 - 6$	34
$40 - 6$	
34	TRUE

The solution is 10.

7.
$$3x - 9 = 33$$
$3x - 9 + 9 = 33 + 9$
$3x = 42$
$\dfrac{3x}{3} = \dfrac{42}{3}$
$x = 14$

Check:

$3x - 9 = 33$	
$3\cdot 14 - 9$	33
$42 - 9$	
33	TRUE

The solution is 14.

9.
$$7x + 2 = -54$$
$7x + 2 - 2 = -54 - 2$
$7x = -56$
$\dfrac{7x}{7} = \dfrac{-56}{7}$
$x = -8$

Check:

$7x + 2 = -54$	
$7(-8) + 2$	-54
$-56 + 2$	
-54	TRUE

The solution is -8.

11.
$$-45 = 6y + 3$$
$-45 - 3 = 6y + 3 - 3$
$-48 = 6y$
$\dfrac{-48}{6} = \dfrac{6y}{6}$
$-8 = y$

Check:

$-45 = 6y + 3$	
-45	$6(-8) + 3$
	$-48 + 3$
	-45 TRUE

The solution is -8.

13.
$$-4x + 7 = 35$$
$$-4x + 7 - 7 = 35 - 7$$
$$-4x = 28$$
$$\frac{-4x}{-4} = \frac{28}{-4}$$
$$x = -7$$

Check:
$$\begin{array}{c|c} -4x + 7 = 35 \\ \hline -4(-7) + 7 & 35 \\ 28 + 7 & \\ 35 & \quad \text{TRUE} \end{array}$$

The solution is -7.

15.
$$-7x - 24 = -129$$
$$-7x - 24 + 24 = -129 + 24$$
$$-7x = -105$$
$$\frac{-7x}{-7} = \frac{-105}{-7}$$
$$x = 15$$

Check:
$$\begin{array}{c|c} -7x - 24 = -129 \\ \hline -7 \cdot 15 - 24 & -129 \\ -105 - 24 & \\ -129 & \quad \text{TRUE} \end{array}$$

The solution is 15.

17. $5x + 7x = 72$
$$12x = 72 \qquad \text{Collecting like terms}$$
$$\frac{12x}{12} = \frac{72}{12} \qquad \text{Dividing by 12 on both sides}$$
$$x = 6$$

Check:
$$\begin{array}{c|c} 5x + 7x = 72 \\ \hline 5 \cdot 6 + 7 \cdot 6 & 72 \\ 30 + 42 & \\ 72 & \quad \text{TRUE} \end{array}$$

The solution is 6.

19. $8x + 7x = 60$
$$15x = 60 \qquad \text{Collecting like terms}$$
$$\frac{15x}{15} = \frac{60}{15} \qquad \text{Dividing by 15 on both sides}$$
$$x = 4$$

Check:
$$\begin{array}{c|c} 8x + 7x = 60 \\ \hline 8 \cdot 4 + 7 \cdot 4 & 60 \\ 32 + 28 & \\ 60 & \quad \text{TRUE} \end{array}$$

The solution is 4.

21. $4x + 3x = 42$
$$7x = 42$$
$$\frac{7x}{7} = \frac{42}{7}$$
$$x = 6$$

Check:
$$\begin{array}{c|c} 4x + 3x = 42 \\ \hline 4 \cdot 6 + 3 \cdot 6 & 42 \\ 24 + 18 & \\ 42 & \quad \text{TRUE} \end{array}$$

The solution is 6.

23. $-6y - 3y = 27$
$$-9y = 27$$
$$\frac{-9y}{-9} = \frac{27}{-9}$$
$$y = -3$$

Check:
$$\begin{array}{c|c} -6y - 3y = 27 \\ \hline -6(-3) - 3(-3) & 27 \\ 18 + 9 & \\ 27 & \quad \text{TRUE} \end{array}$$

The solution is -3.

25. $-7y - 8y = -15$
$$-15y = -15$$
$$\frac{-15y}{-15} = \frac{-15}{-15}$$
$$y = 1$$

Check:
$$\begin{array}{c|c} -7y - 8y = -15 \\ \hline -7 \cdot 1 - 8 \cdot 1 & -15 \\ -7 - 8 & \\ -15 & \quad \text{TRUE} \end{array}$$

The solution is 1.

27. $10.2y - 7.3y = -58$
$$2.9y = -58$$
$$\frac{2.9y}{2.9} = \frac{-58}{2.9}$$
$$y = -20$$

Check:
$$\begin{array}{c|c} 10.2y - 7.3y = -58 \\ \hline 10.2(-20) - 7.3(-20) & -58 \\ -204 + 146 & \\ -58 & \quad \text{TRUE} \end{array}$$

The solution is -20.

29.
$$x + \frac{1}{3}x = 8$$
$$\left(1 + \frac{1}{3}\right)x = 8$$
$$\frac{4}{3}x = 8$$
$$\frac{3}{4} \cdot \frac{4}{3}x = \frac{3}{4} \cdot 8$$
$$x = 6$$

Check:
$$\begin{array}{c|c} x + \dfrac{1}{3}x = 8 \\ \hline 6 + \dfrac{1}{3} \cdot 6 & 8 \\ 6 + 2 & \\ 8 & \quad \text{TRUE} \end{array}$$

The solution is 6.

31. $8y - 35 = 3y$
$$8y = 3y + 35 \qquad \text{Adding 35 and simplifying}$$
$$8y - 3y = 35 \qquad \text{Subtracting } 3y \text{ and simplifying}$$
$$5y = 35 \qquad \text{Collecting like terms}$$
$$\frac{5y}{5} = \frac{35}{5} \qquad \text{Dividing by 5}$$
$$y = 7$$

Check: $\dfrac{8y - 35 = 3y}{\begin{array}{c|c} 8 \cdot 7 - 35 & 3 \cdot 7 \\ 56 - 35 & 21 \\ 21 & \end{array}}$ TRUE

The solution is 7.

33. $8x - 1 = 23 - 4x$

$8x + 4x = 23 + 1$ Adding 1 and $4x$

$12x = 24$ Collecting like terms

$\dfrac{12x}{12} = \dfrac{24}{12}$ Dividing by 12

$x = 2$

Check: $\dfrac{8x - 1 = 23 - 4x}{\begin{array}{c|c} 8 \cdot 2 - 1 & 23 - 4 \cdot 2 \\ 16 - 1 & 23 - 8 \\ 15 & 15 \end{array}}$ TRUE

The solution is 2.

35. $2x - 1 = 4 + x$

$2x - x = 4 + 1$ Adding 1 and subtracting x

$x = 5$ Collecting like terms

Check: $\dfrac{2x - 1 = 4 + x}{\begin{array}{c|c} 2 \cdot 5 - 1 & 4 + 5 \\ 10 - 1 & 9 \\ 9 & \end{array}}$ TRUE

The solution is 5.

37. $6x + 3 = 2x + 11$

$6x - 2x = 11 - 3$

$4x = 8$

$\dfrac{4x}{4} = \dfrac{8}{4}$

$x = 2$

Check: $\dfrac{6x + 3 = 2x + 11}{\begin{array}{c|c} 6 \cdot 2 + 3 & 2 \cdot 2 + 11 \\ 12 + 3 & 4 + 11 \\ 15 & 15 \end{array}}$ TRUE

The solution is 2.

39. $5 - 2x = 3x - 7x + 25$

$5 - 2x = -4x + 25$ Collecting like terms

$4x - 2x = 25 - 5$

$2x = 20$

$\dfrac{2x}{2} = \dfrac{20}{2}$

$x = 10$

Check: $\dfrac{5 - 2x = 3x - 7x + 25}{\begin{array}{c|c} 5 - 2 \cdot 10 & 3 \cdot 10 - 7 \cdot 10 + 25 \\ 5 - 20 & 30 - 70 + 25 \\ -15 & -40 + 25 \\ & -15 \end{array}}$ TRUE

The solution is 10.

41. $4 + 3x - 6 = 3x + 2 - x$

$3x - 2 = 2x + 2$ Collecting like terms

$3x - 2x = 2 + 2$

$x = 4$

Check: $\dfrac{4 + 3x - 6 = 3x + 2 - x}{\begin{array}{c|c} 4 + 3 \cdot 4 - 6 & 3 \cdot 4 + 2 - 4 \\ 4 + 12 - 6 & 12 + 2 - 4 \\ 16 - 6 & 14 - 4 \\ 10 & 10 \end{array}}$ TRUE

The solution is 4.

43. $4y - 4 + y + 24 = 6y + 20 - 4y$

$5y + 20 = 2y + 20$ Collecting like terms

$5y - 2y = 20 - 20$

$3y = 0$

$\dfrac{3y}{3} = \dfrac{0}{3}$

$y = 0$

Check: $\dfrac{4y - 4 + y + 24 = 6y + 20 - 4y}{\begin{array}{c|c} 4 \cdot 0 - 4 + 0 + 24 & 6 \cdot 0 + 20 - 4 \cdot 0 \\ 0 - 4 + 0 + 24 & 0 + 20 - 0 \\ 20 & 20 \end{array}}$ TRUE

The solution is 0.

45. $\dfrac{7}{2}x + \dfrac{1}{2}x = 3x + \dfrac{3}{2} + \dfrac{5}{2}x$

The least common multiple of all the denominators is 2.
We multiply by 2 on both sides.

$2\left(\dfrac{7}{2}x + \dfrac{1}{2}x\right) = 2\left(3x + \dfrac{3}{2} + \dfrac{5}{2}x\right)$

$2 \cdot \dfrac{7}{2}x + 2 \cdot \dfrac{1}{2}x = 2 \cdot 3x + 2 \cdot \dfrac{3}{2} + 2 \cdot \dfrac{5}{2}x$

$7x + x = 6x + 3 + 5x$

$8x = 11x + 3$

$8x - 11x = 3$

$-3x = 3$

$\dfrac{-3x}{-3} = \dfrac{3}{-3}$

$x = -1$

Check: $\dfrac{7}{2}x + \dfrac{1}{2}x = 3x + \dfrac{3}{2} + \dfrac{5}{2}x$

$\begin{array}{c|c} \dfrac{7}{2}(-1) + \dfrac{1}{2}(-1) & 3(-1) + \dfrac{3}{2} + \dfrac{5}{2}(-1) \\[2mm] -\dfrac{7}{2} - \dfrac{1}{2} & -3 + \dfrac{3}{2} - \dfrac{5}{2} \\[2mm] -\dfrac{8}{2} & -\dfrac{3}{2} - \dfrac{5}{2} \\[2mm] -4 & -\dfrac{8}{2} \\[2mm] & -4 \end{array}$ TRUE

The solution is -1.

47. $\dfrac{2}{3} + \dfrac{1}{4}t = \dfrac{1}{3}$

The least common multiple of all the denominators is 12.
We multiply by 12 on both sides.

$$12\left(\dfrac{2}{3} + \dfrac{1}{4}t\right) = 12 \cdot \dfrac{1}{3}$$

$$12 \cdot \dfrac{2}{3} + 12 \cdot \dfrac{1}{4}t = 12 \cdot \dfrac{1}{3}$$

$$8 + 3t = 4$$

$$3t = 4 - 8$$

$$3t = -4$$

$$\dfrac{3t}{3} = \dfrac{-4}{3}$$

$$t = -\dfrac{4}{3}$$

Check:

$$\dfrac{2}{3} + \dfrac{1}{4}t = \dfrac{1}{3}$$

$$\begin{array}{c|c} \dfrac{2}{3} + \dfrac{1}{4}\left(-\dfrac{4}{3}\right) & \dfrac{1}{3} \\[2ex] \dfrac{2}{3} - \dfrac{1}{3} & \\[2ex] \dfrac{1}{3} & \text{TRUE} \end{array}$$

The solution is $-\dfrac{4}{3}$.

49. $\dfrac{2}{3} + 3y = 5y - \dfrac{2}{15}, \qquad$ LCM is 15

$$15\left(\dfrac{2}{3} + 3y\right) = 15\left(5y - \dfrac{2}{15}\right)$$

$$15 \cdot \dfrac{2}{3} + 15 \cdot 3y = 15 \cdot 5y - 15 \cdot \dfrac{2}{15}$$

$$10 + 45y = 75y - 2$$

$$10 + 2 = 75y - 45y$$

$$12 = 30y$$

$$\dfrac{12}{30} = \dfrac{30y}{30}$$

$$\dfrac{2}{5} = y$$

Check:

$$\dfrac{2}{3} + 3y = 5y - \dfrac{2}{15}$$

$$\begin{array}{c|c} \dfrac{2}{3} + 3 \cdot \dfrac{2}{5} & 5 \cdot \dfrac{2}{5} - \dfrac{2}{15} \\[2ex] \dfrac{2}{3} + \dfrac{6}{5} & 2 - \dfrac{2}{15} \\[2ex] \dfrac{10}{15} + \dfrac{18}{15} & \dfrac{30}{15} - \dfrac{2}{15} \\[2ex] \dfrac{28}{15} & \dfrac{28}{15} \quad \text{TRUE} \end{array}$$

The solution is $\dfrac{2}{5}$.

51. $\dfrac{5}{3} + \dfrac{2}{3}x = \dfrac{25}{12} + \dfrac{5}{4}x + \dfrac{3}{4}, \qquad$ LCM is 12

$$12\left(\dfrac{5}{3} + \dfrac{2}{3}x\right) = 12\left(\dfrac{25}{12} + \dfrac{5}{4}x + \dfrac{3}{4}\right)$$

$$12 \cdot \dfrac{5}{3} + 12 \cdot \dfrac{2}{3}x = 12 \cdot \dfrac{25}{12} + 12 \cdot \dfrac{5}{4}x + 12 \cdot \dfrac{3}{4}$$

$$20 + 8x = 25 + 15x + 9$$

$$20 + 8x = 15x + 34$$

$$20 - 34 = 15x - 8x$$

$$-14x = 7x$$

$$\dfrac{-14}{7} = \dfrac{7x}{7}$$

$$-2 = x$$

Check:

$$\dfrac{5}{3} + \dfrac{2}{3}x = \dfrac{25}{12} + \dfrac{5}{4}x + \dfrac{3}{4},$$

$$\begin{array}{c|c} \dfrac{5}{3} + \dfrac{2}{3}(-2) & \dfrac{25}{12} + \dfrac{5}{4}(-2) + \dfrac{3}{4} \\[2ex] \dfrac{5}{3} - \dfrac{4}{3} & \dfrac{25}{12} - \dfrac{5}{2} + \dfrac{3}{4} \\[2ex] \dfrac{1}{3} & \dfrac{25}{12} - \dfrac{30}{12} + \dfrac{9}{12} \\[2ex] & \dfrac{4}{12} \\[2ex] & \dfrac{1}{3} \qquad \text{TRUE} \end{array}$$

The solution is -2.

53. $2.1x + 45.2 = 3.2 - 8.4x$

Greatest number of decimal places is 1

$$10(2.1x + 45.2) = 10(3.2 - 8.4x)$$

Multiplying by 10 to clear decimals

$$10(2.1x) + 10(45.2) = 10(3.2) - 10(8.4x)$$

$$21x + 452 = 32 - 84x$$

$$21x + 84x = 32 - 452$$

$$105x = -420$$

$$\dfrac{105x}{105} = \dfrac{-420}{105}$$

$$x = -4$$

Check:

$$2.1x + 45.2 = 3.2 - 8.4x$$

$$\begin{array}{c|c} 2.1(-4) + 45.2 & 3.2 - 8.4(-4) \\ -8.4 + 45.2 & 3.2 + 33.6 \\ 36.8 & 36.8 \qquad \text{TRUE} \end{array}$$

The solution is -4.

55. $1.03 - 0.62x = 0.71 - 0.22x$

Greatest number of decimal places is 2

$$100(1.03 - 0.62x) = 100(0.71 - 0.22x)$$

Multiplying by 100 to clear decimals

$$100(1.03) - 100(0.62x) = 100(0.71) - 100(0.22x)$$

$$103 - 62x = 71 - 22x$$

$$32 = 40x$$

$$\dfrac{32}{40} = \dfrac{40x}{40}$$

$$\dfrac{4}{5} = x, \text{ or}$$

$$0.8 = x$$

Check:
$$\begin{array}{c|c}1.03 - 0.62x = 0.71 - 0.22x \\ \hline 1.03 - 0.62(0.8) & 0.71 - 0.22(0.8) \\ 1.03 - 0.496 & 0.71 - 0.176 \\ 0.534 & 0.534 \qquad \text{TRUE}\end{array}$$

The solution is $\frac{4}{5}$, or 0.8.

57. $\frac{2}{7}x - \frac{1}{2}x = \frac{3}{4}x + 1$, LCM is 28

$28\left(\frac{2}{7}x - \frac{1}{2}x\right) = 28\left(\frac{3}{4}x + 1\right)$

$28 \cdot \frac{2}{7}x - 28 \cdot \frac{1}{2}x = 28 \cdot \frac{3}{4}x + 28 \cdot 1$

$8x - 14x = 21x + 28$

$-6x = 21x + 28$

$-6x - 21x = 28$

$-27x = 28$

$x = -\frac{28}{27}$

Check:
$$\frac{2}{7}x - \frac{1}{2}x = \frac{3}{4}x + 1$$
$$\begin{array}{c|c} \frac{2}{7}\left(-\frac{28}{27}\right) - \frac{1}{2}\left(-\frac{28}{27}\right) & \frac{3}{4}\left(-\frac{28}{27}\right) + 1 \\ -\frac{8}{27} + \frac{14}{27} & -\frac{21}{27} + 1 \\ \frac{6}{27} & \frac{6}{27} \qquad \text{TRUE}\end{array}$$

The solution is $-\frac{28}{27}$.

59. $3(2y - 3) = 27$

$6y - 9 = 27$ 　　Using a distributive law

$6y = 27 + 9$ 　　Adding 9

$6y = 36$

$y = 6$ 　　Dividing by 6

Check:
$$\begin{array}{c|c}3(2y - 3) = 27 \\ \hline 3(2 \cdot 6 - 3) & 27 \\ 3(12 - 3) & \\ 3 \cdot 9 & \\ 27 & \text{TRUE}\end{array}$$

The solution is 6.

61. $40 = 5(3x + 2)$

$40 = 15x + 10$ 　　Using a distributive law

$40 - 10 = 15x$

$30 = 15x$

$2 = x$

Check:
$$\begin{array}{c|c}40 = 5(3x + 2) \\ \hline 40 & 5(3 \cdot 2 + 2) \\ & 5(6 + 2) \\ & 5 \cdot 8 \\ & 40 \qquad \text{TRUE}\end{array}$$

The solution is 2.

63. $2(3 + 4m) - 9 = 45$

$6 + 8m - 9 = 45$ 　　Collecting like terms

$8m - 3 = 45$

$8m = 45 + 3$

$8m = 48$

$m = 6$

Check:
$$\begin{array}{c|c}2(3 + 4m) - 9 = 45 \\ \hline 2(3 + 4 \cdot 6) - 9 & 45 \\ 2(3 + 24) - 9 & \\ 2 \cdot 27 - 9 & \\ 54 - 9 & \\ 45 & \text{TRUE}\end{array}$$

The solution is 6.

65. $5r - (2r + 8) = 16$

$5r - 2r - 8 = 16$

$3r - 8 = 16$ 　　Collecting like terms

$3r = 16 + 8$

$3r = 24$

$r = 8$

Check:
$$\begin{array}{c|c}5r - (2r + 8) = 16 \\ \hline 5 \cdot 8 - (2 \cdot 8 + 8) & 16 \\ 40 - (16 + 8) & \\ 40 - 24 & \\ 16 & \text{TRUE}\end{array}$$

The solution is 8.

67. $6 - 2(3x - 1) = 2$

$6 - 6x + 2 = 2$

$8 - 6x = 2$

$8 - 2 = 6x$ 　　Adding $6x$ and subtracting 2

$6 = 6x$

$1 = x$

Check:
$$\begin{array}{c|c}6 - 2(3x - 1) = 2 \\ \hline 6 - 2(3 \cdot 1 - 1) & 2 \\ 6 - 2(3 - 1) & \\ 6 - 2 \cdot 2 & \\ 6 - 4 & \\ 2 & \text{TRUE}\end{array}$$

The solution is 1.

69. $5(d + 4) = 7(d - 2)$

$5d + 20 = 7d - 14$

$20 + 14 = 7d - 5d$

$34 = 2d$

$17 = d$

Check:
$$\begin{array}{c|c}5(d + 4) = 7(d - 2) \\ \hline 5(17 + 4) & 7(17 - 2) \\ 5 \cdot 21 & 7 \cdot 15 \\ 105 & 105 \qquad \text{TRUE}\end{array}$$

The solution is 17.

71. $8(2t + 1) = 4(7t + 7)$

$16t + 8 = 28t + 28$

$16t - 28t = 28 - 8$

$-12t = 20$

$t = -\frac{20}{12}$

$t = -\frac{5}{3}$

Check:
$$\frac{8(2t+1) = 4(7t+7)}{}$$

$$\begin{array}{c|c} 8\left(2\left(-\dfrac{5}{3}\right)+1\right) & 4\left(7\left(-\dfrac{5}{3}\right)+7\right) \\[2mm] 8\left(-\dfrac{10}{3}+1\right) & 4\left(-\dfrac{35}{3}+7\right) \\[2mm] 8\left(-\dfrac{7}{3}\right) & 4\left(-\dfrac{14}{3}\right) \\[2mm] -\dfrac{56}{3} & -\dfrac{56}{3} \quad \text{TRUE} \end{array}$$

The solution is $-\dfrac{5}{3}$.

73. $3(r-6)+2 = 4(r+2)-21$
$3r-18+2 = 4r+8-21$
$3r-16 = 4r-13$
$13-16 = 4r-3r$
$-3 = r$

Check:
$$\frac{3(r-6)+2 = 4(r+2)-21}{}$$

$$\begin{array}{c|c} 3(-3-6)+2 & 4(-3+2)-21 \\ 3(-9)+2 & 4(-1)-21 \\ -27+2 & -4-21 \\ -25 & -25 \quad \text{TRUE} \end{array}$$

The solution is -3.

75. $19-(2x+3) = 2(x+3)+x$
$19-2x-3 = 2x+6+x$
$16-2x = 3x+6$
$16-6 = 3x+2x$
$10 = 5x$
$2 = x$

Check:
$$\frac{19-(2x+3) = 2(x+3)+x}{}$$

$$\begin{array}{c|c} 19-(2\cdot 2+3) & 2(2+3)+2 \\ 19-(4+3) & 2\cdot 5+2 \\ 19-7 & 10+2 \\ 12 & 12 \quad \text{TRUE} \end{array}$$

The solution is 2.

77. $2[4-2(3-x)]-1 = 4[2(4x-3)+7]-25$
$2[4-6+2x]-1 = 4[8x-6+7]-25$
$2[-2+2x]-1 = 4[8x+1]-25$
$-4+4x-1 = 32x+4-25$
$4x-5 = 32x-21$
$-5+21 = 32x-4x$
$16 = 28x$
$$\frac{16}{28} = x$$
$$\frac{4}{7} = x$$

The check is left to the student.
The solution is $\dfrac{4}{7}$.

79. $0.7(3x+6) = 1.1-(x+2)$
$2.1x+4.2 = 1.1-x-2$
$10(2.1x+4.2) = 10(1.1-x-2)$
$\qquad\qquad\qquad\qquad$ Clearing decimals
$21x+42 = 11-10x-20$
$21x+42 = -10x-9$
$21x+10x = -9-42$
$31x = -51$
$$x = -\frac{51}{31}$$

The check is left to the student.
The solution is $-\dfrac{51}{31}$.

81. $a+(a-3) = (a+2)-(a+1)$
$a+a-3 = a+2-a-1$
$2a-3 = 1$
$2a = 1+3$
$2a = 4$
$a = 2$

Check:
$$\frac{a+(a-3) = (a+2)\quad(a+1)}{}$$

$$\begin{array}{c|c} 2+(2-3) & (2+2)-(2+1) \\ 2-1 & 4-3 \\ 1 & 1 \quad \text{TRUE} \end{array}$$

The solution is 2.

83. Do the long division. The answer is negative.

$$3.4_{\wedge}\overline{\smash{)}2\,2.1_{\wedge}\,0}$$
$$\begin{array}{r} 6\,.\,5 \\ \hline 2\,0\,4 \\ \hline 1\,7\,0 \\ 1\,7\,0 \\ \hline 0 \end{array}$$

$-22.1 \div 3.4 = -6.5$

85. Since -15 is to the left of -13 on the number line, -15 is less than -13, so $-15 < -13$.

87. $\dfrac{y-2}{3} = \dfrac{2-y}{5}$, LCM is 15

$$15\left(\frac{y-2}{3}\right) = 15\left(\frac{2-y}{5}\right)$$
$$5(y-2) = 3(2-y)$$
$$5y-10 = 6-3y$$
$$5y+3y = 6+10$$
$$8y = 16$$
$$y = 2$$

The solution is 2.

89. $\dfrac{5+2y}{3} = \dfrac{25}{12} + \dfrac{5y+3}{4}$, LCM is 12

$$12\left(\frac{5+2y}{3}\right) = 12\left(\frac{25}{12} + \frac{5y+3}{4}\right)$$
$$4(5+2y) = 25+3(5y+3)$$
$$20+8y = 25+15y+9$$
$$20+8y = 34+15y$$
$$-7y = 14$$
$$y = -2$$

The solution is -2.

91. $\frac{2}{3}(2x - 1) = 10$

$3 \cdot \frac{2}{3}(2x - 1) = 3 \cdot 10$

$2(2x - 1) = 30$

$4x - 2 = 30$

$4x = 32$

$x = 8$

The solution is 8.

Exercise Set 12.5

1. $b + 6$, or $6 + b$

3. $c - 9$

5. $q + 6$, or $6 + q$

7. $a + b$, or $b + a$

9. $y - x$

11. $w + x$, or $x + w$

13. $r + s$, or $s + r$

15. $2x$

17. $5t$

19. Let x represent the number. Then we have $97\%x$, or $0.97x$.

21. *Familiarize*. Let $x =$ the number. Then "what number added to 85" translates to $x + 85$.

***Translate*.**

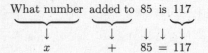

$$x \quad + \quad 85 = 117$$

***Solve*.** We solve the equation.

$$x + 85 = 117$$
$$x + 85 - 85 = 117 - 85 \qquad \text{Subtracting 85}$$
$$x = 32$$

***Check*.** 32 added to 85, or $32 + 85$, is 117. The answer checks.

***State*.** The number is 32.

23. *Familiarize*. Let $h =$ the number of hours the consultant worked in order to earn \$53,400. Then \$80 $\cdot\, h$ represents this amount.

***Translate*.**

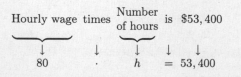

$$80 \qquad \cdot \qquad h \quad = 53,400$$

***Solve*.** We solve the equation.

$$80 \cdot h = 53,400$$
$$\frac{80 \cdot h}{80} = \frac{53,400}{80} \qquad \text{Dividing by 80}$$
$$h = 667.5$$

***Check*.** \$80 $\cdot$ 667.5 $=$ \$53,400, so the answer checks.

***State*.** The consultant worked 667.5 hr.

25. *Familiarize*. Let $c =$ the cost of one 12-ox box of oat flakes. Then four boxes cost $4c$.

***Translate*.**

$$\underbrace{\text{The cost of four boxes}}_{\downarrow} \quad \underbrace{\text{was}}_{\downarrow} \quad \underbrace{\$9.56}_{\downarrow}$$
$$4c \qquad\qquad = \qquad 9.56$$

***Solve*.** We solve the equation.

$$4c = 9.56$$
$$c = 2.39$$

***Check*.** If one box cost \$2.39, then four boxes cost 4(\$2.39), or \$9.56. The result checks.

***State*.** One box cost \$2.39.

27. *Familiarize*. Let $x =$ the number. Then "four times the number" translates to $4x$, and "17 subtracted from four times the number" translates to $4x - 17$.

***Translate*.** We reword the problem.

$$\text{Four} \quad \text{times} \quad \underbrace{\text{a number}}_{\downarrow} \quad \text{less 17 is} \quad 211$$
$$4 \qquad \cdot \qquad x \qquad - \quad 17 = 211$$

***Solve*.** We solve the equation.

$$4x - 17 = 211$$
$$4x = 228 \qquad \text{Adding 17}$$
$$\frac{4x}{4} = \frac{228}{4} \qquad \text{Dividing by 4}$$
$$x = 57$$

***Check*.** Four times 57 is 228. Subtracting 17 from 228 we get 211. The answer checks.

***State*.** The number is 57.

29. *Familiarize*. Let $y =$ the number.

***Translate*.** We reword the problem.

$$\underbrace{\text{Two times a number}}_{\downarrow} \quad \text{plus} \quad 16 \quad \text{is} \quad \underbrace{\frac{2}{5} \text{ of the number}}_{\downarrow}$$
$$2 \cdot y \qquad + \quad 16 = \quad \frac{2}{5} \cdot y$$

Solve. We solve the equation.

$$2y + 16 = \frac{2}{5}y$$

$$5(2y + 16) = 5 \cdot \frac{2}{5}y \qquad \text{Clearing the fraction}$$

$$10y + 80 = 2y$$

$$80 = -8y \qquad \text{Subtracting 10y}$$

$$-10 = y \qquad \text{Dividing by -8}$$

Check. We double -10 and get -20. Adding 16, we get -4. Also, $\frac{2}{5}(-10) = -4$. The answer checks.

State. The number is -10.

31. Familiarize. First draw a picture.

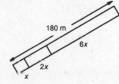

We let $x = $ the first length. Then $2x = $ the second length, and $3 \cdot 2x$, or $6x = $ the third length.

Translate. The lengths of the three pieces add up to 180 m. This gives us the equation.

Length of 1st piece	plus	Length of 2nd piece	plus	Length of 3rd piece	is 180
x	$+$	$2x$	$+$	$6x$	$= 180$

Solve. We solve the equation.

$$x + 2x + 6x = 180$$
$$9x = 180$$
$$x = 20$$

Check. If the first piece is 20 m long, then the second is $2 \cdot 20$ m, or 40 m, and the third is $6 \cdot 20$ m, or 120 m. The lengths of these pieces add up to 180 m ($20 + 40 + 120 = 180$). This checks.

State. The first piece measures 20 m. The second measures 40 m, and the third measures 120 m.

33. Familiarize. We draw a picture. Let $w = $ the width of the rectangle. Then $w + 60 = $ the length.

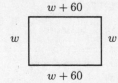

The perimeter P of a rectangle is given by the formula $2l + 2w = P$, and the area A of a rectangle is given by the formula $A = l \cdot w$, where $l = $ the length and $w = $ the width in each case.

Translate. We substitute $w + 60$ for l and 520 for P in the formula for perimeter.

$$2l + 2w = P$$
$$2(w + 60) + 2w = 520$$

Solve. We solve the equation.

$$2(w + 60) + 2w = 520$$
$$2w + 120 + 2w = 520$$
$$4w + 120 = 520$$
$$4w = 400$$
$$w = 100$$

Possible dimensions are $w = 100$ ft and $w + 60 = 160$ ft. Then the area is given by

$$A = 160 \text{ ft} \cdot 100 \text{ ft} = 16,000 \text{ ft}^2.$$

Check. The length is 60 ft more than the width. The perimeter is $2 \cdot 160$ ft $+ 2 \cdot 100$ ft, or 520 ft. We can redo our computation of the area also. The result checks.

State. The width of the rectangle is 100 ft; the length is 160 ft; and the area is 16,000 ft^2.

35. Familiarize. We draw a picture. Let $l = $ the length of the paper. Then $l - 6.3 = $ the width.

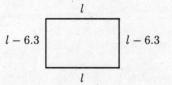

The perimeter P of a rectangle is given by the formula $2l + 2w = P$, where $l = $ the length and $w = $ the width.

Translate. We substitute $l - 6.3$ for w and 99 for P in the formula for perimeter.

$$2l + 2w = P$$
$$2l + 2(l - 6.3) = 99$$

Solve. We solve the equation.

$$2l + 2(l - 6.3) = 99$$
$$2l + 2l - 12.6 = 99$$
$$4l - 12.6 = 99$$
$$4l = 111.6$$
$$l = 27.9$$

Possible dimensions are $l = 27.9$ cm and $l - 6.3 = 21.6$ cm.

Check. The width is 6.3 cm less than the length. The perimeter is $2(27.9$ cm$) + 2(21.6$ cm$)$, or 99 cm. The answer checks.

State. The length is 27.9 cm, and the width is 21.6 cm.

37. Familiarize. The total cost is the daily charge plus the mileage charge. The mileage charge is the cost per mile times the number of miles driven. Let $m = $ the number of miles that can be driven for \$80.

Translate. We reword the problem.

Daily rate	plus	Cost per mile	times	Number of miles driven	is	Amount
↓	↓	↓	↓	↓	↓	↓
34.95	+	0.10	·	m	=	80

Solve. We solve the equation.

$$34.95 + 0.10m = 80$$
$$100(34.95 + 0.10m) = 100(80) \qquad \text{Clearing decimals}$$
$$3495 + 10m = 8000$$
$$10m = 4505$$
$$m = 450.5$$

Check. The mileage cost is found by multiplying 450.5 by \$0.10 obtaining \$45.05. Then we add \$45.05 to \$34.95, the daily rate, and get \$80.

State. The businessperson can drive 450.5 mi on the car-rental allotment.

39. **Familiarize**. Let s = one score. Then four score = $4s$ and four score and seven = $4s + 7$.

Translate. We reword .

1776	plus	four score and seven	is	1863
↓	↓	↓	↓	↓
1776	+	$(4s + 7)$	=	1863

41. **Familiarize**. Let l = the length of the rectangle. Then the width is $\frac{3}{4}l$. When the length and width are each increased by 2 cm they become $l + 2$ and $\frac{3}{4}l + 2$, respectively. We will use the formula for perimeter, $2l + 2w = P$.

Translate. We substitute $l + 2$ for l, $\frac{3}{4}l + 2$ for w, and 50 for P in the formula.

$$2(l + 2) + 2\left(\frac{3}{4}l + 2\right) = 50$$

Solve. We solve the equation.

$$2(l + 2) + 2\left(\frac{3}{4}l + 2\right) = 50$$

$$2l + 4 + \frac{3}{2}l + 4 = 50$$

$$2\left(2l + 4 + \frac{3}{2}l + 4\right) = 2(50)$$
$$\qquad\qquad \text{Clearing the fraction}$$
$$4l + 8 + 3l + 8 = 100$$
$$7l + 16 = 100$$
$$7l = 84$$
$$l = 12$$

Possible dimensions are $l = 12$ cm and $w = \frac{3}{4} \cdot 12$ cm = 9 cm.

Check. If the length is 12 cm and the width is 9 cm, then they become 12 + 2, or 14 cm, and 9 + 2, or 11 cm,

respectively, when they are each increased by 2 cm. The perimeter becomes $2 \cdot 14 + 2 \cdot 11$, or $28 + 22$, or 50 cm. This checks.

State. The length is 12 cm, and the width is 9 cm.

43. **Familiarize**. Let x = the number of half dollars. Then

$$2x = \text{number of quarters,}$$
$$4x = \text{number of dimes } (2 \cdot 2x = 4x), \text{ and}$$
$$12x = \text{number of nickels } (3 \cdot 4x = 12x).$$

The value of x half dollars is $0.50(x)$.

The value of $2x$ quarters is $0.25(2x)$.

The value of $4x$ dimes is $0.10(4x)$.

The value of $12x$ nickels is $0.05(12x)$.

Translate. The total value is \$10.

$$0.50(x) + 0.25(2x) + 0.10(4x) + 0.05(12x) = 10$$

Solve.

$$0.50(x) + 0.25(2x) + 0.10(4x) + 0.05(12x) = 10$$
$$0.5x + 0.5x + 0.4x + 0.6x = 10$$
$$2x = 10$$
$$x = 5$$

Possible answers for the number of each coin:

Half dollars = $x = 5$

Quarters = $2x = 2 \cdot 5 = 10$

Dimes = $4x = 4 \cdot 5 = 20$

Nickels = $12x = 12 \cdot 5 = 60$

Check. The value of

$$
\begin{aligned}
5 \text{ half dollars} &= \$2.50 \\
10 \text{ quarters} &= \$2.50 \\
20 \text{ dimes} &= \$2.00 \\
60 \text{ nickels} &= \$3.00
\end{aligned}
$$

The total value is \$10. The numbers check.

State. The storekeeper got 5 half dollars, 10 quarters, 20 dimes, and 60 nickels.